ACTA BIOLOGICA ET MEDICA GERMANICA

Zeitschrift für funktionelle Biowissenschaften

VIIIth Symposium Structure and Function of Erythrocytes Part II

Band 36
Heft 5/6 · 1977
Seite 611–940

EVP 48,– M

ABMGAJ 36 (5/6) 611–940 (1977)

31002

AKADEMIE-VERLAG
BERLIN

Aufnahmebedingungen

1. Die ACTA BIOLOGICA ET MEDICA GERMANICA, Zeitschrift für funktionelle Biowissenschaften, publiziert Arbeiten aus den Fachgebieten Biochemie, Molekular- und Zellbiologie, Physiologie (einschließlich Pathophysiologie), Pharmakologie und Immunbiologie. Es werden nur Arbeiten angenommen, die nicht an anderer Stelle mit demselben Inhalt veröffentlicht oder zur Veröffentlichung angeboten werden. Der Autor verpflichtet sich nach Annahme, die Arbeiten an keiner anderen Stelle zu veröffentlichen.

2. Die Arbeit muß wissenschaftlich wertvoll sein. Bestätigungen bekannter Tatsachen, Versuche und Beobachtungen ohne positives Ergebnis werden, wenn überhaupt, nur in kürzester Form aufgenommen. Nicht aufgenommen werden Polemiken und spekulative Arbeiten, falls sie nicht wesentlich neue Gesichtspunkte enthalten.
   Die Arbeiten sollen den Charakter wissenschaftlicher Originalarbeiten haben. Als solche gelten alle Mitteilungen, die zur vorwärtsführenden Erweiterung des Erkenntnisstandes auf den genannten Fachgebieten führen. Originalarbeiten sollen 20 Manuskriptseiten nicht überschreiten. Kurzmitteilungen werden bei der Drucklegung zeitlich bevorzugt; sie dürfen 5 Manuskriptseiten nicht überschreiten. Als Kurzmitteilung gelten solche Arbeiten, in denen über neue Ergebnisse berichtet wird, ohne Details einer Originalarbeit zu enthalten.
   Besonders aktuelle Untersuchungsergebnisse können in kurzer Form (bis 4 Seiten) im Offsetverfahren publiziert werden, wofür reproduktionsreife Manuskripte erforderlich sind. In Form von Übersichtsarbeiten (Reviews) werden Artikel entgegengenommen, die zu aktuellen Gebieten einen Überblick geben, in dem Fakten dargestellt, besprochen und kritisch bewertet werden.

3. Die Arbeiten müssen so kurz als möglich abgefaßt werden und in einem druckreifen Zustand geschrieben sein. Einleitung (Problematik), Methodik, Befunde und Diskussion sollen deutlich in Erscheinung treten. Der Arbeit ist eine Zusammenfassung der wesentlichsten Ergebnisse voranzustellen, wobei bei deutschsprachigen Manuskripten auch eine englische Übersetzung notwendig ist.
   Arbeiten werden in Deutsch, Englisch und Russisch angenommen. Die Manuskripte sind in zweifacher Ausfertigung einzureichen. Bei Manuskripten in deutscher Sprache ist die Schreibweise des „Duden" verbindlich; bei eingedeutschten Wörtern ist die „K-Z"-Schreibweise anzuwenden.
   Von den Abbildungen sind 2 Kopien sowie 1 Satz reproduktionsreife Vorlagen beizufügen. Genaue Hinweise zur Manuskriptgestaltung sind von der Redaktion der Zeitschrift anzu fordern und unbedingt einzuhalten. Manuskripte, die diesen Bedingungen nicht entsprechen, gehen unbearbeitet zur Revision an den Autor zurück.

4. Die Arbeiten werden im Sofortumbruch gesetzt; Korrekturen in Form von Streichungen bzw. Zusätzen sind daher in der Umbruchkorrektur nicht zulässig.

5. Manuskripte sind an die Redaktion der ACTA BIOLOGICA ET MEDICA GERMANICA, DDR-1115 Berlin-Buch, Lindenberger Weg 70, zu senden.

6. Von jeder Originalarbeit werden kostenlos 80 Sonderdrucke geliefert.

Chefredaktion/Herausgeber

Zeitschrift ACTA BIOLOGICA ET MEDICA GERMANICA

Verlag: Akademie-Verlag, DDR-108 Berlin, Leipziger Straße 3—4; Fernruf 2236229 odeo 2236221; Telex-Nr. 114420; Postscheckkonto: Berlin 35021. Bank: Staatsbank der DDR, Berlin, Kto.-Nr.: 6836-26-20712.

Chefredaktion: Prof. Dr. Heinz Bielka, Prof. Dr. Werner Scheler.

Anschrift der Redaktion: DDR-1115 Berlin-Buch, Lindenberger Weg 70. Fernruf: 569 78 51, App. 222.

Veröffentlicht unter der Lizenznummer 1318 des Presseamtes beim Vorsitzenden des Ministerrates der Deutschen Demokratischen Republik.

Gesamtherstellung: VEB Druckerei „Thomas Müntzer", DDR-582 Bad Langensalza.

Erscheinungsweise: Die Zeitschrift erscheint monatlich. Die 12 Hefte eines Jahrganges bilden einen Band. Bezugspreis je Heft 35,— M (Preis für die DDR: 24,— M); Bandpreis 420,— M zuzüglich Versandspesen (Preis für die DDR 288,— M).

Bestellnummer dieses Heftes: 1053/XXXVI/5/6.

A N (EDV) 50117

ACTA BIOLOGICA ET MEDICA GERMANICA

Herausgeber:
R. Baumann, H. Dutz,
A. Graffi, F. Jung,
O. Prokop, S. M. Rapoport

Chefredaktion:
H. Bielka, W. Scheler

Band 36 1977 Heft 5—6

Zeitschrift für funktionelle Biowissenschaften

Acta biol. med. germ., Band 36, Seite 611—619 (1977)

Physiologisch-Chemisches Institut der Friedrich-Schiller-Universität, 69 Jena, DDR

The effects of calcium on glycolysis and ATP concentration in complete and membrane-poor hemolyzates of human erythrocytes

D. Brox, B. Petermann, and H. Frunder

Summary

Hemolyzates prepared from packed human red cells with 30 µM total calcium were employed as a means to examine the relationship between ATP consumption and lactate formation. Hemolyzates exposed to ultracentrifugation accumulate membrane fragments in the top layer yielding membrane-poor fractions in the bottom layers of the centrifuge tube. Lactate formation accompanied by ATP depletion amounts to 12 µmoles per ml and hour in complete hemolyzates fortified with NAD. Complexation of calcium results in about 50% inhibition of the lactate formation with a concomitant increase of ATP. Lactate formation is reduced in membrane-poor hemolyzates approximately concurrently to the extent of membrane removal which produces no discernible change in the glyceraldehydephosphate dehydrogenase activity. 50—200 µM total calcium has no effect on the membrane-independent lactate formation which amounts to 1—2 µmoles per ml and hour. Triton X-100 seems to solubilise also the membrane components responsible for the high calcium-dependent ATP consumption which governs the lactate formation.

Introduction

Human erythrocytes normally contain small amounts of calcium. Excess intracellular calcium initiates a series of structural and metabolic alterations which include loss of potassium, spheroechinocytic transformation, deterioration of cell membrane deformability, activation of (Ca^{2+} + Mg^{2+})-ATPase and of glycolysis [1—6]. The connected paper shows that erythrocytes exposed to the divalent cation ionophore A 23187, rapidly increase glycolysis and become ATP depleted [5].

Inhibition of the (Ca^{2+} + Mg^{2+})-ATPase by ruthenium red or lanthanum inhibits glycolysis and simultaneously speeds up ATP depletion. This suggests that calcium-dependent ATP consuming reactions not related to (Ca^{2+} + Mg^{2+})-ATPase are responsible for ATP depletion.

As a means to characterise these unknown ATP consuming processes and to examine direct effects of low calcium concentrations on glycolysis, complete and membrane-poor hemolyzates were applied. The majority of previous hemolyzate studies have utilized diluted hemolyzates which have inherent limitations by alterations of all intracellular constituent concentrations [7—9]. Thus, in the studies reported here, "thick" hemolyzates were employed to approach "in vivo" conditions according to Reimann et al. [10].

The experimental procedure is described for the removal of membrane fragments from "thick" hemolyzates avoiding gross changes of cellular metabolite concentrations. The results obtained with complete and membrane-poor hemolyzates demonstrate that a moderate excess calcium influences glycolysis and ATP consumption exclusively by membrane processes.

Materials and methods

Preparation of erythrocytes and hemolyzates

Freshly drawn human blood was either defibrinized or mixed with EGTA (final concentration 5 mM). Erythrocytes were separated from the serum (plasma) and buffy coat layer and washed twice with standard medium containing 140 mM NaCl, 5 mM KCl, 1 mM P_i and 5.5 mM glucose, *p*H 7.4. The last centrifugation at 20000 × *g* and 5 °C for 15 min yielded red cell sediments with hematocrits between 95 and 98%. The total calcium content in the cell sediment, measured by atomic absorption spectrophotometry (AAS 1, VEB Carl Zeiss, Jena) was in the range of 30—50 μM.

Ultra sound hemolyzates: 15 ml packed cells were gently sonicated in a sonifier (VEB Elektromat, Dresden) at maximum output and 3 °C. 60—100% hemolysis was obtained by 3—15 min sonication.

Triton X-100 and saponin hemolyzates: 12.5 ml packed cells were rigorously mixed with 0.1 ml 20% Triton X-100 (v/v) or 0.2 ml 10% saponin (w/v) at room temperature. 90—100% hemolysis was obtained within 10 min. The degree of hemolysis was estimated by hemoglobin measurements in supernatants of 1:10 with water or saline diluted and centrifuged hemolyzates by the cyanmethemoglobin method.

Removal of membrane fragments

The hemolyzate was centrifuged in an ultracentrifuge (VAC 601, VEB Zentrifugenbau Engelsdorf) at 200000 × *g* and 10 °C for 150 min. Then, the contents of the centrifuge tube were divided into 5—10 fractions from top to bottom by gentle suction. An aliquot of each fraction was diluted (1:40) with 0.3% (v/v) Triton X-100 for the determination of the acetylcholine esterase activity according to Weber [11]. In some experiments membrane-poor hemolyzate fractions (No. 4—7, cf. Fig. 1) and hemolyzates diluted with saline (1:40) were submitted to an ultrafiltration through a set of millipore filters with 3.0, 0.8, 0.3 and 0.12 μm pore diameter and finally through an Amicon XM 300 membrane.

Abbreviations: GAPD: Glyceraldehydephosphate dehydrogenase (EC 1.2.1.12); D-Glyceraldehyde-3-phosphate: NAD oxidoreductase (phosphorylating); PGK: Phosphoglycerate kinase (EC 2.7.2.3); ATP: 3-phospho-D-glycerate 1-phosphotransferase; GAP: D-Glyceraldehyde-3-phosphate; DAP: Dihydroxyacetone phosphate; EGTA: ethyleneglycol-bis-(β-aminoethylether)-N',N'-tetraacetic acid

Fig. 1 shows the distribution of the acetylcholine esterase activity after high speed centrifugation of the hemolyzate. The lowest activity was found near the bottom of the centrifuge tube. From bottom to top there is a gradient with a fourfold increase of the activity in the top layer. About 10% of the acetylcholine esterase pass the XM 300 membrane (Fig. 2). This can be taken as a measure of the solubilisation of acetylcholine esterase by the preparation procedure. The activities retained by millipore filters are assumed to belong to small membrane fragments not floating during ultracentrifugation.

Incubation conditions

1 or 2 ml of hemolyzate were mixed vigorously with 0.01 or 0.02 ml 50 mM NAD solution in a prewarmed test tube, stopped and incubated at 37 °C for 35 min. Other substances were added in the same way, i. e. in a volume corresponding to 1% of the hemolyzate volume. The *p*H-value of the hemolyzate was 7.2 ± 0.005 at 37 °C. It did not change noticeably during incubation.

The results in Fig. 3 show that cellular NAD is rapidly destroyed, which limits glycolysis; added NAD is destroyed, too. Nicotinamide inhibits the NAD destruction indicating the activation of NAD nucleosidases by gentle sonication of packed cells. NAD above 0.05 mM has no noticeable influence on glycolysis. 0.35—0.5 mM NAD chosen as regular addition ensures that NAD does not decrease below 0.05 mM during the incubation period at 37 °C. For the maintenance of a constant lactate formation with a 2:1 to 3:1 ratio of lactate formation and glucose consumption in complete hemolyzates it is sufficient, too.

Biochemical assays

After 5 and 35 min of incubation approximately 0.5 ml of hemolyzate was withdrawn, weighted and deproteinized with 3 ml 6% (w/v) perchloric acid. The supernatant was neutralized with K_2CO_3, centrifuged and aliquots taken for metabolite determinations with standard methods [12]. The method of Rose and Liebowitz [13] was used to measure 2,3-biphosphoglycerate. Measurements of P_i were performed with the method of Martin and Doty, modified as described in [14]. GAPD was determined according to Delbrück et al. [15] and PGK according to Bergmeyer [12] after appropriate dilution with 0.3% Triton X-100.

Results

Hemolyzates

Hemolyzates preparad by gentle sonication of packed cells prewashed in saline show a high lactate formation (Table 1), which approximately corresponds to V_{max} of hexokinase of human erythrocytes. ATP decreases and P_i increases during the incubation, indicating an imbalance between ATP production and consumption. To examine the dependence of this imbalance on calcium, hemolyzates were prepared from packed cells prewashed with EGTA. The amount of EGTA added corresponded to the calcium impurities of the wash solutions (30—50 μM) and was thought not to be in excess. Lactate formation is significantly lower and the ATP level more stable in these hemolyzates. Addition of EGTA in excess over cellular calcium (0.1—1.0 mM) has only a small supplementary effect on lactate formation. However, the ATP level is much more stable during the incubation period with the tendency to increase.

Lactate formation increases in the experiment shown in Table 1 from 7.8 to 11.0 by addition of calcium in excess over EGTA. This is regularly connected with a rapid ATP depletion.

It can be seen from the P_i values given in Table 1 and from 2,3-biphosphoglycerate determinations (not shown in Table 1) that the 2,3-bisphosphoglycerate store contributes only little to the lactate formation. Then, it may be concluded that gentle sonication activates ATP consumption and, in consequence, glycolysis. Even

Table 1

Lactate formation, ATP and P_i concentrations in hemolyzates

Hemolyzate	Additions	Lactate formation [µmoles/ml · hour]	ATP [mM] before incubation	ATP [mM] after incubation	P_i [mM] before incubation	P_i [mM] after incubation
Ultrasound (*s*)	none	12.0 ± 4.9 (11)	1.1 ± 0.3 (9)	0.8 ± 0.5 (9)	1.2 ± 0.5 (18)	3.7 ± 1.3 (18)
Ultrasound (*e*)	none	8.3 ± 2.4 (9)	1.5 ± 0.4 (6)	1.3 ± 0.6 (6)	1.3 ± 0.5 (17)	2.1 ± 1.2 (17)
Ultrasound (*e*)	EGTA	7.2 ± 0.5 (5)	1.9 ± 0.3 (4)	2.0 ± 0.2 (4)	1.0 (1)	1.5 (1)
Ultrasound (*e*)	calcium	11.0 (1)	1.5 (1)	0.1 (1)	—	—
Triton (*e*)	none	6.9 ± 1.3 (8)	1.4 ± 0.5 (8)	1.5 ± 0.4 (9)	—	—
Triton (*e*)	EGTA (1 mM)	6.0 (1)	—	—	—	—

Ultra sound hemolyzates (70—90% hemolysis) and Triton X-100 hemolysates (90—98% hemolysis) prepared from packed cells either prewashed with saline (= *s*) or with saline plus 50 µM EGTA (= *e*). Final concentrations of added substance: EGTA = 0.1 mM; calcium = 0.15 mM. 30 min incubation at 37 °C and *p*H 7.2. Means ± S.D. refered to hemolyzate water; *n* in ( ).

Table 2

Lactate formation, ATP and P_i concentrations in membrane-poor hemolyzates

Hemolyzate	Additions	Acetylcholine-esterase [%]	Lactate formation [µmoles/ml · hour]	ATP [mM] before incubation	ATP [mM] after incubation
Ultrasound (*s*)	none	37 ± 22 (11)	3.9 ± 1.5 (11)	1.2 ± 0.2 (9)	1.5 ± 0.2 (9)
Ultrasound (*e*)	none	20 ± 7 (9)	2.0 ± 1.4 (6)	1.6 ± 0.2 (6)	1.6 ± 0.3 (6)
Ultrasound (*e*)	EGTA	24 ± 6 (3)	42 ± 21 (3)*	1.3 ± 0.1 (3)	1.5 ± 0.1 (3)
Ultrasound (*e*)	calcium	20 ± 5 (6)	15 ± 35 (6)**	1.3 ± 0.2 (6)	1.4 ± 0.2 (6)
Ultrasound (*e*)	ruthenium red	26 ± 1 (6)	33 ± 48 (6)*	1.4 ± 0.2 (6)	1.6 ± 0.1 (6)
Triton (*e*)	none	60	6.0 ± 1.9 (4)	1.5 ± 0.5 (3)	1.3 ± 0.3 (3)

The conditions of incubation are those indicated in Table 1. The results are the mean ± S.D. refered to hemolyzate water; *n* in ( ). Acetylcholine esterase is given as percentage of the activity in the complete hemolyzate. Final concentration of ruthenium red was 0.025 mM.

*% Inhibition when compared with parallel experiments without addition;

**% activation when compared with parallel experiments without calcium no differences of the ATP concentrations in parallel experiments

maximum glycolysis does not keep pace with the ATP consumption and the ATP level decreases. Complexation of calcium by EGTA clearly inhibits ATP consumption and glycolysis, which now maintains a normal ATP level. Triton X-100 hemolyzates show a very rapid ATP depletion during incubation when prepared from cells prewashed with saline. However, complexation of extracellular calcium impurities by EGTA yields hemolyzates which respond like ultrasound hemolyzates with respect to lactate formation and ATP concentration.

Membrane-poor hemolyzates

Membrane fragments migrate to the top of the centrifuge tube during high speed centrifugation of hemolyzates prepared by gentle sonication (or saponin treatment) of packed cells. This leads to hemolyzate fractions poor in membrane fragments (Fig. 1). The pooled fractions 4—7 used in this series of experiments contained between 15 and 35% of the acetylcholine esterase activity of the initial hemolyzate. Approximately 10% of the membrane-bound acetylcholine esterase are solubilised during the hemolyzate preparation (Fig. 2). Thus, the amount of membrane fragments in membrane-poor fractions is less than indicated by the percentage of acetylcholine esterase. Taking this into account it may be concluded from a comparison of the results shown in Tables 1 and 2 that the decrease of lactate formation in membrane-poor hemolyzates approximately corresponds to the extent of membrane removal from ultra sound hemolyzates.

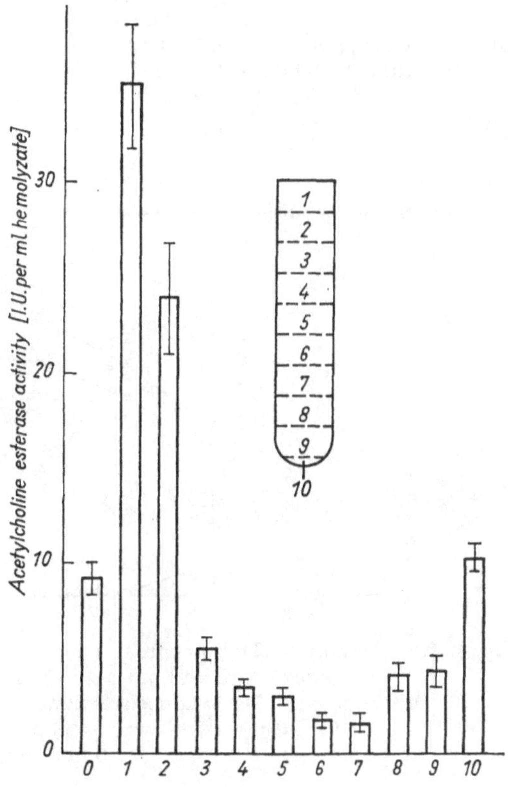

Fig. 1. Distribution of the acetylcholine esterase activity in ultrasound hemolyzates after 150 min centrifugation at 200000 × *g* and 10 °C rotor temperature. 1—10 markes the fractions examined from top to bottom of the centrifuge tube. The results are the mean ± S. D. of 6 experiments. Acitivity in I.U. per ml hemolyzate

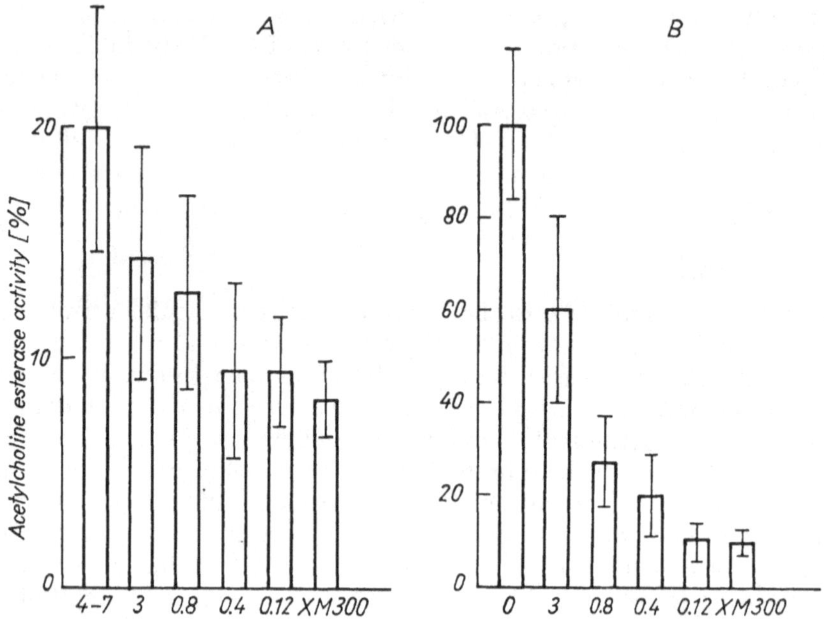

Fig. 2. A) Acetylcholine esterase activity of the mixed fractions 4—7 after ultrafiltration through a set of millipore filters with 3.0, 0.8, 0.4 and 0.12 μm pore diameter and finally through a XM 300 Amicon membrane. The results are the mean ± S.D. of 5 experiments. Activity as percentage of the activity in the complete hemolyzate.

B) Acetylcholine esterase activity of 1:40 with saline diluted hemolyzates (0) after ultrafiltration (like A). The results are the mean ± S.D. of 3 experiments. Activity as percentage of the activity in the complete hemolyzates (0)

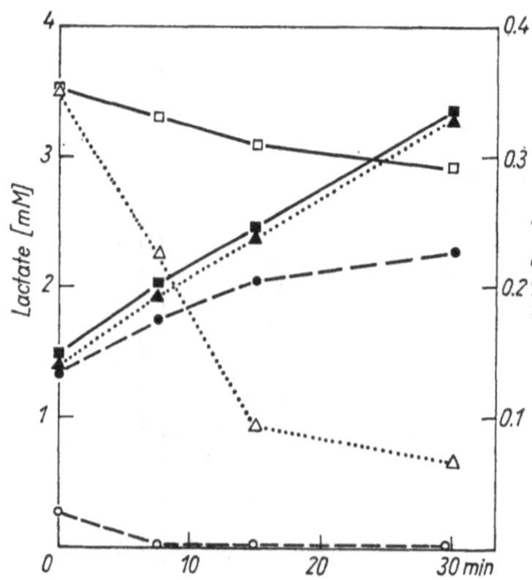

Fig. 3. Lactate and NAD concentrations in ultrasound hemolyzates during 30 min incubation at 37 °C and *p*H 7.2 in μmoles per ml hemolyzate water.

Open symbols: NAD concentrations, black symbols: lactate concentrations.

– – – no addition; ······· with 0.35 mM NAD; ——— with 0.35 mM NAD plus 10 mM nicotinamide

The ATP concentrations either slightly increase or remain unchanged during incubation. Correspondingly, the P_i levels are low and constant in all series (0.9 $\pm$ $\pm$ 0.4 (30) before and 1.0 $\pm$ 0.2 (30) after incubation). Thus, ATP consuming reactions are membrane-associated either exclusively or predominantly.

To further define the relationship between the rate of lactate formation and calcium, excess EGTA or calcium was added to membrane-poor hemolyzates. The addition of EGTA results in 42% inhibition of the lactate formation with a concomitant slight increase of the ATP concentration. Excess calcium augments the lactate formation only at an average of 15%. In 5 parallel experiments of this series there is practically no effect of excess calcium. The exception is one parallel experiment included in this series where calcium activates the lactate formation. No discernible differences in ATP and P_i concentrations are found in parallel experiments.

These results are in accordance with the assumption that small amounts of membrane fragments remaining in membrane-poor hemolyzates are responsible for the effects of EGTA and calcium on lactate formation. However, there remains a lactate formation of about 1—2 μmoles per ml and hour, which cannot be explained by the presence of membrane fragments.

It has to be examined whether this remaining glycolysis is due to a futile cycle via 2,3-bisphosphoglycerate without net production and consumption of ATP.

According to Anand and Dalla [16] ruthenium red acts as an inhibitor of the (Ca^{2+} + Mg^{2+})- and (Mg^{2+})-dependent ATPases. It causes only a slight inhibition of the high lactate formation in complete hemolyzates. However, it decreases the small lactate formation in membrane-poor hemolyzates by about 30%. This sug-

Table 3

Concentrations of triosephosphates and fructose-1,6-bisphosphate in complete and membrane-poor hemolyzates

	GAP + DAP [mM]	F-1,6-P_2 [mM]
Complete hemolyzates	<0.1	<0.01
Membrane-poor hemolyzates	0.7 $\pm$ 0.33 (13)	0.2 $\pm$ 0.15 (12)

Samples were withdrawn after 30 min incubation at 37 °C and *p*H 7.2. The hemolyzates used were ultrasound hemolyzates prepared from packed cells prewashed with saline plus EGTA. Means $\pm$ S.D. refered to hemolyzate water; *n* in ().

Table 4

GAPD, PGK and acetylcholine esterase in membrane-poor hemolyzates

Acetylcholine esterase [%]	GAPD [%]	PGK [%]
28 $\pm$ 9 (10)	98 $\pm$ 19 (19)	110 $\pm$ 46 (3)

Activity in complete hemolyzates taken as 100; means $\pm$ S.D.; *n* in ().

gests that ruthenium red (used without purification) influences the lactate formation by membrane-independent side effects.

In membrane-poor hemolyzates there is a marked accumulation of triosephosphates and fructose-1,6-bisphosphate, which is not observed in complete hemolyzates (Table 3). Thus, the activities of GAPD and PGK were determined in membrane-poor and complete hemolyzates. The results shown in Table 4 demonstrate that at least under the conditions described for the preparation and separation of the hemolyzates, no significant amounts of these enzymes are removed together with the membrane fragments from the hemolyzates. However, it will be reported in a forthcoming paper that under other conditions a remarkable binding of glycolytic enzymes to membrane fragments can be found. The reasons remain to be elucidated for these metabolite accumulations or why these metabolites do not accumulate in complete hemolyzates with much higher lactate concentrations.

The main part of the acetylcholine esterase is solubilized in Triton X-100 hemolyzates and cannot be removed by high speed centrifugation. Therefore, the rate of lactate formation in the corresponding hemolyzate fractions is approximately as high as in complete Triton X-100 hemolyzates. The membrane-bound proteins responsible for the ATP consumption are either solubilized together with the acetylcholine esterase by Triton X-100 or the delipidated membrane remnant which contains these proteins in a bound form migrates neither to the top nor to the bottom of the centrifuge tube during high speed centrifugation. Preliminary data point to a solubilisation.

Discussion

The results of the connected paper [5] demonstrate that there is a calcium dependent ATP consuming reaction of high capacity in human red cells which cannot be related to the well known (Ca^{2+} + Mg^{2+})-dependent ATPase. The experiments described in this paper were carried out in order to further characterize this reaction in hemolyzates prepared from packed cells.

Hemolysis induces a rapid destruction of NAD and ATP which limits glycolysis. When NAD is added in excess the hemolyzates prepared in the presence of calcium impurities in the micromolar range show maximum glycolysis which probably is limited by the hexokinase capacity. However, ATP consuming reactions are much more stimulated. In consequence, the ATP level decreases more or less rapidly which again limits glycolysis.

Complexation of extracellular calcium impurities by EGTA reduces the lactate formation of the hemolyzates from 12 to 8 μmoles per ml and hour. Now, glycolysis keeps pace with ATP consumption which must be reduced, too. Thus, the ATP level remains approximately constant during 30 min of incubation (or even longer). Addition of EGTA in excess over cellular calcium reduces the lactate formation only a little (from 8 to 7 μmoles per ml and hour) and the ATP level increases. Taking into account that the ratio lactate formation/glucose consumption is in the range of 2—3, it may be concluded from the results that more than 5 μmoles ATP per ml hemolyzate and hour are consumed by reactions dependent on micromolar total calcium concentrations.

Ruthenium red and ouabain inhibit the total and EGTA insensitive lactate formation only to a small extent which can be taken as a hint that the known ATP-ases contribute only little to the EGTA insensitive ATP consumption.
EGTA sensitive and insensitive ATP consumption, which apparently govern the rate of lactate formation, are almost completely bound to the membrane compartment. This can be derived from the results obtained with membrane-poor hemolyzates. Here, the lactate formation is reduced approximately concurrently to the extent of membrane removal. A concomitant removal of possibly membrane-bound glycolytic enzymes which could also explain the reduced lactate formation, was ruled out with respect to GAPD and PGK.
Excess EGTA reduces the low lactate formation of membrane-poor hemolyzates possibly a bit more than attributable to the amount of membrane fragments present when compared with the effect of EGTA in complete hemolyzates. A small part of calcium sensitive ATP consuming membrane constituents may be detached from the membrane during the preparation of membrane-poor hemolyzates. On the other hand, excess calcium stimulates the lactate formation at an average of only 15%, in accordance with the small amount of membrane fragments present. From this result it may be concluded that total calcium up to 0.2 mM has no direct effect on the membrane-independent lactate formation.
Triton X-100 hemolyzates qualitatively respond towards calcium as ultra sound hemolyzates. However, the sensitivity towards calcium seems to be higher which may be attributed to a partial solubilisation of the corresponding membrane proteins. This offers a path for the future characterization of the calcium dependent ATP consuming reaction which works after membrane disruption in hemolyzates with a capacity far above V_{max} of the $(Ca^{2+} + Mg^{2+})$-dependent ATPase [5].

Acknowledgements

The authors express their gratitude to Mrs. Ch. Kaschlik, Mrs. W. Hallmeyer and Miss K. Behrens for technical assistance.

References

[1] Harrison, D. G., and C. Long: J. Physiol., Lond. **199**, 367 (1968)
[2] Sarkadi, B., J. Szasz, and G. Gardos: J. Membrane Biol. **26**, 357 (1976)
[3] Palek, J., W. A. Curby, and F. J. Lionetti: Am. J. Phyisol. **220**, 1028 (1971)
[4] Schatzmann, H. J.: J. Physiol., Lond. **235**, 551 (1973)
[5] Till, U., H. Petermann, I. Wenz, and H. Frunder: Acta biol. med. germ. **36**, 597 (1977)
[6] Edmondson, J. W., and Li, Ting-Kai: Biochim. biophys. Acta **443**, 106 (1976)
[7] Chapman, R. G., F. M. Hennessy, and B. W. Gabrio: J. clin. Invest. **41**, 1245 (1962)
[8] Oski, F. A., and C. Smith: Pediatrics **41**, 473 (1968)
[9] Baskan, N., S. Moses, and A. Livue: FEBS Lett. **36**, 213 (1973)
[10] Reimann, B., G. Küttner, D. Maretzki, and S. Rapoport: Acta biol. med. germ. **34**, 1777 (1975)
[11] Weber, H.: Dt. med. Wschr. **91**, 1927 (1966)
[12] Bergmeyer, H. U. (Ed.): Methoden der enzymatischen Analyse. 2. Ed. Akademie-Verlag, Berlin 1970
[13] Rose, E. B., and J. Liebowitz: Anal. Biochem. **35**, 177 (1970)
[14] Till, U., E. Blume, J. Günther, J. G. Reich, D. Zahn, R. Klinger, K. Jaroszewicz, and H. Frunder: Eur. J. Biochem. **6**, 373 (1968)
[15] Delbrück, A., E. Zebe, and Th. Bücher: Biochem. Z. **331**, 273 (1959)
[16] Anand, M. B., and N. S. Dhalla: Fedn Proc. Fedn Am. Socs exp. Biol. **35**, 535 (1976)

Acta biol. med. germ., Band 36, Seite 621—624 (1977)

Zakład Genetyki Człowieka, Polska Akademia Nauk, ul. Święcickiego 6, 60—781 Poznań, Poland

Effect of ligands of ferric hemes on interaction between ferric and ferrous chains in partially oxidized hemoglobin A

A. L. Pawlak

Summary

Normal values of Bohr effect of oxygenation of partially oxidized hemoglobin A with ferrihemes liganded either with H_2O and OH or with CN have been found in the range of *p*H values from 6.8 to 7.6 in 45 μM (Fe)-hemoglobin containing 36—38% of ferrihemes. As the changes of oxygen affinity of Hb A induced by changes of *p*H are due to the modifications of R state, this quaternary conformation is thought to be unchanged in the studied forms of partially oxidized hemoglobin.
It is suggested that interactions between ferric and ferrous hemes leading to the increased affinity of ferrous hemes to oxygen occur in deoxygenated form of partially oxidized hemoglobin. In partially oxidized hemoglobin with ferric hemes liganded with H_2O asymmetry of oxygen binding curves has been noted, which is not observed in forms with ferric hemes liganded with OH ot CN. This shows the effect of ligands of ferric hemes on interactions between chains containing ferric and ferrous hemes.

Introduction

The asymmetry of oxygen binding curves of partially oxidized hemoglobin A (Hb(II/III)) with ferrihemes in aquoform (Hb(II/III H_2O)), but not of Hb(II/III) with ferrihemes in OH or CN forms (Hb(II/III OH) and Hb(II/III CN)) indicates that ligands of ferrihemes change the interaction between ferric and ferrous chains in Hb(II/III) [1, 2].
Here these results are complemented with determinations of changes of p_{50} and n_{50} of oxygenation of Hb(II/III H_2O-OH) and of Hb(II/III CN) induced by changing *p*H. The results obtained suggest that the effect of ligands of ferrihemes on the symmetry of oxygen binding to ferrohemes is due to the differences in conformations between deoxygenated Hb(II/III H_2O) and Hb(II/III CN).

Materials and methods

Normal human adult blood was collected in heparin. Red blood cells (rbc) were washed in 0.9% NaCl and depleted of 2,3-diphosphoglycerate by overnight incubation at room temperature without glucose. After incubation, rbc were once more suspended in 0.9% NaCl-solution and separated by centrifugation at 150 × *g* for 10 min. Subsequently, rbc were hemolyzed with 2 vol. of cold H_2O. The hemolysate was centrifuged at 10000 × *g* for 25 min at 4 °C. The clear supernatant was stored in separate samples at −8 °C. The measurements were performed during 3 weeks from the preparation of the samples. The hemolysates prepared as described did not contain 2,3-diphosphoglycerate, as assessed by enzymatic assay. The concentration of methemoglobin was less than 4%.
The affinity of hemoglobin to oxygen was determined by spectrophotometric method [3] in Thunberg type cuvette with optical path 1.4 cm and a volume of 74 ml. The change of absorption at 560 nm was taken as a basis for estimation of oxygen binding to ferrohemes, and absorption at 522 nm was monitored as an isosbestic point.

Partially oxidized hemoglobin (Hb(II/III)) was obtained from deoxygenated hemoglobin by addition of ferricyanide solution (ca. 10 mM) from the side arm of the Thunberg type cuvette 5 min before the start of measurement of oxygenation equilibrium. Cyan-met form of ferrihemes in Hb(II/III) was obtained from oxygenated Hb(II/III) by addition of diluted solution of KCN in 10% excess in relation to the amount of ferric hemes.

Results

Hb(II/III) displays asymmetry in oxygen binding curves at *p*H from 6.8 to 7.1. It is characterized by the increase of interaction of hemes in the course of oxygen binding (Fig. 1). This effect is not seen at higher *p*H and in Hb(II/III CN) at *p*H 6.8.

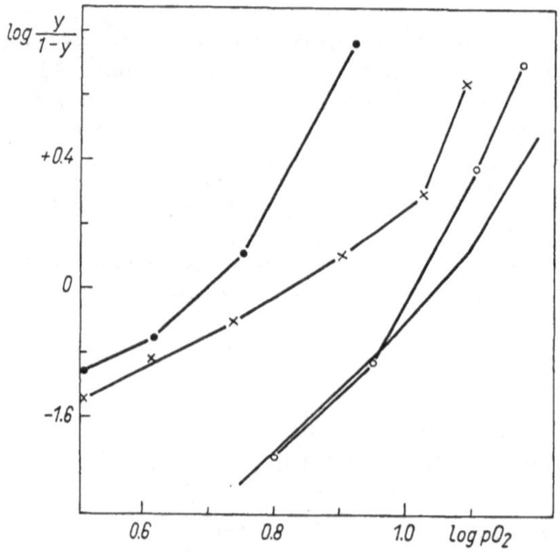

Fig. 1. The effect of different concentrations of ferrihemes in Hb(II/III) on Hill plots of oxygen binding to ferrohemes.
Hb(II/III) 45 µM (Fe), phosphate buffer 67 mM, *p*H 6.8; concentrations of ferrihemes: ——— 0%, o———o 16%, ×———× 50%, •———• 75 %

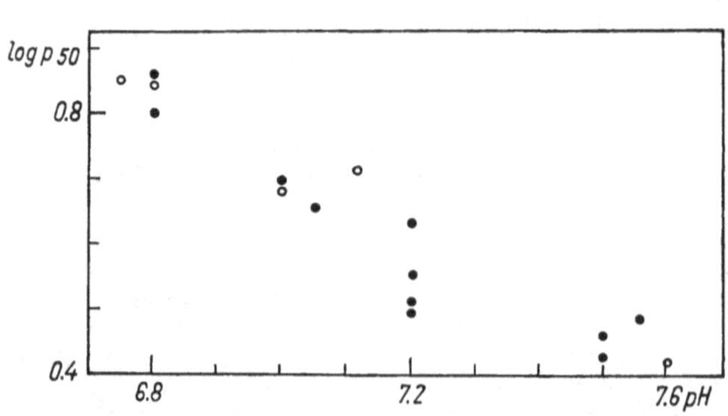

Fig. 2. The effect of *p*H on p_{50} of oxygenation of ferrohemes in Hb(II/III) with different ligands of ferrihemes. (o) Hb(II/III H_2O) and Hb(II/III OH), (•) Hb(II/III CN); Hb(II/III) 45 µM (Fe), concentration of ferrihemes 36—38%, phosphate buffers 67 mM

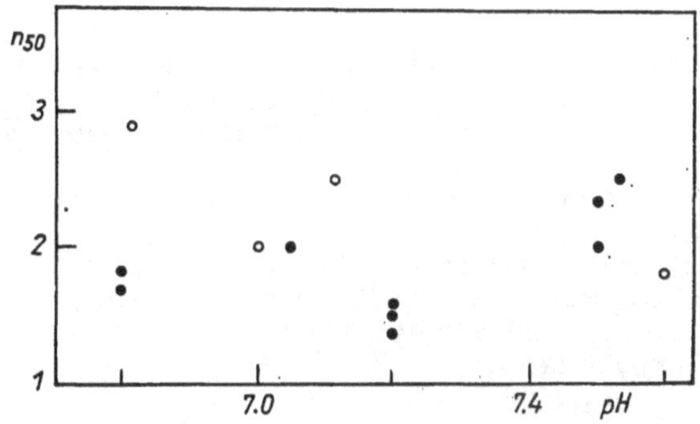

Fig. 3. The effect of pH on n_{50} of oxygenation of ferrohemes in Hb(II/III) with different ligands of ferrihemes. (○) Hb(II/III H_2O) and Hb(II/III OH), (●) Hb(II/III CN); Hb(II/III) 45 μM (Fe), concentration of ferrihemes 36—38%, phosphate buffers 67 mM

The oxygenation Bohr effect of both (H_2O-OH) and (CN) forms of Hb(II/III) is close to normal value in the range of pH from 6.8 to 7.6 (Fig. 2): $\Delta \log p_{50}/\Delta pH = 0.51$. Changes of pH (in the range 6.8—7.6) do not influence the value of the coefficient of cooperativity of oxygenation of ferrohemes (n_{50}) in Hb(II/III H_2O-OH) and Hb(II/III CN) (Fig. 3).

Discussion

Homotropic interactions between ferrous chains during oxygenation of Hb A are due to the release of constraints which are present in quaternary T conformation. The relaxed R quaternary conformation of Hb A is characterized by lack of constraints of quaternary structure on the tertiary conformation of chains [4].

The interactions between ferric and ferrous chains in Hb(II/III) were thought to resemble homotropic interactions in Hb A [5]. The asymmetry of oxygen binding by Hb(II/III H_2O) shows that interaction between aquoferrichains and ferrochains is different from normal homotropic mechanisms in Hb A. These differences could be due to modification of either T or R state of Hb(II/III). The change of oxygen affinity of Hb A induced by pH changes (Bohr effect) is due to modification of R state [6]. A normal Bohr effect in Hb(II/III H_2O-OH) has been observed by Enoki et al. [7] in 5.6 mM (Fe) Hb(II/III) solutions using the manometric method. The findings presented here of normal Bohr effect in Hb(II/III H_2O-OH) and in Hb(II/III CN) suggest that the conformation R in these hemoglobins is not changed as compared to that in Hb A.

It may be concluded that interactions between ferric and ferrous chains in Hb(II/III) occur when ferrous hemes are deoxygenated and quaternary conformation resembles T state of Hb A. In this state some differences in quaternary conformation are also conceivable between Hb(II/III H_2O) and Hb(II/III CN). The interactions producing assymmetry of oxygen binding curves, which are seen only in Hb(II/III) with ferric hemes in high spin form, probably are linked to the change of tertiary conformation of ferric chains during oxygenation of ferrous chains, which can be detected by accompanying decrease of spin of ferrihemes [8, 9].

References

[1] Pawlak, A. L.: Abstr. Commun. 9 th FEBS Meet., Budapest, 1974, p. 351
[2] Pawlak, A. L.: Abstr. Commun. 10th FEBS Meet., Paris, 1975, abstr. 642
[3] Rossi-Fanelli, A., and E. Antonini: Archs Biochem. Biophys. **77**, 478 (1958)
[4] Perutz, M. F.: Nature, Lond. **228**, 726 (1970)
[5] Benesch, R., R. E. Benesch, and I. Tyuma: Proc. natn. Acad. Sci. U.S.A. **56**, 1268 (1966)
[6] McDonald, M. J., and R. W. Noble: J. biol. Chem. **247**, 4282 (1972)
[7] Enoki, Y., H. Tokui, and I. Tyuma: Respiration Physiol. **7**, 300 (1969)
[8] Perutz, M. F.: Nature, Lond. **237**, 495 (1972)
[9] Pawlak, A. L.: unpublished data

Address: Andrezj L. Pawlak, Institute of Human Genetics, Polish Academy of Sciences, ul. Swiecickiego 6, 60–781 Poznan, Poland

Acta biol. med. germ., Band. 36, Seite 625—629 (1977)

Institut für Physiologische und Biologische Chemie des Bereichs Medizin (Charité) der Humboldt-Universität Berlin, 104 Berlin, DDR

Glykolyse und ATP-Verbrauch in membranfreien Hämolysaten

D. Maretzki, M. Brenneis, Zs. Schwarz, I. Lange und S. M. Rapoport

Zusammenfassung

Es wird die Präparation eines membranfreien Hämolysats mit sehr kleiner ATPase-Aktivität beschrieben. Solche annähernd ATPase-freien Hämolysate haben eine Glykolyse entsprechend der Harden-Young-Gleichung: Glukose + anorganisches Phosphat → Laktat + C_3-Phosphat (Fruktose-1,6-bis-phosphat und Triosephosphat).
Der 2,3-Glyzerat-bis-phosphat-Weg hat in den membranfreien Hämolysaten wahrscheinlich eine sehr niedrige Aktivität. Die Glykolysegeschwindigkeit entspricht bei Berücksichtigung der Hämolyseverdünnung (1:4) den Verhältnissen in intakten Zellen *p*H 7,2. Durch Erhöhung der Mg-ATP-Konzentration (1,5 bzw. 2,5 mM) treten Übergangszustände bei der Laktat- und Fruktose-1,6-bis-phosphat/Triosephosphat-Bildung auf, die durch die ATP-Hemmung der Phosphofruktokinase verursacht sind.
Es wird geschlußfolgert, daß die basale ATP-Produktion nicht von der ATPase-Aktivität beeinflußt wird. Bei glukosefreien Inkubationen wurde eine niedrige ATPase-Aktivität im Hämolysat nachgewiesen (0,1—0,2 mM ATP-Abbau/l Hämolysat/l bei 37 °C).
Das membranfreie Hämolysatsystem ist geeignet, den Einfluß von ATPasen auf die Glykolysegeschwindigkeit zu untersuchen.

Einleitung

In konzentrierten Zellsuspensionen von Erythrozyten gelingt durch Ultraschallhämolyse die Zerstörung der Membranen ohne eine stark erhöhte ATPase-Aktivität. Solche Hämolysate sind für Inkubationen bis zu 1 Std. geeignet, weil der Glukoseverbrauch und die Laktat-Bildung mit den Ergebnissen aus Inkubationen mit intakten Zellen übereinstimmen [1]. Diese Hämolysate haben den Nachteil, daß die Ultraschallbehandlung nicht ausreichend gut reproduzierbar ist und daher sehr unterschiedliche ATPase-Aktivitäten resultieren.
Ein Ziel der vorliegenden Arbeit war, möglichst ein ATPase-freies Hämolysat zu präparieren, um den Einfluß von ATPasen auf die Glykolyse untersuchen zu können. Anfängliche Untersuchungen induzierten zwei Fragen:

1. Wie läuft die Erythrozytenglykolyse in membranfreien Hämolysaten ab?
2. Existiert eine zytosolische ATPase?

Material und Methoden

Hoffmann et al. [2] und Schwoch und Passow [3] haben gezeigt daß Erythrozytenmembranen nach hypotoner Hämolyse, in Gegenwart von Mg^{++}-Ionen, durch Inkubation in isotoner Alkalisalzlösung „resealed ghosts" bilden. Diese Tatsache wurde zur Abtrennung der Membranen ausgenutzt.
Erythrozyten des Menschen wurden bei 5000 × *g* für 10 min aus frischem, heparinisiertem Blut abzentrifugiert und dreimal bei 37 °C gewaschen. Die Waschlösung enthielt 125 mM NaCl, 25 mM $NaHCO_3$, 1 mM KH_2PO_4, 300 µM Pyruvat und 5 mM Glukose. Der *p*H-Wert wurde auf 7,4 eingestellt. Die gewaschenen Zellen wurden bei 0 °C hämolysiert. 1 Volumen-

teil Zellen (Hämatokrit ca. 90%) wurde zu 3 Volumenteilen Hämolysemedium gegeben, das 1 mM KH_2PO_4, 1 mM $MgCl_2$, 1 mM Glutathion, 150 µM NAD und 2,5 mM Glukose enthielt (*p*H 7,2). Nach 5 min Hämolyse wurde durch Zugabe von festem KCl (130 mM) und NaCl (20 mM) Isotonie in der Lösung eingestellt, die zur Bildung von „resealed ghosts" führt. Diese rekonstruierten Membranen wurden durch Zentrifugation bei 100000 × *g* und 4 °C innerhalb von 30 min abgetrennt. 1/3 bis 1/4 des Überstandes wurde gesammelt. Vor der Inkubation bei 37 °C wurden nochmals 150 µM NAD zum Hämolysat zugegeben und entsprechende Konzentrationen von Mg-ATP eingestellt.
Für Inkubationen mit glukosefreien Hämolysaten wurden die Wasch- und Hämolyselösungen ohne Glukose verwendet. Die Hämolysate waren gegenüber intakten Zellen im Volumen 1:4 verdünnt und hatten einen Hämoglobingehalt von 6—7 g%. Die Angaben zu den Metabolitkonzentrationen wurden daher vereinfacht auf l Hämolysat bezogen. Die Hämolysate wurden bis zu 4 Std. bei 37 °C inkubiert und der *p*H-Wert wurde bei 7,25 ± 0,1 durch Zugabe von 0,1 N NaOH aufrechterhalten. Zur Metabolitanalyse wurden 1 ml Proben für die Perchlorsäurefällung (5%) entnommen. Die Neutralisation der Proben erfolgte mit 1 M Triäthanolamin/1,3 M K_2CO_3.
Laktat, Adeninnukleotide, Triosephosphate und Fruktose-1,6-bis-phosphat wurden durch enzymatische Analyse bestimmt [4]. Glukose wurde mit dem Glukoseoxydase/Peroxydase-Testansatz vom Arzneimittelwerk Dresden nach $ZnSO_4$/NaOH Enteiweißung bestimmt.

Ergebnisse

Inkubation ohne Glukose

Nach dreimaligem Waschen der Zellen ohne Glukosezusatz wurde bei den nachfolgenden Hämolysatinkubationen keine Laktatbildung gemessen. Um den ATP-Verbrauch in den membranfreien Hämolysaten zu bestimmen (ATP-Konzentration 0,45 ± 0,1 mM), wurden glukosefreie Inkubationsansätze mit 1 mM und 2 mM Mg-ATP untersucht (Endkonzentration 1,5 bzw. 2,5 mM). Abb. 1 zeigt, daß die ATP-Konzentration bei etwa 0,5 mM ATP konstant bleibt. Bei erhöhter ATP-Konzentration beginnt nach etwa 2 Std. Inkubation ein langsamer ATP-Abbau. Aus Versuchen mit 7stündiger Inkubation (nicht dargestellt) wurde eine ATP-Abbaugeschwindigkeit von ca. 0,1 mM/l Hämolysat/Std. ermittelt, wobei die Summe der Adeninnukleotide um den gleichen Betrag abfällt. Dieser ATP-Abbau tritt erst nach 2 Std. Inkubation auf. Die stabile Anfangsphase der ATP-Konzentration ist reproduzierbar und läßt auf das Vorkommen einer reaktivierten ATPase

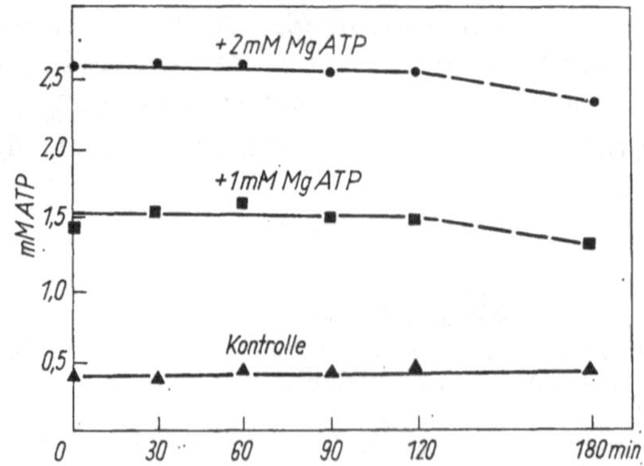

Abb. 1. ATP-Konzentrationen im membranfreien Hämolysat bei Inkubationen ohne Glukose

im verwendeten Hämolysatsystem schließen. Kontrollversuche mit einem gekochten Hämolysat, dem nachfolgend 2 mM Mg–ATP zugesetzt wurde, ergaben, daß der beobachtete ATP-Abbau nicht auf eine einfache Hydrolyse des ATP unter Inkubationsbedingungen zurückzuführen ist.

Inkubationen mit Glukose

Die Abb. 2—5 zeigen die Ergebnisse aus drei Inkubationen von membranfreien Hämolysaten in Gegenwart von Glukose. Bei *p*H 7,2 und einer stationären Mg–ATP-Konzentration von 0,45 mM betrug der Glukoseverbrauch 0,2—0,3 mM/l Hämolysat/Std. bei 37 °C (Abb. 2). Bei Erhöhung des Mg-ATP auf 1,5 mM wurden 0,3—0,4 mM Glukose/l Hämolysat/Std. verbraucht. Diese Werte entsprechen, bei Berücksichtigung der 1 : 4 Verdünnung der Hämolysate gegenüber dem Zytosol der Erythrozyten, der Glykolysegeschwindigkeit intakter Zellen bei *p*H 7,2 (1—1,2 mM Glukose-Verbrauch/l Erythrozyten/Std. bei 37 °C). Die Laktatbildung ist in diesen Hämolysaten annähernd dem Glukoseverbrauch im Verhältnis 1:1 äquivalent (0,3 mM Laktatbildung/l Hämolysat/Std. bei 0,45 mM Mg-ATP bzw. 0,4 mM Laktat bei 1,5 mM Mg–ATP) (Abb. 3).

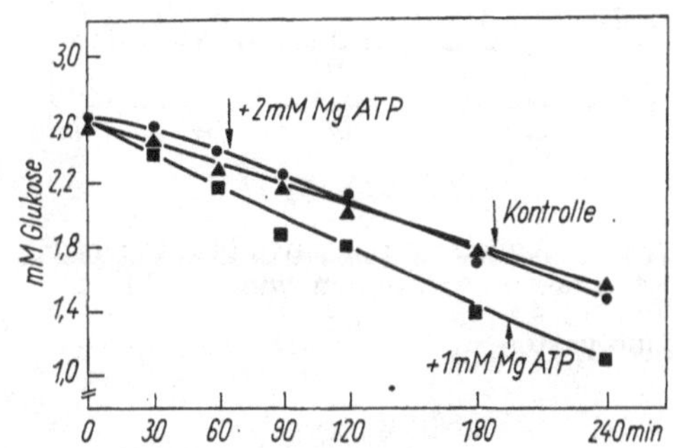

Abb. 2

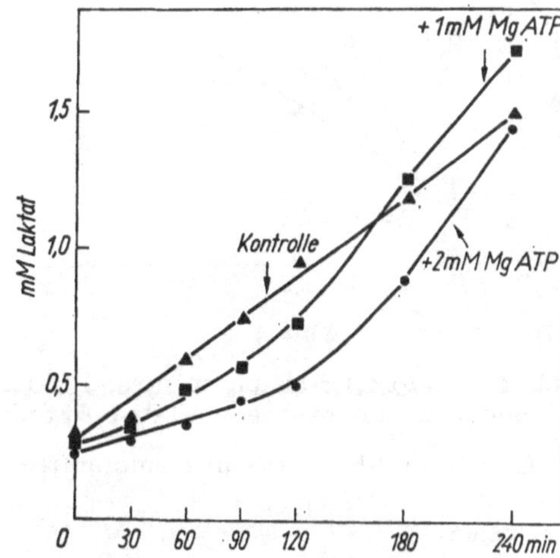

Abb. 3

Abb. 2. Glukoseverbrauch in membranfreien Hämolysaten bei verschiedenen ATP-Konzentrationen. (Kontrolle = 0,45 mM Mg-ATP.
+ 1 mM MgATP = 1,5 mM Mg-ATP (Endkonzentration);
+ 2 mM Mg ATP = 2,5 mM Mg-ATP (Endkonzenration)

Abb. 3. Laktatbildung bei verschiedenen ATP-Konzentrationen (ATP-Konzentrationen wie in Abb. 2)

Bei 2,5 mM Mg–ATP trat im Vergleich zu 1,5 mM Mg–ATP eine größere Hemmung der Glykolysegeschwindigkeit in der Anfangsphase der Inkubation auf, nach deren Überwindung aber vergleichbare Änderungen im Laktat und C_3-Phosphat beobachtet wurden (Abb. 3 und 4). In Abb. 4 sind die Änderungen von Fruktose-1,6-bis-phosphat und Triosephosphat als Summe C_3-Phosphat aufgetragen. Aus der

Bilanzierung des Glukoseverbrauchs im Verhältnis zur Laktat- und C_3-Phosphat-Bildung ergibt sich, daß die Glykolyse in membranfreien und damit annähernd ATPase-freien Hämolysaten prinzipiell einer Harden-Young-Gleichung entspricht Glukose + anorg. Phosphat → Laktat + C_3-Phosphat (C_3-Phosphat = Fruktose-1,6-bis-phosphat + Triosephosphat).

Bei Erhöhung der Mg–ATP-Konzentration (1,5 bzw. 2,5 mM) traten in den ersten beiden Stunden der Inkubation Übergangszustände in der C_3-Phosphat- und Laktatbildung auf, die aber annähernd der Bilanzgleichung entsprechen.

Die ATP-Konzentration ist in den Hämolysatinkubationen mit Glukose im untersuchten Zeitraum (4 Std.) annähernd konstant (Abb. 5). Voraussetzung ist, daß die Phosphatkonzentration entsprechend der Harden-Young-Gleichung nicht begrenzend wird. Deshalb wurde nach 2—3 Std. Inkubation 1 mM Phosphat zugegeben.

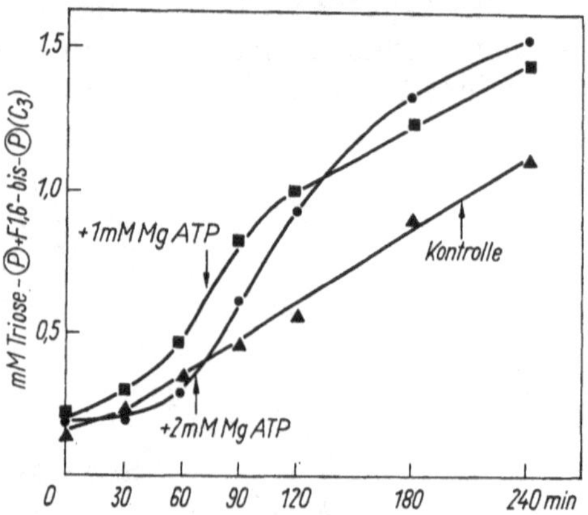

Abb. 4

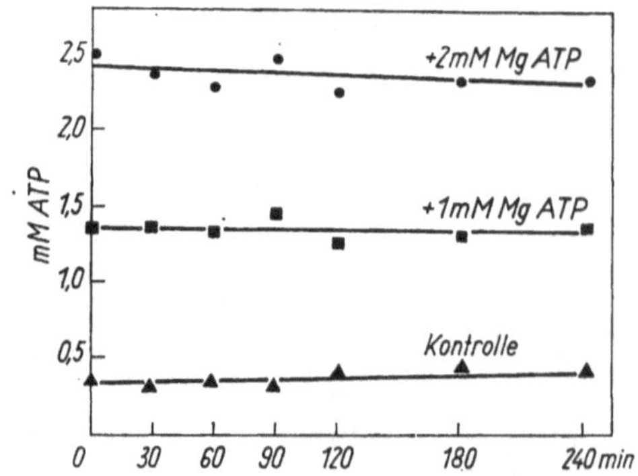

Abb. 5

Abb. 4. Fruktose-1,6-bis-phosphat-und Triosephosphat-Anhäufung in membranfreien Hämolysaten mit niedriger ATPase-Aktivität. (ATP-Konzentrationen wie in Abb. 2)

Abb. 5. ATP-Konstanz in membranfreien Hämolysaten in Gegenwart von Glukose

Diskussion

Durch die Abtrennung der Membranen nach dem beschriebenen Verfahren wurde ein Hämolysat mit einer sehr niedrigen, langsam reaktivierten ATPase-Aktivität gewonnen. Diese ATPase-Aktivität wurde in glukosefreien Inkubationsansätzen bei 2,5 und 1,5 mM Mg–ATP nachgewiesen, während bei 0,45 mM MgATP kein ATP-Abbau erfolgt. Daraus ist zu schließen, daß es sich um eine ATPase mit niedriger Mg–ATP-Affinität handelt. Über die Lokalisation dieser ATPase in Erythrozyten kann bisher keine Aussage getroffen werden. Es ist denkbar, daß eine Membran-ATPase während der Hämolysatpräparation von der Membran abgelöst wurde oder aber, daß im Zytosol eine ATPase mit geringer Aktivität vorkommt (0,1—0,2 mM Mg–ATP-Abbau/l Hämolysat/Std. bei 37 °C). In derartigen annähernd ATPase-freien Hämolysaten läuft die Glykolyse entsprechend der bereits von HARDEN [5] für Hefe-Zytolysate beschriebenen Gleichung ab. Pro Mol umgesetzter Glukose entstehen 1 Mol Laktat und äquivalente Mengen C_3-Phosphat (berechnet aus der Bildung von Fruktose-1,6-bis-phosphat und Triosephosphat). Die Änderung des C_3-Phosphats entspricht der ATP-Produktion der Glykolyse, wenn anorganisches Phosphat und NAD nicht begrenzende Faktoren sind. ADP wird durch die Hexokinase- und Phosphofruktokinase-Reaktionen regeneriert und ist sekundär über diese Enzyme mit der Glykolysegeschwindigkeit verknüpft. Bedingt durch die ATP-Hemmung der Phosphofruktokinase trat in der 1. Std. der Inkubation bei erhöhter Mg–ATP-Konzentration (2,5 bzw. 1,5 mM) sowohl eine Hemmung der Laktat- als auch der Fruktose-1,6-bis-phosphat/Triosephosphat-Bildung auf (Abb. 3 und 4).

Im weiteren Verlauf der Inkubationen wird diese Hemmung überwunden, so daß bei Berücksichtigung der Verdünnung der Hämolysate eine mit intakten Zellen vergleichbare Glykolysegeschwindigkeit resultiert.

Zunehmende ATPase-Aktivität sollte in diesen Hämolysaten zu einer Abnahme des C_3-Phosphat-Anteils mit äquivalenter Erhöhung der Laktatbildung führen.

Die getroffenen Aussagen sind für die Hämolysate nur gültig, wenn der 2,3-Glyzerat-bis-phosphat-Weg nicht aktiv bzw. sehr klein ist. Da in glukosefreien Hämolysaten nach Entfernung der Membranen keine Laktatbildung gemessen wurde, ist es wahrscheinlich, daß in den verwendeten Hämolysaten bei *p*H 7,2 kein 2,3-Glyzerat-bis-phosphat-Abbau stattfindet. Die Bilanzierung von Glukose, Laktat und C_3-Phosphat ist aus methodischen Gründen mit 10—20% Fehler behaftet, so daß 2,3-Glyzerat-bis-phosphat-Änderungen in dieser Höhe am Glukoseumsatz beteiligt sein könnten. Zur Rolle des 2,3-Glyzerat-bis-phosphat-Weges im Hämolysat sind weitere Untersuchungen notwendig.

Unsere Ergebnisse aus den bisherigen Untersuchungen an membranfreien Hämolysaten bestätigen, daß eine basale Energieproduktion in Erythrozyten durch das Hexokinase-Phosphofruktokinase-System und nicht durch ATPasen kontrolliert wird. Bei der Ausschaltung der Membran-ATPasen läuft im Hämolysat eine mit intakten Zellen vergleichbare ATP-Bildung ab, die zur äquivalenten Phosphorylierung von Glukose zu Fruktose-1,6-bis-phosphat und Triosephosphat und damit zu einer ADP-Regenerierung führt. Bei erhöhter Mg–ATP-Konzentration (2,5 mM) tritt nach Überwindung einer Hemmung der Glykolyse keine wesentlich höhere Glykolysegeschwindigkeit auf. Aber auch bei erniedrigter Mg–ATP-Konzentration (0,45 mM) besitzen die Hämolysate eine annähernd gleiche Glykolysegeschwindigkeit.

Das beschriebene Hämolysesystem sollte daher zur Untersuchung des Einflusses ATP-verbrauchender Prozesse auf die Glykolysegeschwindigkeit geeignet sein.

Literatur

[1] REIMANN, B., G. KÜTTNER, D. MARETZKI u. S. RAPOPORT: Acta biol. med. germ. **34**, 1777 (1975)
[2] HOFFMANN, J. F., D. C. TOSTESON u. R. WHITTAM: Nature, Lond. **185**, 186 (1960)
[3] SCHWOCH, G., u. H. PASSOW: Molec. Cell. Biochem. **2**, 197 (1973)
[4] BERGMEYER, U. (Hrsg.): Methoden der enzymatischen Analyse. Verlag Chemie, Weinheim 1972
[5] HARDEN, A.: Alcoholic Fermentation. Longmans Green and Co., London 1932

Acta biol. med. germ., Band 36, Seite 631—638 (1977)

Cátedra de Bioquímica, Facultad de Veterinaria, Universidad Complutense, Madrid-3, Spain

Comparative activation by AMP and cyclic-AMP of rat erythrocyte and reticulocyte glycolysis

J. LUQUE, P. RONCALÉS, C. TEJERO, and M. PINILLA

Summary

Incubation of rat erythrocytes and reticulocytes in Tris-Ringer's medium with 5 mM cyclic-AMP or AMP increased lactate formation and glucose utilization. The glycolysis-stimulating effect of cyclic-AMP is very similar to that of AMP and, in both cases, it seems to be higher in reticulocytes than in erythrocytes. 0.5 mM norepinephrine produced a much higher lactate formation in reticulocytes than in erythrocytes, suggesting a greater adenylate cyclase activity in younger cells.

300 μM cyclic-AMP and AMP reverse inhibition produced by ATP (up to 1.5 mM) on phosphofructokinase from rat reticulocyte haemolysates. Both nucleotides are positive allosteric effectors of the enzyme as shown by displacement of F6P-saturation curve to hyperbolic kinetics. Similar results were previously obtained with rat erythrocytes. This deinhibitory effect is suggested to be responsible of the above glycolysis-stimulating effect.

Introduction

Cyclic adenosine-3',5'-monophosphate (cAMP) has an evident role in cellular growth and development [1, 2]. The addition of exogenous cAMP or the application of agents which increase endogenous cAMP produce a stimulation of erythropoiesis [3—5]. Since erythrocyte ghosts of mammals (mouse and rat mainly), contrary to previous reports, contain a β-catecholamine stimulated adenylate cyclase [6, 7] and this activity is greater in younger cells [8], we became interested in the problem of the age-dependent changes which might be occurring during the process of erythroid differentiation and maturation, under the influence of (1) exogenous cAMP (5 mM) which is 10^4 times higher than the intracellular level to mimic the effect of a hormone [9]; (2) enough amount of a β-catecholamine hormone, 0.5 mM norepinephrine, to activate experimentally adenylate cyclase. Glycolysis is the biochemical pathway here studied since it has been shown that these agents stimulate lactate formation in human erythrocytes [10].

A description of the influence of any extracellular agent during the maturation process would require the physical separation of cells of different ages. Although we are presently using a counter-current distribution technique for this purpose [11, 12], we have first approached the problem by studying the glycolysis-stimulating effect produced by 5 mM cAMP or AMP and 0.5 mM norepinephrine on rat erythrocytes and reticulocytes.

In order to localize the site of action in the glycolytic pathway of the above glycolysis-stimulating effect, we have studied the ability of 300 μM cAMP and AMP to overcome the ATP-inhibition of rat reticulocyte phosphofructokinase. This "deinhibitory effect" was previously demonstrated in rat erythrocytes [13].

Material and methods

Chemicals

cAMP was from Sigma Chemical, USA. AMP, ATP, NAD^+, NADH, $NADP^+$, glucose-6-phosphate (G6P), fructose-6-phosphate (F6P) and coupling enzymes were from Boehringer, Mannheim. Dithiothreitol (DTT), Tris, EDTA, cresyl-brilliant blue and Giemsa's solution were from E. Merck Darmstadt. Phenylhydrazine was from Hopkins and Williams, England. All other chemicals were of analytical grade.

Incubations experiments

Anesthesized Wistar rats weighing 200–300 g were decapitated and whole blood collected in heparinized tubes. After centrifuging at $1100 \times g$ for 10 min to separate plasma, cells were washed three times with 0.12 M NaCl. Top layer was removed after each centrifugation. Packed cells were adjusted to a 16% hematocrit with Tris-Ringer medium containing (in mM): NaCl, 120; KCl, 6; $MgCl_2$, 2; CaCl, 2; Na_2HPO_2, 1; glucose, 10; Tris, 100 and enough HCl to obtain the final *p*H indicated in results. To obtain reticulocytes, rats were made anemic by i.p. injecting during six days with 0.5 ml of HCl-neutralized solution of phenylhydrazine (20 mg/kg body weight). The animals were killed on the third day after cessation of injections and blood was collected and cells were washed as described for erythrocytes. To calculate the number or reticulocytes, 5 µl of cell suspension was mixed (2 min) with an equal volume of cresyl-brilliant blue (10 mg/ml in 0.12 M NaCl), smeared on a glass slide, air dried, fixed with methanol (2 min) and, when dry, stained with Giemsa's solution (30 min). After examination of smears more than 90% of the total number of cells consisted of reticulocytes.

To 1 ml of erythrocyte or reticulocyte suspension enough amount of the different agents (dissolved in Tris-Ringer medium of appropriate *p*H) were added to obtain, in a final volume of 2 ml, the adequate concentration. The *p*H was adjusted and the suspension was incubated at 37 °C in an Unitronic-320 shaker bath (Selecta, Spain) cycled at 80/min. A 20 min period was allowed to equilibrate, during which *p*H values were corrected with 0.1 N NaOH. Perchloric acid extracts were prepared at 20 min and at 80 min of incubation by taking 0.5 ml aliquots into 1 ml of cold $HClO_4$. During the incubation period hemoglobin determinations were made by the cyanmethaemoglobin method [14].

Determinations of glucose were carried out by an enzymatic procedure [15]. The assay mixture contained per ml: 100 mM Tris-HCl, *p*H 7.6, 4 mM $MgCl_2$, 0.6 mM $NADP^+$, 0.7 mM ATP, 25 µl/ml perchloric acid extract, 4 units/ml glucose-6-phosphate dehydrogenase and 2 units/ml hexokinase. Lactate was determined enzymatically [16]. The assay mixture contained, per ml: 200 mM $NaAsO_2$-HCl *p*H 9.6, 4 mM EDTA, 5 mM $HONH_3Cl$, 6 mM NAD^+, 100 µl/ml perchloric acid extract and 20 units/ml lactate dehydrogenase.

Kinetic experiments

For enzyme assays, the packed reticulocytes obtained after the third washing in 0.12 M NaCl-solution, were lysed and then, frozen and thawed twice, centrifuged at $30{,}000 \times g$ for 20 min to remove stroma, and dialyzed for 24 hrs against water. All these steps were carried out at 4 °C. A coupled optical test was used [17, 18]. The assay mixture contained (final volume 2 ml): 100 mM Tris-HCl *p*H 7.1, 0.3 mM EDTA, 3 mM $MgCl_2$, 90 mM KCl, 0 to 1.5 mM ATP, 0.15 mM NADH, 5 mM DTT, 20 to 50 µl of haemolysate, 0 to 5 mM F6P, 3×(F6P) mM G6P, 0.1 unit/ml fructose-diphosphate aldolase, 0.5 units/ml glycerol-3-phosphate dehydrogenase and 6 units/ml triosephosphate isomerase. The reaction was initiated by addition of ATP. The decrease in absorption at 340 nm was followed with a Shimadzu UV-200 double beam recording spectrophotometer. Cyclic-AMP and AMP were added to the assay mixture in the cuvette and the change in absorption of the following linear period was recorded.

Results and discussion

The incubation of rat erythrocytes and reticulocytes in Tris-Ringer medium of different *p*H have been made, and lactate determined, in order to establish in our conditions the pH-dependence of glycolysis in both kinds of cells. As shown in Table 1, an increase in pH up to 8.2 induces a parallel elevation in the amount of lactate formed. An optimal pH-value of around 8.0 was found for erythrocytes [19] and reticulocytes. It must be noted that the capacity for lactate formation is always higher in reticulocytes than in erythrocytes, probably as a consequence of the much greater activity of some of the enzymes of glycolysis at all pH studied [20].

Table 1
Effect of pH on lactate formation in rat erythrocytes and reticulocytes

pH	6.8	7.0	7.2	7.4	7.8	8.2
Erythrocytes	0.171 (2)	0.298 ± 0.047 (11)	0.577 ± 0.056 (9)	0.859 ± 0.040 (5)	1.156 (2)	0.956 (2)
Reticulocytes	0.977 (2)	1.236 ± 0.058 (6)	1.574 ± 0.064 (6)	1.891 (2)	2.752 (2)	

Mean ± S.E. are given as μmoles of lactate formed/μmoles of hemoglobin/hour. (n) indicates number of experiments.

The stimulation of glycolysis by the exogenous addition of 5 mM cAMP or AMP is shown in Table 2. cAMP increased the amount of lactate formed after 1 hr of incubation at the two pH-values studied (7.0 and 7.2). This increase (around 20—40%), which can be equally observed in both kinds of cells, was significantly greater in reticulocytes ($p < 0.01$) than in erythrocytes ($p < 0.1$ or $p < 0.05$). On the other hand, the % change in lactate production is not only significantly greater ($p < 0.01$ or $p < 0.001$) but also higher in the presence of AMP: The elevation seems to be more pronounced for reticulocytes (between 79% and 122%) than for erythrocytes (between 61% and 78%). It is interesting that our results about the cAMP effect on lactate output are rather similar to those obtained by Ford and Omachi [10] with human erythrocytes, while these authors contrarily found that AMP caused no significant change in lactate production.

Since rat erythrocytes contain both a soluble and a membrane-bound cAMP phosphodiesterase [7, 21], it would appear that a high degradation of cAMP added might have been occurred during incubation of cells with this agent. To check this point, additional experiments were made in the presence of 5 mM caffeine as phosphodiesterase inhibitor. As shown in Table 2, the highly significant lactate elevations produced when 5 mM cAMP and 5 mM caffeine were simultaneously present ($p < 0.001$) nearly reached the same percentage change (between 44% and 102%) as that obtained when only AMP was present, indicating that caffeine is a potent inhibitor of rat erythrocyte and reticulocyte phosphodiesterase. The lower values of lactate formation found in the presence of cAMP alone were then a consequence of the partial degradation of cAMP during the incubation period. cAMP effect is, therefore, not mediated through formation of AMP.

In general, it may be suggested from these experiments that the glycolysis-stimulating effect of cAMP is very similar to that of AMP and, in both cases, it seems to be higher in reticulocytes than in erythrocytes.

Table 2

Increase in lactate formation in the presence of cAMP, AMP, norepinephrine (NE) and cAMP plus caffeine

	Control	5 mM cAMP	5 mM AMP	0.5 mM NE	cAMP + 5 mM caffeine
Erythrocytes pH 7.0	0.298 ±0.047 (11)	0.39° (+31) 0.39° (+31) 0.40* (+64) 0.42* (+41)	0.48** (+61) 0.49** (+64) 0.53*** (+78) 0.52*** (+74)	0.40° (+35) 0.36 (+21)	0.53*** (+78) 0.53*** (+78)
pH 7.2	0.577 ±0.056 (9)	0.68° (+18) 0.69° (+20) 0.69° (+20) 0.77** (+34)		0.92*** (+60) 0.91*** (+58) 0.81** (+40)	10.5*** (+81) 1.06*** (+84) 1.08*** (+87)
Reticulocytes pH 7.0	1.236 ±0.058 (6)	1.53** (+24) 1.50** (+21) 1.63** (+32) 1.51** (+22)	2.74*** (+122) 2.59*** (+110)	6.10*** (+394) 5.47*** (+343)	2.22*** (+80) 1.78*** (+44) 2.17*** (+76)
pH 7.2	1.574 ±0.064 (6)	1.97** (+25) 1.98** (+26) 1.98** (+26) 1.95** (+24)	3.05*** (+94) 2.94*** (+87) 2.82*** (+79)	5.93*** (+277) 4.58*** (+191)	2.27*** (+44) 2.60*** (+65) 3.18*** (+102)

Results are given as μmoles of lactate formed/μmoles hemoglobin/hr. The figures in parentheses refer to % change from control. (n) indicates number of experiments. The paired t-test was used to evaluate statistical significance: ° $p < 0.1$; * $p < 0.05$; ** $p < 0.01$; *** $p < 0.001$.

The increase in glucose consumption was also checked. As can be seen in Table 3, both cAMP and AMP increased the amount of glucose utilized by erythrocytes or reticulocytes, in the absence of phosphodiesterase inhibitor. The elevation in glucose consumption is greater in reticulocytes (39—60%) than in erythrocytes (19—40%) while the increase in lactate production, also in the absence of caffeine, was found to be the same for both cells (Table 2). This discrepancy may be a consequence of the simultaneous exhibition by reticulocytes of anaerobic and aerobic route of glucose degradation. In any case, the higher glucose degradation shown by reticulocytes is in favour of the greater glycolysis-stimulating effect of cAMP and AMP on these cells.

The stimulation of lactate production exerted by 0.5 mM norepinephrine is shown in Table 2, in comparison with that obtained with cAMP or AMP. There is a significant elevation in most of the experiments made with erythrocytes (40—60%). This is due to the β-catecholamine stimulated adenylate cyclase present in rat erythrocytes membranes [6, 7]. Very interesting is the large increase in lactate formation observed when reticulocytes are incubated with 0.5 mM norepinephrine: they are highly significant ($p < 0.001$) and the change from control varied from 190% to 390%. This may be an indirect demonstration of a greater adenylate cyclase activity suggested by SHEPPARD and BURGHARDT [8] for younger red cells.

The second part of this work is an attempt to localize phosphofructokinase as the main site of action in the glycolytic pathway of the glycolysis-stimulating effect

Table 3
Glucose consumption in the presence of cAMP and AMP

	Control	5 mM cyclic AMP	5 mM AMP
Erythrocytes			
*p*H 7.0	0.271 ± 0.040	0.321 (+19)	0.377° (+39)
	(5)	0.328 (+21)	0.376° (+40)
*p*H 7.2	0.350 ± 0.019	0.444** (+27)	0.415* (+18)
	(5)	0.467** (+33)	0.472** (+35)
Reticulocytes			
*p*H 7.0	0.325 ± 0.025	0.455** (+40)	0.452** (+39)
	(5)	0.487** (+50)	0.520*** (+60)

Results are given in µmoles of glucose utilized/µmoles of hemoglobin/hr. The figures in parentheses refer to % change from control. (*n*) indicates number of experiments. The paired *t*-test was used to evaluate statistical significance: ° $p < 0.1$; * $p < 0.05$; ** $p < 0.01$; *** $p < 0.001$.

exerted a) by cAMP or AMP, which penetrated cell membranes after their exogenous addition, b) by norepinephrine, which stimulated adenylate cyclase. Activation of phosphofructokinase appears to be a reasonable explanation for this effect since the enzyme has been reported to be activated by cAMP and AMP in this and other systems [6, 10, 17, 19, 22, 23]. We have previously shown the deinhibitory effect of 300 µM cAMP or AMP on the ATP-inhibition of rat erythrocyte phosphofructokinase [13]. Therefore, in order to obtain comparative results we have studied here the influence of cAMP and AMP on the ATP-inhibition of phosphofructokinase from reticulocyte haemolysates. The concentrations of cAMP used were in magnitude one (3 µM) and three orders (300 µM) higher than the physiological level normally found inside the cells (in the general range of 0.1 µM [9]). The concentrations of AMP are nearly physiological (300 µM) or lower (3 µM).

That ATP is both a substrate and a powerful inhibitor of the enzyme is first shown in Fig. 1 by the ATP-saturation control curves obtained at physiological *p*H (7.1) and two relatively low concentrations of F6P. The enzyme activities gave a hyperbolic control curve for the catalytic part (ATP being used as substrate) followed by a deinhibition at all ATP levels beyond 0.25—0.50 mM (ATP acting here as allosteric inhibitor). As can be seen in the same figure, the higher concentration of both nucleotides (300 µM) are able to overcome, in a great extent and in a similar fashion, the ATP-inhibition of phosphofructokinase at both fixed levels of F6P studied (0.1 mM and 0.5 mM). The low concentration of cAMP or AMP (3 µM) did not significantly relieve the ATP-inhibition of the enzyme.

This deinhibitory effect of 300 µM cAMP or AMP is further supported by studying their influence on the F6P-saturation curves at physiological *p*H (7.1), where the regulatory properties of the enzyme can be demonstrated. The higher concentration (300 µM), but not the lower one (3 µM), shifted the sigmoidal control curve to an hyperbolic kinetics (Fig. 2) and lowered Hill values (*n*). It may be concluded from these findings that 300 µM cAMP and AMP are positive allosteric effectors which increase the affinity of phosphofructokinase for F6P. Similar properties

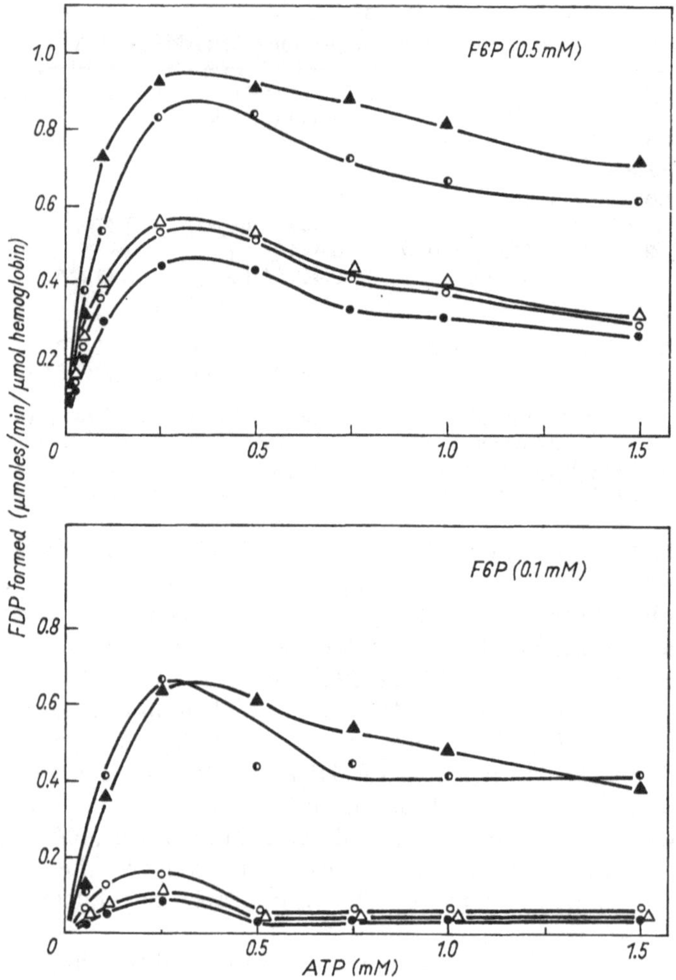

Fig. 1. Influence of cAMP and AMP on the ATP-saturation curves for reticulocyte phosphofructokinase at *p*H 7.1 and two different concentrations of F6P. ●, control; △, 3 μM AMP; ○, 3 μM cAMP; ▲, 300 μM AMP; ◐, 300 μM cAMP

were found in rat erythrocytes [13] as well as in the purified enzyme from human erythrocytes [22, 23].

Since the intracellular concentration of ATP is around 1.5 mM the in vitro deinhibition of phosphofructokinase by cAMP or AMP, at concentrations (300 μM) which are higher than physiological for cAMP while they are near physiological for AMP, suggest that AMP must be the only intracellular modulator of the enzyme in the normal situations of cells.

The deinhibitory effect of cAMP may be considered significant only in those in vitro (experimental) or in vivo (pathological) situations of augmented cAMP production: a) Assuming that the high levels here used (300 μM) to deinhibit the enzyme could be reached after incubation of both kinds of cells with 5 mM

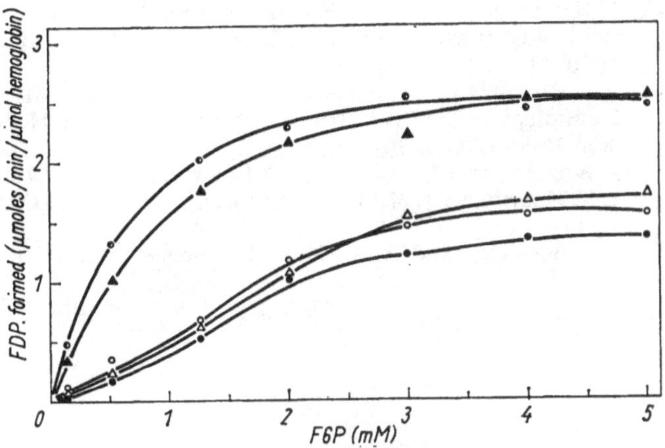

Fig. 2. Comparative deinhibitory effect of cAMP and AMP on the F6G-saturation curves for reticulocyte phosphofructokinase at *p*H 7.1 and inhibitory concentrations of ATP (1.5 mM). ●, control; △, 3 μM AMP; ○, 3 μM cAMP; ▲, 300 μM AMP, ◐, 300 μM cAMP

cAMP, the deinhibitory effect of 300 μM cAMP reported here in reticulocytes (Fig. 1 and 2) and previously in erythrocytes [13], may serve to explain the increased glycolysis rate (Tables 2 and 3). Since rat erythrocyte membranes contain a norepinephrine stimulated adenylate cyclase [6, 7] and this activity is probably greater in younger cells [8], the phosphofructokinase activation by 300 μM cAMP could equally serve to give an explanation to the high increase in lactate formation observed in reticulocytes incubated in the presence of norepinephrine (Table 2).

References

[1] Pastan, I. H., G. S. Johnson, and W. B. Andersen: A. Rev. Biochem. **44**, 491 (1975)
[2] Strada, S. J., and W. J. Pledger in: Cyclic Nucleotides in Disease. 3. B. Weiss (Ed.). University Park Press, Baltimore 1975
[3] Gidari, A. S., E. D. Zanjani, and A. S. Gordon: Life Sci. **10**, 895 (1971)
[4] Scholley, J. C., and L. J. Mahlmann: Proc. Soc. exp. Biol. Med. **150**, 215 (1975)
[5] Golwasser, E.: Fedn Proc. Fedn Am. Scs exp. Biol. **34**, 2285 (1975)
[6] Sheppard, H., and C. R. Burghardt: Biochem. Pharmac. **18**, 2576 (1969)
[7] Sheppard, H., and C. R. Brughardt: Molec. Pharmac. **6**, 425 (1970)
[8] Sheppard, H., and C. R. Brughardt: Biochem. Pharmac. **22**, 427 (1973)
[9] Robison, G. A., R. W. Butcher, and E. W. Sutherland: Cyclic AMP. Academic Press, New York 1971, p. 17
[10] Ford, D. L., and A. Omachi: Biochim. biophys. Acta **279**, 587 (1972)
[11] Albertsson, P. A.: Partition of Cell Particles and Macromolecules. Almqvist and Wiksell, Stockholm 1971
[12] Walter, H.: Meth. Cell. Biol. **9**, 25 (1975)
[13] Pinilla, M., and L. Luque: Submitted for publication
[14] Van Kampen, E. J., and W. G. Zijlstra: Clinica chim. Acta **6**, 538 (1961)
[15] Bergmeyer, H. U., E. Bernt, F. Schmidt, and H. Stork in: Methods of Enzymatic Analysis. **3**, 1196. 2. Ed. H. U. Bergmeyer (Ed.). Academic Press, New York 1974
[16] Gerchen, G.: Z. physiol. Chem. **320**, 180 (1960)
[17] Wenzel, K. W., G. Zimmermann, J. Gauer, W. Diezel, S. Liebe, and E. Hoffmann: FEBS Lett. **19**, 285 (1971)
[18] Layzer, R. S.: Meth. Enzymol. **42**, 110 (1975)

[19] JACOBASCH, G., S. MINAKAMI, and S. M. RAPOPORT in: Cellular and Molecular Biology of Erythrocytes. H. YOSHIKAWA and S. M. RAPOPORT (Eds.). University of Tokyo Press, Tokyo 1974, p. 55
[20] RAPOPORT S. M., S. ROSENTHAL, T. SCHEWE, M. SCHULTZE, and M. MÜLLER in: Cellular and Molecular Biology of Erythrocytes. H. YOSHIKAWA and S. M. RAPOPORT (Eds.). University of Tokyo Press, Tokyo 1974, p. 93
[21] SHEPPARD, H., G. WIGGAN, and W. H. TSIEN: Adv. cycl. Nucleot. Res. **1**, 103 (1972)
[22] STAAL, G. E, J., J. F. KOSTER, C. J. M. BANZINGER, and L. MILLIGEN-BOERSMA: Biochim. biophys. Acta **276**, 113 (1972)
[23] LEE, L. M. Y., R. M. KRUPKA, and R. A. COOK: Biochemistry **12**, 3503 (1973)

Acta biol. med. germ., Band 36, Seite 639—643 (1977)

Institut für Physiologische Chemie, Lehrstuhl II der Universität Erlangen-Nürnberg, BRD

Carbon balance studies with various substrates in human erythrocytes

K.-H. Quadflieg and K. Brand

Summary

Carbon metabolism in erythrocytes has been found to be in balance with a variety of substrates studied. The contribution of the 2,3-bisphosphoglycerate pathway to carbon metabolism depends on the rates of carbon utilization and increases with increasing metabolic rates. Net decrease of the 2,3-bisphosphoglycerate level acts as a NAD regenerating system thus facilitating uptake of polyalcohols such as xylitol.

Introduction

Erythrocytes are known to utilize a variety of sugars and polyalcohols [1, 2] for their energy supply which in part are of clinical use for parenteral nutrition. In this we report on carbon balance studies with various substrates which provide information about complete intracellular metabolism. These investigations lead us to the problem of the flow rates via the 2,3-bisphosphoglycerate bypass, which is still under discussion [3].

Material and methods

2,5 ml of packed erythrocytes were incubated in parallel experiments in Krebs-Ringer phosphate buffer, free of calcium ions at *p*H 7.4 or 7.6 for 120 min at 37 °C with the following substrates: D-ribose, xylitol, D-xylulose, D-fructose, D-glucose and mixtures of these compounds. Uptake of these compounds as well as the end products and the intermediate levels of the glycolytic, the 2,3-bisphosphoglycerate and the pentose phosphate pathway have been determined enzymatically or by isotopic techniques.

Results and discussion

A balance of carbon uptake from the substrates and the formation of intermediates and products was constructed. As can be seen from Table 1 the lowest rate of uptake was obtained with D-ribose as substrate followed by xylitol, D-fructose, D-xylulose and D-glucose. Most of the carbon atoms taken up were found in the products lactate, pyruvate and CO_2. The accumulation of pyruvate is negatively correlated to the amount of carbon atoms taken up. In case of xylitol as substrate, however, very little pyruvate was formed although the rate of uptake was very low. This is due to the fact that NADH formed in the D-xylulose reductase reaction has been reoxidized by converting pyruvate to lactate.
D-xylulose accumulated when xylitol was applied while with glucose as substrate at *p*H 7.6 glyceraldehyde-3-phosphate, dihydroxyaceton-phosphate and fructose 1,6-bisphosphate accumulated. The uptake of carbon atoms and the formation of products, however, was balanced only when the change in the level of 2,3-bisphosphoglycerate was taken into account.

Table 1

Balance of carbon metabolism from various substrates in human erythrocytes.
Mean values of 1—4 experiments with duplicate samples each are expressed as ng atom C/ml packed erythrocytes per 120 min of incubation at *p*H 7.4 or 7.6.

Substrates	mM	*p*H	No. of exp.	Utilization of C-atoms from:		$\sum C$	Recovery of C-atoms as:				$\sum C$	% recovery
				substrate	2,3-P_2-glycerate		lactate	pyruvate	$^{14}CO_2$	others		
—	—	7.4	3	—	3132	3132	494	2945	—	—	3439	110
	—	7.6	3	—	3072	3072	483	2630	—	—	3113	101
D-Ribose	2	7.4	2	2434	2652	5086	2592	2599	28	—	5219	103
Xylitol	4	7.4	3	5438	2424	7862	6732	119	35	994*	7880	100
D-Xylulose	4	7.4	1	6935	1539	8474	7458	1776	N.D.	N.D.	9234	109
D-Fructose	2	7.4	2	6497	2100	8597	6933	1534	88	—	8555	100
	1.6	7.6	1	5979	1767	7746	5973	1584	57	—	7614	98
D-Glucose	2	7.4	3	9431	1119	10550	9064	932	105	—	10101	96
	1.6	7.6	1	26232	—	26232	20616	—	N.D.	4860**	25476	97

* Formation of D-xylulose; ** formation of dihydroxyaceton-phosphate and D-fructose-1,6-bisphosphate

As shown in Fig. 1 at low rates of carbon uptake from various substrates the 2,3-bisphosphoglycerate level decreased during the incubation period. At rates of substrate uptake within the range of 2—13 μg atom C/ml packed cells/120 min the change of the 2,3-bisphosphoglycerate level was positively correlated with the rates of carbon utilization. At higher metabolic rates steady state levels of 2,3-bisphosphoglycerate were obtained. Because of the constant level of this compound accumulation of NADH is avoided. Only in the presence of electron acceptors like pyruvate or phenazine methosulphate the 2,3-bisphosphoglycerate levels were found to be increased without decreasing the rate of carbon utilization. Because of the observed interrelationship between the metabolic rates and the change in the 2,3-bisphosphoglycerate content under various metabolic conditions it was possible to calculate the flow rates via the 2,3-bisphosphoglycerate bypass and the glycolytic pathway. The procedure for calculation and the conclusions drawn from these results are discussed in detail in the preceding paper (K. Brand and K.-H. Quadflieg).

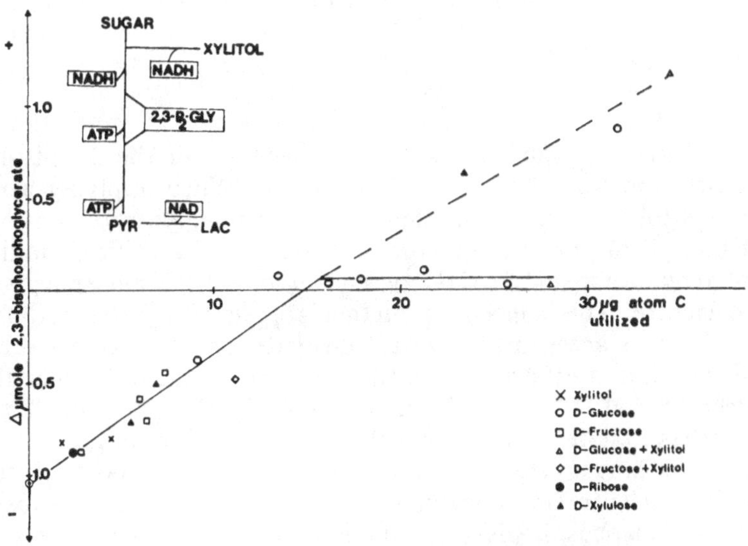

Fig. 1. Effect of carbon utilization from various substrates on the level of 2,3-bisphosphoglycerate in human erythrocytes.

2.5 ml packed cells were incubated with various substrates in different concentrations at *p*H 7.4 or 7.6 for 120 min at 37 °C. The points of the dashed line represent the results obtained in the presence of pyruvate or phenazine methosulphate at *p*H 7.6. Values are given as μg atom C or μmole/ml packed cells/120 min. For the rates of carbon utilization only those carbon atoms are considered which have been utilized for ATP and 2,3-bisphosphoglycerate formation.

Xylitol utilization in erythrocytes strongly depends on the ability of the cells to remove „extra NADH" formed during the D-xylulose reductase reaction. Our results indicate the xylitol metabolism in red blood cells to be coupled with the decrease of 2,3-bisphosphoglycerate which in turn is converted to lactate thus regenerating NAD.

Table 2

Dependence of xylitol utilization on NADH reoxidation.
Values are expressed as μmole/ml packed cells/120 min of incubation at *p*H 7.6

Substrate	mM	Δ Xylitol	Δ D-Xylulose	Δ[2,3-P_2-glycerate]	Δ[F-1,6-bis P + DAP + GAP]
Xylitol	1.6	−828	+466	−819	–
Xylitol PMS	1.6 0.02	−679	+466	−722	–
D-Xylulose	0.4	+78	−1229	−709	–
Glucose + Xylitol	1.6 1.6	−7	–	+153	+1212
Glucose + Xylitol + PMS	1.6 1.6 0.02	−1067	+311	+1194	–
Glucose + Xylitol + Pyruvate	1.6 0.4 0.4	−266	+19	+818	–
Glucose + D-Xylulose	1.6 0.4	+536	−1318	+521	+443

As shwon in Table 2 uptake of xylitol and decrease of the 2,3-bisphosphoglycerate concentration was almost stoichiometrical. When applying a mixture of glucose and xylitol no decrease in the 2,3-bisphosphoglycerate content was observed and no xylitol but only glucose was found to be utilized, indicating the necessity of reoxidation of NADH by converting 2,3-bisphosphoglycerate via pyruvate to lactate. This concept is further supported by the finding that the addition of electron acceptors such as pyruvate or phenazine methosulphate led to the utilization of xylitol also in the presence of glucose. Since the 2,3-bisphosphoglycerate concentration decreased also in the presence of xylitol and phenazine methosulphate it can be concluded that net decomposition of 2,3-bisphosphoglycerate during the incubation period primarily occured in order to keep the ATP concentration constant. When small amounts of substrate were metabolized, 2,3-bisphosphoglycerate obvioulsy acts as an "energy source", yielding ATP in the pyruvate kinase step. Breakdown of 2,3-bisphosphoglycerate to lactate, however, concomitantly provides the reoxidation of the "extra NADH" thus facilitating uptake and metabolism of polyalcohols like xylitol in erythrocytes.

The rate limiting step for xylitol metabolism in red blood cells was shown to be the D-xylulose kinase reaction. This is proven by the accumulation of D-xylulose, observed with xylitol as a substrate even when regeneration of NAD was sufficient either by 2,3-bisphosphoglycerate breakdown or by the addition of electron acceptors (Table 2).

From the balance studies it can be concluded that erythrocytes are able of metabolizing a variety of substrates as a means of energy but with significantly different rates which strongly affect the 2,3-bisphosphoglycerate level as well as the flow rates via the glycolytic pathway and the 2,3-bisphosphoglycerate bypass.

References

[1] BÄSSLER, K. H., and W. V. REIMOLD: Klin. Wschr. **43**, 169 (1965)

[2] ASAKURA, T., K. ADACHI, S. MINAKAMI, and H. YOSHIKAWA: J. Biochem., Tokyo **62**, 184 (1967)

[3] JACOBASCH, G., S. MINAKAMI, and S. M. RAPOPORT in: Cellular and Molecular Biology of Erythrocytes. H. YOSHIKAWA and S. M. RAPOPORT (eds.). Urban and Schwarzenberg, München, Berlin, Wien, 1974, p. 55

Acta biol. med. germ., Band 36, Seite 645—649 (1977)

Department of Occupational Diseases, Institute of Internal Diseases, School of Medicine, Wrocław, Poland

The effect of CS_2 on red cells metabolism

W. Sidorowicz, W. Zatoński, R. Andrzejak, and R. Smolik

Summary

The erythrocyte metabolism was studied in 118 workers exposed for a long time to CS_2 at concentrations in excess of accepted standards, and in a control group of 44 healthy men. The erythrocyte count, hemoglobin concentration in 1 ml erythrocytes, reticulocyte count, MCHC, MCV, MCH, MCD, MCT and minimum, maximum and medium osmotic resistance were determined. The activity of certain erythrocyte anaerobic glycolytic cycle enzymes was studied including phosphohexoisomerase, phosphofructokinase, aldolase, pyruvate kinase, and those of pentose cycle: glucose-6-phosphate dehydrogenase. The contents of free adenine nucleotides (AMP, ADP, ATP) and 2,3-diphosphosphoglyceric acid (2,3-DPG) in acid-soluble fraction of erythrocytes were studied using ion exchange chromatography in Dowex columns (1 + 8, 200—400 mesh).

Statistically significant decrease in hemoglobin concentration in 1 ml erythrocytes, osmotic resistance and the activity change of the studied enzymes were found to occur. The AMP, ADP and 2,3-DPG levels were increased and that of ATP decreased.

These results corroborate the hemolytic effect of CS_2. The disturbance in activity of key enzymes of glycolytic cycle and ATP level decrease seem to be responsible for hemolysis increase. Considerable increase in 2,3-DPG level is indicative of profound hypoxia of the tissues and is liable to be a compensating mechanism. The observed dysfunction of anaerobic glycolysis reduces the erythrocyte function and can be significantly associated with clinically manifest neurogenic disturbances and premature vascular atheromatosis in the patients exposed to long-term CS_2 effect.

Introduction

The mechanism of action of carbon disulphide (CS_2) on the human organism is still under discussion. At present, opinions on the action of CS_2 deal mainly with enzymatic disturbances, lipids regulation, metabolism of catecholamines and vitamin B_6 [1—4]. In the literature concerned with the toxicology of CS_2, papers on the influence of CS_2 on red blood cell are of frequent occurrence [5—10].

Binet and Bourlier [5] demonstrated that inhalation of CS_2 caused anemia in dogs. Vidacović et al. [10] have demonstrated a decrease in hemoglobin and erythrocyte level. Pilarska et al. [8] observed changes in peripheral blood, namely the decrease in hemoglobin and hematocrit level and red cell count, in rats after peritoneal administration of CS_2. No change in red cell count has been observed in this experimental anemia.

Investigations on workers exposed to CS_2 have shown cases of anemia, which normally were attributed to individual sensitiveness [9]. It was supposed that the damage of the blood system or a hemolysing mechanism was the cause of anemia [6—9]. The above mentioned reports and also our own clinical observations led to investigation about the state of red blood cell system and its metabolism [11, 12].

Materials and methods

The investigations covered 118 persons working in viscose fibre plant, exposed to chronic action of CS_2 on concentration ranges 30—50 mg/m^3. The mean age of patients was 35 years, and the period of service ranged from 1 to 20 years. The control group comprised 44 healthy men, working in the same plant but without occupational contact with CS_2. The patients were divided into three groups depending on age and period of service, according to generally accepted criteria. The reticulocyte count, erythrocyte count, hemoglobin concentration in 1 ml of erythrocyte mass used for further determinations, mean concentration of hemoglobin in erythrocyte (MCHC), mean volume of erythrocyte (MCV), mean weight of hemoglobin (MCH), mean diameter (MCD) and mean thickness (MCT) were determined by the generally accepted method in all patients. The osmotic resistance of red blood cells was also determined by the spectrophotometric method, defining its minimum, mean and maximum value [13].

The activity of key enzymes of anaerobic glycolytic cycle in erythrocytes (phosphohexoisomerase, phosphofructukinase, aldolase) and of pentose cycle (glucose-6-phosphate dehydrogenase) and the amount of lactate were measured by standard methods [14, 15]. Free adenine nucleotides in acid-soluble fraction of erythrocyte: AMP, ADP, ATP were determined by ion exchange chromatography on columns according to Hurlbert et al. [16] in the modification of Mills et al. [16, 17].

2,3-diphosphoglyceric acid (2,3-DPG) and the concentration of total phosphorus were determined according to Bartlett [18].

Statistically significant differences of arithmetic mean values were estimated by the Student's t-test.

Results

A statistically significant decrease in hemoglobin level in 1 ml of red blood cells was found in workers exposed to CS_2 in comparison to the control group (Fig. 1).

No significant deviations in other elements of red blood cell morphogram were found; therefore they are not included in this paper.

In all workers exposed to CS_2 a statistically significant decrease in osmotic resistance was found. Maximal osmotic resistance was lowered in workers with up to 5 and 10 years of work (Fig. 2).

Statistically significant elevation in mean values of AMP was found in all men exposed as compared with the control group. Also the ADP level was elevated in workers exposed to CS_2. A statistically significant decrease in ATP level was found in workers of 1 to 5 years of work period (Fig. 3, group 1), and a statistically significant decrease was found in the remaining groups working more than 5 or 10 years (Fig. 3, groups 2 and 3).

Moreover, a positive correlation has been found between the decrease of ATP level in workers with above 10 years of working period (group 3) and the decrease of osmotic resistance (minimal + 0.10, mean + 0.09, and maximal + 0.11).

A statistically significant increase in the 2,3-DPG level in all workers exposed to CS_2 was found as compared with the control group. The highest level of 2,3-DPG was observed in workers of above 10 years of working period (Fig. 4, group 3). A statistically significant increase in the activities of aldolase, phosphofructokinase and glucose-6-phosphate dehydrogenase was observed, but the activity of phosphohexoisomerase was normal or slightly increased (Fig. 5).

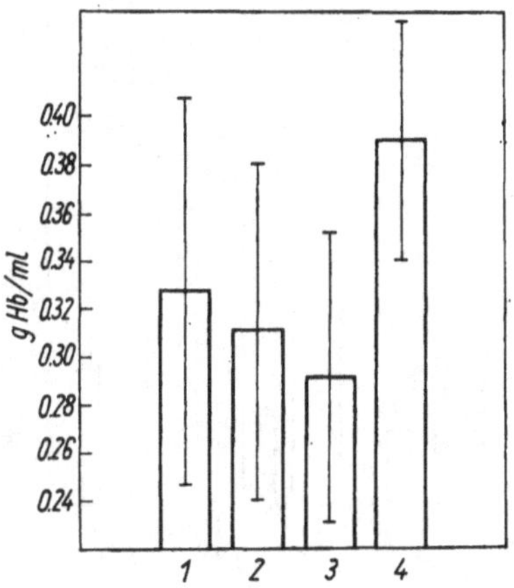

Fig. 1

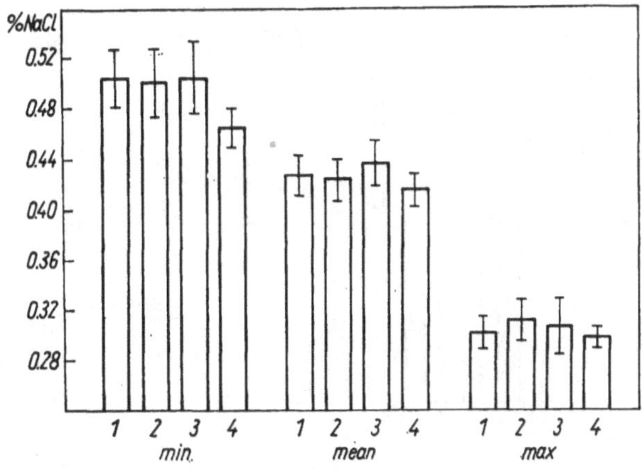

Fig. 2

Fig. 1. Mean values and standard deviations of hemoglobin level (as grams of hemoglobin/ml of red cells) in workers exposed to CS_2: 1 to 5 years of work 1, 5 to 10 years of work 2, above 10 years of work 3, and in the control group 4.

Fig. 2. Mean values and standard deviations of minimum, mean and maximum osmotic resistance in exposed workers 1, 2, 3, and in control group 4

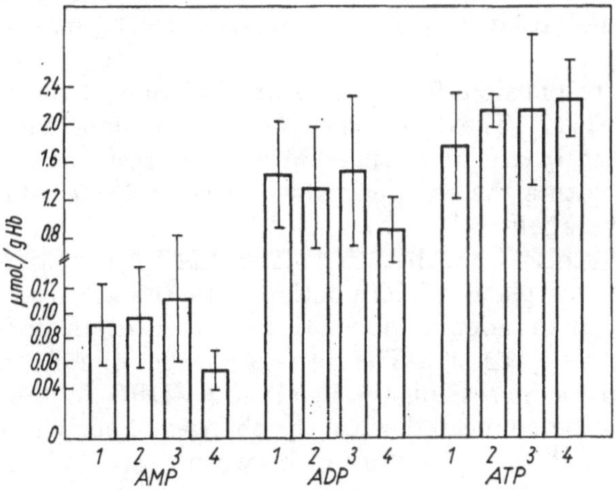

Fig. 3. Mean values and standard deviations of AMP, ADP, ATP levels in exposed workers 1, 2, 3 and in the control group 4

Discussion

In workers exposed to CS_2 the red cell in pulmonary circulation comes directly into contact with CS_2 and transports it to all tissues in the organism. On the grounds of the red blood cell function in the human organism, the disorders in its metabolism have an essential influence in the whole organism and may also occur in advance of some organ changes [19—21]. Essential decrease in osmotic resistance found in workers exposed to CS_2 may be attributable to impair-

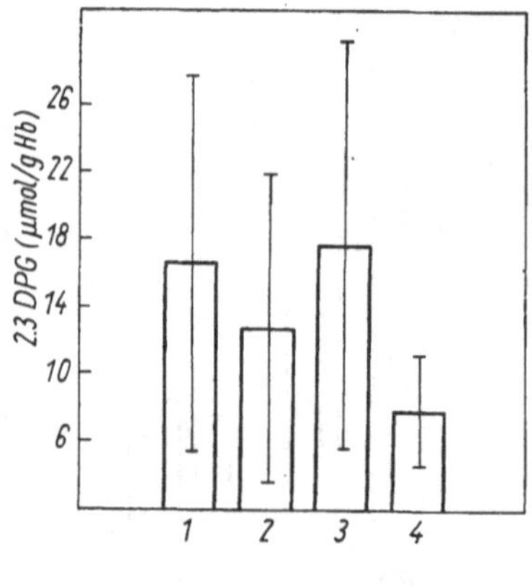

Fig. 4

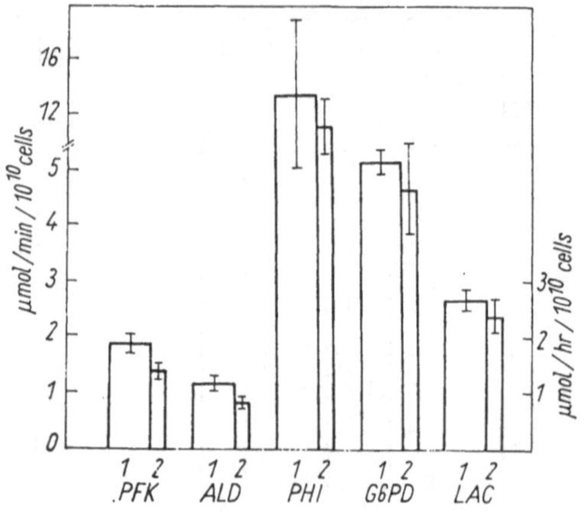

Fig. 5

Fig. 4. Level of 2,3-DPG and standard deviations in exposed workers (1, 2, 3) and in control group (4)

Fig. 5. Red blood enzymes activity (mean values and standard deviations) in exposed (1) and in control group (2).
Phosphohexoisomerase (PHI), phosphofructokinase (PFK), aldolase (ALD), glucose-6-phosphate dehydrogenase (= G-6-PD); amount of lactate (LAC).

ment of cell membrane, which may cause early disintegration of red cells. The decrease of hemoglobin level in workers exposed to CS_2 may be associated with this phenomenon.

The main factors responsible for appropriate function of red blood cells are the adenine nucleotides [22]. The decrease in ATP level enhances a tendency to hemolysis and consequently affects changes in osmotic resistance [23, 24]. The ratio of AMP/ATP indicating the energetic state of red cell simultaneously designate the course of its metabolism [25]. Increased levels of ATP act as a negative effector of anaerobic glycolysis; on the other hand AMP is acting as a positive effector. This action takes place in conjunction with the key enzymes of anaerobic glycolysis, for which the adenine nucleotide constitutes an important agent controling the metabolism [22, 25]. The decreased level of ATP may be due to increased utilisation for increasing the level of 2,3-DPG in the Rapoport-Luebering cycle, or may be connected with the changes of activity of some anaerobic glycolytic enzymes. Positive correlation between decrease of ATP and decrease of osmotic resistance points to a possible cause of increased red cell hemolysis dependent on the decrease of ATP level. This is confirmed by the hemolytic action of CS_2 on red blood cells. The most important role of 2,3-DPG in red cell is its contribution in the oxygen transportation to the tissues. The deficiency of 2,3-DPG affects the oxygen transmission of hemoglobin and also influences the course of glycolysis in erythrocytes [20].

A significant increase in 2,3-DPG level in workers exposed to CS_2 may point to increased oxygen demand due to toxic action or may be an indication of increased metabolism in tissues. Considering previous clinical observations the question arises whether the increase in 2,3-DPG level is sufficient to meet fully the oxygen demand of the tissues. Assuming the intoxication with CS_2 as a model for changes

in the central nervous system or premature atherosclerosis, the increase in 2,3-DPG level in chronic exposure should be considered as insufficient to meet fully the oxygen demand of the organs. The disturbances in red blood cell metabolism found in patients exposed to chronic actions of CS_2 give evidence for a profound multi-directional influence on the cell and allows it to be classed among compounds of hemolytic action. The decrease of ATP level in red blood cells should be considered as one of the reasons of increased hemolysis. Multidirectional changes found in the clinical picture of workers exposed to CS_2 point to a heterogenous mechanism of action, where the influence on the red blood cells and disturbance of its metabolism constitute the main factor of toxic action.
In order to explain the relation between the appearance of changes in the nervous system, premature arteriosclerosis and disturbances in red cell metabolism further observations might be necessary.

References

[1] MAGOS, L.: Ann. occup. Hyg. **15**, 303 (1972)
[2] MOESCHLIN, S.: Klinik und Therapie der Vergiftungen. Thieme, Stuttgart 1972, p. 196
[3] SMOLIK, R.: Zaburzenia biochemiczne w doswiadczalnym zatruciu dwusiarczkiem węgla z uwzglednieniem oddziaływania na hydrokortyzon. Prace naukowe Akademii Medycznej we Wrocławiu 1968
[4] TEISINGER, J.: Am. ind. Hyg. Ass. J. **55** (1974)
[5] BINET, L., and F. BOURLIER: Archs Mal. prof. Méd. trav. **6**, 12 (1944)
[6] KICZAK, J., T. ZAJACZKOWSKI, K. SZYSZKA, and A. ABUCEWICZ: Medycyna Pr. **23**, 265 (1975)
[7] KLECZEŃSKI, A., J. MATOLSPSZY, M. WRZYSZEZ, and M. SZOTT: Medycyna Pr. **17**, 99 (1966)
[8] PILARSKA, K., H. OWAJDA, and M. WOYKE: Acta Haemat. Pol. **4**, 34 (1973)
[9] WOYKE, M., and K. PILARSKA: Medycyna Pr. **22**, 22 (1971)
[10] VIDACOVIĆ, A., E. VISCONTI, V. VISNIC, S. DODIC, and J. REZMAN: Les études hematologiques de l'exposition par le sulfure de carbone. Toxicology of carbon disulphide. Proceedings of a Symposium, Prague 1966. H. BRIEGER and J. TEISINGER (eds.). Excerpta Medica Foundation, Amsterdam 1967
[11] ZATOŃSKI, W.: Pr. nauk. Akad. Med., Wrocławiu, **8**, 83 (1974)
[12] SIDOROWICZ, W.: Medycyny Pr. Kraków, 210 (1975)
[13] KRAWCZYŃSKI, J., and T. OSIŃSKI: Laboratoryjne metody diagnostyczne. PZWL, Warszawa 1967
[14] SZCZEKLIK, E.: Enzymologia kliniczna. PZWL, Warszawa 1974
[15] RICHTERICH, R.: Chemia kliniczna, PZWL, Warzawa 1971
[16] HURLBERT, R. B., H. SMITZ, A. F. BRUMM, and V. R. POTTER: J. biol. Chem. **23**, 209 (1954)
[17] MILLS, G. C., D. O. BURGER, M. SCHNEIDER, and W. C. LEVIN: J. Lab. clin. Med. **58**, 725 (1961)
[18] BARTLETT, G. R.: J. biol. Chem. **466**, 234 (1959)
[19] KWIATKOWSKA, J., W. ZATOŃSKI, and T. BARANOWSKI: Clinica chim. Acta. **40**, 101 (1972)
[20] PAWLAK, A. L.: Post. Biochemii **16**, 221 (1970)
[21] ZDEBSKA, E., S. MAJ, J. DASZYŃSKI and J. RADZISZEWSKI: Acta Haemat. Pol. **2**, 119 (1972)
[22] RAPOPORT, T. A., and R. HEINRICH: Steady state theory of glycolysis. 9th Meeting of the Federation of European Biochemical Societies, Budapest 1974, Abstracts, p. 335
[23] KWIATKOWSKA, J.: Choroby układu krwiotwórczego w Enzymologii Klinicznej pod. red. Szczeklika, PZWL, Warszawa, 527, 1974.
[24] KWIATKOWSKA, J., and E. BOHDANOWICZ: Acta Haemat. Pol. **3**, 3 (1972)
[25] GUMIŃSKA, M.: PWN, Warszawa, **47**, 24 (1971)

The author's address: Dr. med. WLADYSLAW SIDOROWICZ, Department of Occupational Diseases, Institute of Internal Diseases, School of Medicine, 50—367 Wrocław, ul. Pasteura 4, Poland

Acta biol. med. germ., Band 36, Seite 651—656 (1977)

II. Medizinische Klinik und Poliklinik rechts der Isar der Technischen Universität München, München, BRD

Der Einfluß von Thyroxin auf den 2,3-DPG-Gehalt der Erythrozyten in vivo und in vitro

U. SCHWEIGART, A. SCHÄTZL und P. BOTTERMANN

Zusammenfassung

Nach Absetzen einer Dauermedikation von Thyroxin sank 2,3-Diphosphoglycerat im Mittel um 0,4 μMol/ml, T_4 im Mittel um 7,6 μg/100 ml. Einmalige Gabe von 1 mg Thyroxin p.o. führte innerhalb von 24 Stunden zu einem Anstieg des 2,3-DPG von 0,2 μMol/ml; der *p*H-Wert in den Erythrozyten stieg im Mittel um 0,02.

Inkubation von Blut mit Thyroxinzusatz in einer Konzentration von 24 μg/100 ml führte zu keinem Anstieg an 2,3-DPG, *p*H-Wert und Phosphat, während im Kontrollblut ohne T_4 eine deutliche Azidose und ein Phosphatanstieg auftraten.

Die Laktatproduktion war unter T_4-Zusatz signifikant geringer, der Glukoseverbrauch hingegen signifikant höher.

Einleitung

BENESCH und BENESCH [1] sowie CHANUTIN und CURNISH [2] haben 1968 die Bedeutung des 2,3-DPG für die Sauerstoffaffinität des Hämoglobins geklärt. In der Folgezeit wurde bei einer Reihe von Zuständen, die mit einem absoluten oder relativen Sauerstoffmangel einhergehen, Erhöhungen des 2,3-DPGs in den Erythrozyten beschrieben. Bei Anämie [3], im Rahmen einer Höhenanpassung [4] und bei körperlichem Training [5] wurden 2,3-DPG-Anstiege gemessen. Zumindest für Höhenanpassung und Anämie ist der 2,3-DPG-Anstieg nicht sicher von einem *p*H-Effekt, hervorgerufen durch eine respiratorische Alkalose, zu trennen.

Ein gewisser Sauerstoffmangel ist auch bei Hyperthyreose anzunehmen, da Schilddrüsenhormone über die allgemeine Stoffwechselsteigerung auch zu einem erhöhten Sauerstoffbedarf führen. Während GAHLENBECK und BARTELS [6] eine Rechtsverschiebung der Sauerstoffbindungskurve bei Patienten mit Hyperthyreose fanden, konnte MONTI [7] keinen eindeutigen Zusammenhang zwischen dem Abfall der Schilddrüsenhormone unter Therapie und dem Verhalten des 2,3-DPGs beobachten. In vitro fanden SNYDER und REDDY [8] dagegen bereits 10 min nach Zugabe von Thyroxin und Trijodthyronin in Konzentrationen, die weit außerhalb des physiologischen Bereiches lagen, einen 2,3-DPG-Anstieg. Unsere Untersuchungen sollten in vivo den Zusammenhang zwischen Thyroxin, 2,3-DPG-Konzentrationen, Hämoglobinkonzentrationen und *p*H-Wert in Erythrozyten klären.

In einer zweiten Serie sollte geklärt werden, inwieweit diese Ergebnisse in vitro reproduzierbar sind.

Versuchspersonen und Methodik

In einer ersten Serie wurden 10 stoffwechselgesunde Versuchspersonen untersucht. Sie erhielten so lange Thyroxin, bis die TSH-Sekretion im TRH-Test nicht mehr zu stimulieren war; in der Regel dauerte das 4—6 Wochen. Die erforderliche Dosis betrug 150—200 µg/täglich. Nach Absetzen der Dauermedikation wurden in wöchentlichen Abständen bis zur Normalisierung der Thyroxin-Spiegel — d. h. ca. 3 Wochen — folgende Parameter bestimmt: 2,3-DPG, Hämoglobinkonzentration (Hb), Hämatokrit (Hk) und Thyroxin-Konzentration (T_4).
In einer zweiten Serie wurden ebenfalls stoffwechselgesunde Versuchspersonen ($n = 11$) untersucht. Sie erhielten einmal wöchentlich 1 mg T_4 peroral verabreicht. Unmittelbar vor, sowie 24 Std. nach der T_4-Gabe wurden 2,3-DPG-Spiegel im Blut, *p*H im Vollblut und in den Erythrozyten gemessen sowie T_4, Hb und Hk bestimmt.
In der dritten Serie wurde Blut von stoffwechselgesunden Probanden ($n = 6$) bei 37 °C mit 8% O_2, 6% CO_2 und 86% N_2 inkubiert. Jeweils der Hälfte des entnommenen Blutes wurde T_4 zugesetzt; die Konzentration betrug im Mittel 24 µg/100 ml. Das Blut ohne T_4-Zusatz diente als Kontrolle, um systematische Veränderungen, die nicht T_4-bedingt waren, auszuschließen. Die Inkubationszeit betrug 120 min. Nach jeweils 30 min wurden Proben zur Bestimmung von *p*H im Vollblut und in den Erythrozyten, 2,3-DPG, Hk, Hb, Phosphat, Chlorid, Laktat, Pyruvat und Glukose entnommen
2,3-DPG wurde nach der Methode von Rose und Liebowitz [9] mit der kommerziellen Ausrüstung der Fa. Sigma, *p*H in den Erythrozyten nach Kältehämolyse, im Vollblut direkt nach Blutentnahme im Astrup-Mikrogerät BMS 2 MK 2 gemessen. Laktat wurde nach der Methode von Hohorst [10], Pyruvat nach der Methode von Czok [11] jeweils mit einem Testbesteck der Firma Boehringer, Mannheim, bestimmt. Die Bestimmung des anorganischen Phosphats erfolgte nach Gomorri [12] (Test-Kit der Firma Merck). Die TSH-Konzentration wurde radioimmunologisch mit einer Doppelantikörpermethode gemessen, die T_4-Bestimmung als Proteinbindungsverfahren.
Die Chloridbestimmung wurde im Marius-Chlormeßgerät durch Titration mit Ag-Ionen durchgeführt. Die Glukosebestimmung erfolgte mit Hilfe der Glukosedehydrogenase-Methode der Firma Merck.
Als statistische Verfahren dienten dieVarianzanalyse und die einfache und multiple Regression

Ergebnisse

Tab. 1 zeigt das Verhalten der gemessenen Parameter. Es fand sich eine signifikante, negative Korrelation zwischen Hämoglobin und 2,3-DPG-Konzentration ($r = 0{,}488$, $p < 0{,}005$) und eine positive Korrelation zwischen T_4-und 2,3-DPG-Spiegeln ($r = 0{,}405$, $p < 0{,}005$). Berechnet man eine multiple Regression zwischen diesen 3 Variablen, so konnten die gleichen Beziehungen statistisch eindeutiger gesichert werden. Die Ergebnisse der 2. Serie sind in Abb. 1 dargestellt. Die T_4-Konzentration lag 24 Std. nach oraler Gabe von 1 mg T_4 im Mittel und bei 3,2 µg/100 ml höher; sie erreichte bis zur nächsten Blutabnahme eine Woche später wieder die jeweiligen Ausgangswerte. Hb und Hk änderten sich nicht signifikant.

Die in vitro-Versuche der dritten Serie ergaben, daß ohne T_4-Zusatz sowohl in den Erythrozyten als auch im Vollblut eine Azidose auftrat, die in den Erythrozyten jedoch deutlicher als im Vollblut ausgeprägt war (Abb. 2). Es kam zu einem Phosphatanstieg von 0,78 mg% im Verlauf von 120 min; der Glukoseverbrauch betrug 0,6 mMol/l/Std. (Abb. 3). Laktat stieg um 1,2 mMol/l/Std. an. Weder 2,3-DPG noch Pyruvat oder Chlorid in Erythrozyten und Serum änderten sich signifikant (Abb. 4). T_4-Zusatz verhinderte den *p*H-Abfall sowohl im Vollblut als auch in den Erythrozyten; der Phosphatanstieg blieb aus. Der Glukoseverbrauch war mit 0.8 mMol/l/Std. signifikant höher ($p < 0{,}001$), die Laktatproduktion mit 1,0 mMol/l/Std. signifikant geringer ($p < 0{,}001$) als ohne T_4.
Die 2,3-DPG-Konzentration änderte sich nicht signifikant (Abb. 4).

Tabelle 1

	Wochen nach Absetzen der T_4-Medikation		
	1	2	3
T_4 [µg%] $\bar{x}$ SEM	15,29 ±0.49	10,17 ±0,42	7,63 ±0,29
ΔTSH [µE/ml]	<1,50	<1,50	3,71 ±0,95
2,3-DPG [mMol/ml Blut]	2,37 ±0,19	2,16 ±0,10	1,97 ±0,05
Hb [g%]	13,79 ±0,79	13,67 ±0,77	13,27 ±0,75
HK [%]	42,89 ±1,26	41,74 ±1,12	41,54 ±1,09

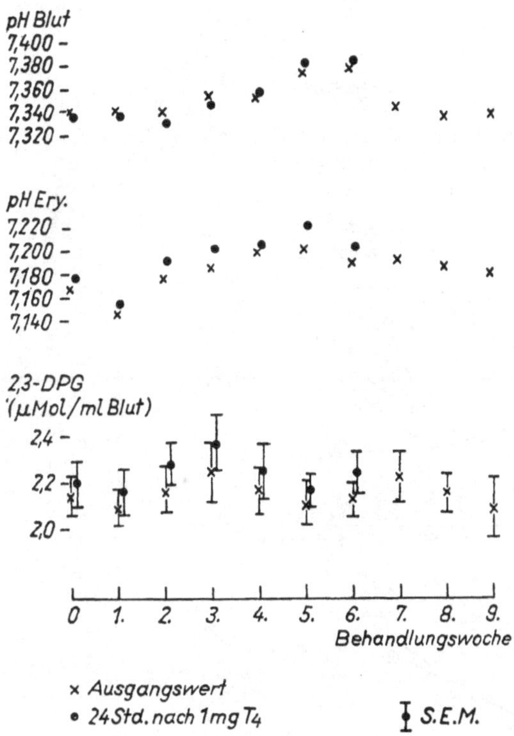

Abb. 1

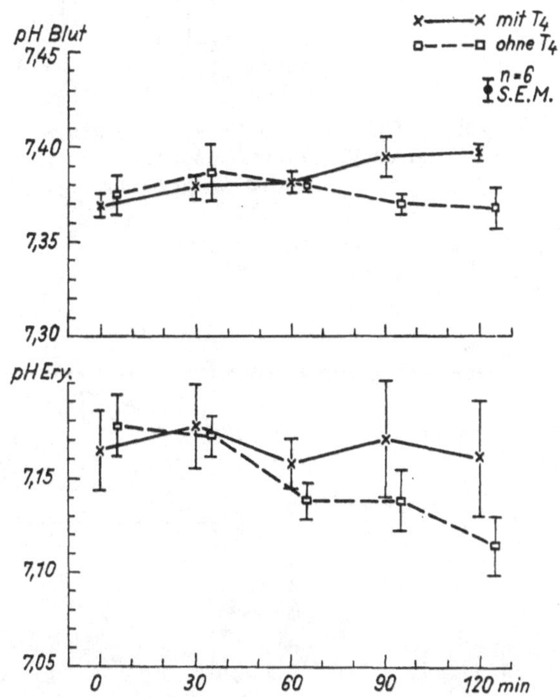

Abb. 2

Abb. 1. Beziehungen zwischen pH_{Ery} und 2,3-DPG-Konzentration. Es besteht eine signifikant positive Korrelation ($r = 0{,}740$, $p < 0{,}001$)

Abb. 2. pH-Werte in Kontrollblut und Blut mit T_4-Zusatz. Die Unterschiede sind ab 60. min signifikant ($p < 0{,}005$)

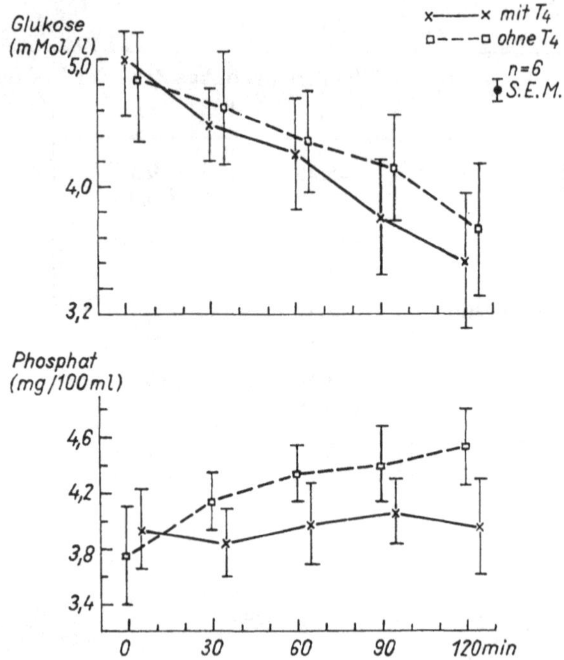

Abb. 3. Glukose- und Phosphatkonzentration in Kontrollblut und Blut mit T_4-Zusatz. Die Unterschiede in der Phosphatkonzentration sind ab 60. min signifikant ($p < 0{,}005$), die der Glukosekonzentration ab 90. min ($p < 0{,}005$)

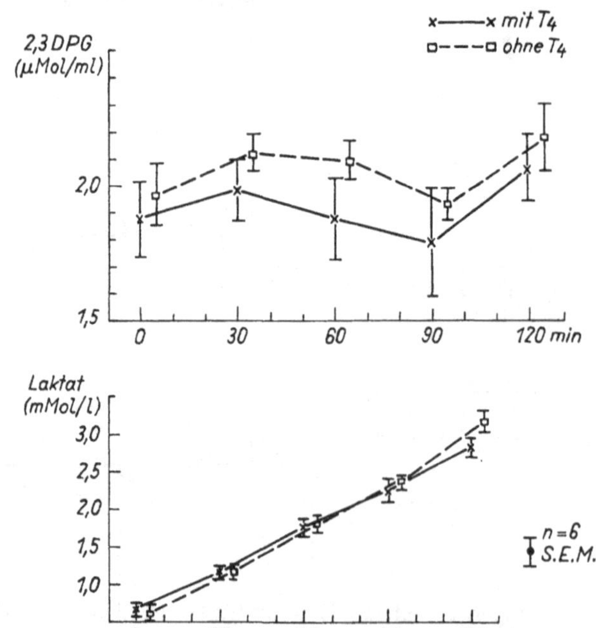

Abb. 4. 2,3-DPG- und Laktatkonzentration.
Die Laktatkonzentration ist 120 min nach Inkubation mit T_4 signifikant geringer ($p < 0{,}005$); die Unterschiede in der 2,3-DPG-Konzentration sind nicht signifikant

Diskussion

Die in vivo-Untersuchungen ließen die Frage offen, worauf der 2,3-DPG-Anstieg nach Thyroxingabe bzw. der Abfall nach Absetzen der Dauermedikation zurückzuführen ist.
Folgende Möglichkeiten standen zur Diskussion:

1. T_4 steigert direkt die Aktivität der 2,3-DPG-Mutase;
2. T_4 verschiebt das Donnan-Gleichgewicht an der Zellmembran;
3. T_4 bewirkt eine Alkalose, die sekundär zu einen 2,3-DPG-Anstieg führt.

Die in vitro-Untersuchungen konnten einen direkten Effekt auf die 2,3-DPG-Mutase ausschließen, da kein 2,3-DPG-Anstieg auftrat. Eine primäre Wirkung auf das Donnan-Gleichgewicht ist auf Grund des gleichbleibenden Quotienten Cl_i/Cl_a bei Inkubation mit T_4 unwahrscheinlich.
Demnach scheint die primäre Wirkung von Thyroxin darin zu bestehen, daß eine Alkalose auftritt, die erst in zweiter Linie zu einem Anstieg des 2,3-DPG-Konzentration in den Erythrozyten führt. Diese Alkalose wird in vitro durch die Laktatbildung verschleiert, läßt sich aber in Relation mit den Proben ohne T_4-Zusatz erkennen. Das Verhalten des Laktats unter Inkubation mit T_4 ist auffallend. Während in den Proben, die ohne Zusatz von Thyroxin inkubiert wurden, aus der verbrauchten Glukose (0,6 mMol/l/Std.) quantitativ Laktat gebildet wurde (1,2 Mol/l/Std.), war die Laktatproduktion in den Proben mit T_4-Zusatz signifikant geringer. Gleichzeitig blieb der Phosphatanstieg aus. Dieses Verhalten kann so gedeutet werden, daß Thyroxin zumindest in vitro einen Seitenweg der Glykolyse stimuliert, in den Phosphat eingeschleust wird. So haben Crevasse et al. [13] berichtet, daß Trijodthyronin den Pentosephosphatshunt beschleunigt. Unsere Untersuchungen könnten diese Ergebnisse auch für Thyroxin bestätigen.

Literatur

[1] Benesch, R., u. R. E. Benesch: Biochem. biophys. Res. Commun. **26**, 162 (1967)
[2] Chanutin, A., u. R. R. Curnish: Archs. Biochem. biophys., **121**, 96 (1967)
[3] Oski, F. A.: New Engl. J. Med., **280**, 1165 (1969)
[4] Lenfant, C.: J. clin. Invest, **47**, 2652 (1968)
[5] Böning, D.: Eur. J. appl. Physiol., **34**, 1 (1975)
[6] Gahlenbeck, H., and H. Bartels: Klin. Wschr. **46**, 547 (1968)
[7] Monti, M.: Acta med. scand. **196**, 263 (1974)
[8] Snyder, L. M., u. W. J. Reddy: J. clin. Invest. **49**, 1993 (1970)
[9] Rose, R. B., u. J. Liebowitz: Anal. Biochem. **35**, 177 (1970)
[10] Hohorst, H. J. in: Methoden der enzymatischen Analyse. **2**, 1425. 2. Aufl. H. U. Bergmeyer (Hrsg.). Verlag Chemie, Weinheim 1970
[11] Czok, R., u. W. Lamprecht in: Methoden der enzymatischen Analyse. **2**, 1407. 2. Aufl. H. U. Bergmeyer (Hrsg.). Verlag Chemie, Weinheim 1970
[12] Gomorri, G.: J. Lab. clin. Med. **27**, 955 (1942)
[13] Crevasse, L.: J. Lab. clin. Med. **65**, 539 (1965)

Summary

U. Schweigart, A. Schätzl and P. Bottermann: The influence of thyroxine on 2,3-DPG-content of erythrocytes in vivo and in vitro

After ending a continous treatment with thyroxine the average dropping of the 2,3 DPG level was 0.4 μmol/ml. T_4 decreased on the average by 7.6 μg/ml. One time application of 1 mg

thyroxine p.o. led within 24 hours to an increase of the 2,3 DPG level of $\bar{x} = 0.2$ µmol/ml, the *p*H in the erythrocytes increased by 0.02 on the average.
Blood incubation with thyroxine added in a concentration of $\bar{x} = 24$ µg/100 ml showed no increase of 2,3 DPG, *p*H and phosphate, while there was a significant acidosis and increase of phosphate in the control blood.
The lactate production was significantly lower and glucose consumption was significantly higher in the blood with thyroxine.

Acta biol. med. germ., Band 36, Seite 657—660 (1977)

Department of Physiology of man and animals, State University, Syktyvkar, USSR

Changes of erythrocytes due to hyperoxygenation

L. I. IRZHAK and V. V. GLADILOV

Summary

After exposure of animals to pure oxygen for an 1—3 hrs daily for 2—8 weeks the reticulocyte concentration increased 1.5—5 times in cats and decreased 2—4 times in rabbits of different age. The hemoglobin affinity to oxygen increased and then returned to normal level in adult rabbits, but only decreased in cats. Blood parameters in newborn rabbits whose mothers were breathing pure oxygen within 4 weeks of gestation up to the parturition were closer to those of adult animals as compared to control newborn rabbits. The hemoglobin affinity to oxygen in hyperoxic animals returned to normal levels after dialysis. This points to the importance of the salt environment within the erythrocytes for changes of the hemoglobin affinity to oxygen in hyperoxic conditions.

Introduction

In recent years much attention has been paid to the problem of the influence of high pO_2 on the functional systems of the organism. The respiratory function of the blood changes in dependence on the content of oxygen in the environment. Hyperoxia was shown to increase the affinity of hemoglobin to oxygen and decrease the content of 2,3-diphosphoglycerate (DPG) in erythrocytes of man, dog, and rat [1—6]. These species are characterised by a relatively high content of DPG in erythrocytes [6—8]. Naturally the question arises how the affinity of hemoglobin to oxygen changes in the species whose erythrocytes have not DPG at all or have it in rather small quantities. This is why the experiments were carried out with adult and newborn rabbits and adult cats. The former have as high as 9 mM DPG in 1 ml of erythrocytes, newborn 10 times less, and in cats it is near to zero [8—10].

Material and methods

Mongrel rabbits and cats of both sexes were held in an atmosphere of pure oxygen for 1—3 hrs daily within an 2—8 week period. The variants of experiment were as follows:

1. group: Adult rabbits, 1 hr exposition daily within 8 weeks;
2. group: Adult rabbits, 3 hrs, 4 weeks;
3. group: Newborn rabbits. Their mothers were held in an atmosphere of pure oxygen for 1 hr daily within a 4 week period before parturition. The blood of newborn animals was investigated;
4. group: Adult cats, 3 hrs, 2 weeks.

The concentration of erythrocytes, hemoglobin and reticulocytes was tested. For measuring of the affinity, $9 \cdot 10^{-5}$ M solutions of hemoglobin in Na-phosphate buffer, *p*H 7.2, were used [11]; dialysis against K-phosphate buffer according to [12].

Results and discussion

In animals of the first group the number of erythrocytes increased by 12.5% as compared to that in control animals, the hemoglobin concentration by 6.7%.

Having increased the duration of exposure up to 3 a day (second group) we obtained such changes at the end of the 4th week of the experiment (Table 1). The blood parameters of newborn rabbits (third group) differed essentially from those of control newborn: the number of erythrocytes was 23% higher, the hemoglobin concentration was 10.4% lower (Table 1). As to the cats (fourth group), their blood picture appears to have no such distinct changes as in rabbits because the blood parameters increased in some cats and decreased in others.

Table 1
Effect of isobaric oxygenation on red blood cells of animals

	Erythrocytes ($10^6/mm^3$)		Hemoglobin (g%)		Reticulocytes (‰)	
Group of animals	Contr.	Exp.	Contr.	Exp.	Contr.	Exp.
1. group ($n = 26$)	4.756 ± 0.210	5.353 ± 0.155	12.7 ± 0.3	13.6 ± 0.3	12.0 ± 2.0	4.0 ± 0.5
2. group ($n = 6$)	4.756 ± 0.210	5.276 ± 0.118	12.7 ± 0.3	12.4 ± 0.2	12.0 ± 2.0	3.0 ± 0.4
3. group ($n = 30$)	2.916 ± 0.158	3.593 ± 0.127	11.6 ± 0.2	10.6 ± 0.3	215.0 ± 19.0	123.0 ± 14.0

The interspecies differences in the changes of concentration of reticulocytes were revealed. These parameters in adult and newborn rabbits became 2 or 3 times less due to oxygen application as compared to the controls (Table 1). At the same time this parameter in cats became higher: 8.0 ± 1.75 in controls, 19.0 ± 3.31 in hyperoxic animals.

It should be stressed that there were no reasons to suggest the development of anemia in cats which have maximal changes of the reticulocyte concentration.

Along with the changes of the blood picture in the conditions of hyperoxygenation, the affinity of hemoglobin to oxygen was considerably changed. The process has a biphasic character in the animals of the first group: The affinity increased by the 4th week of experiment and decreased again to a normal level by the 8th (Fig. 1). But both phases were seen within one month when the exposures were 3 hrs daily. Absolutely opposite results were obtained in cats: The affinity decreased, the level of P_{50} rose by 6.8 mm Hg (Fig. 2). The changes of the affinity are characteristic only of "unstripped" hemoglobin: after dialysis the hemoglobins of control and experimental rabbits reached equal levels of affinity (Fig. 1 and 2). Hence, in these animals, not the hemoglobin but the salt environment within erythrocytes is affected by hyperoxygenation, though the direction of the changes of the affinity is opposite in rabbits and cats. The lack of changes of hemoglobin was also confirmed by the results of electrophoretic investigations.

Usually the hemoglobin of newborn rabbits is characterized by a much higher affinity to oxygen as compared to adult hemoglobin: P_{50} is equal (in our experiments) to 18.2 and 32.9 mm Hg, respectively. After dialysis the difference decreases but does not disappear completely: P_{50} is equal to 16.3 and 18.0 mm Hg, respectively. Evidently, the hemoglobin in newborn rabbits is characterized by somewhat higher affinity to oxygen as compared to the adult hemoglobin in a

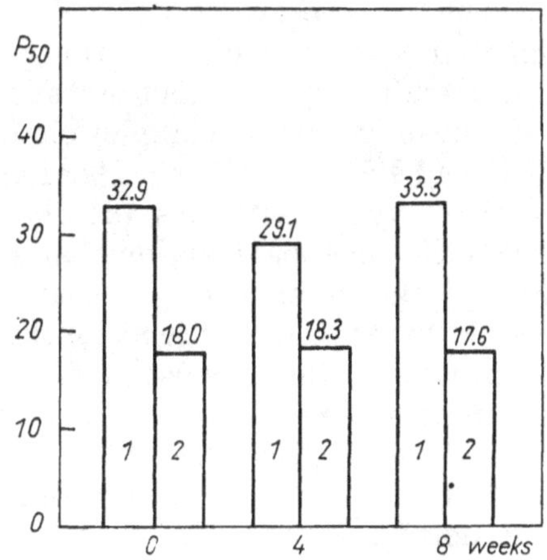

Fig. 1

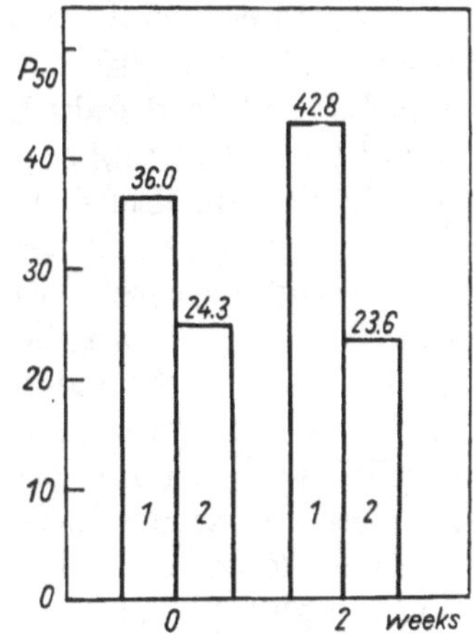

Fig. 2

Fig. 1. P_{50} of the hemoglobin of adult rabbits.
1: before dialysis; 2: after dialysis. Ordinate: P_{50} (mm Hg); Abscissa: weeks of experiment

Fig. 2. P_{50} of the hemoglobin of adult cats

similar salt environment. The affinity in newborn rabbits whose mothers were kept in the hyperoxic conditions throughout the period of pregnancy (third group of animals), was less than in control newborn rabbits (Fig. 3). After dialysis, on the contrary, the affinity of hemoglobin in experimental newborn rabbits was somewhat higher than in control animals of the same age. Thus, in the conditions of hyperoxic exposure the affinity of hemoglobin to oxygen in adult and embryos changes in opposite directions, and this is why differences between P_{50} of both groups appear to be less than in control animals. As it has been shown by the results of dialysis the changes of affinity are mainly due to the salt environment in which hemoglobin functions.

Thus, the different types of reaction to hyperoxia are observed depending on the species and age of animals. It is essential that the maturation of the erythroid cells under the influence of oxygen changes differently in cats and rabbits. Accordingly, the reticulocytes concentration either increased (in cats), or decreased (in rabbits). The changes of the affinity hemoglobin to oxygen are also different.

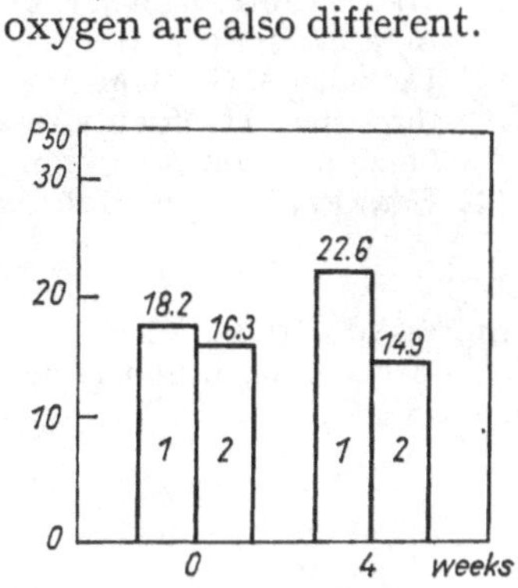

Fig. 3. P_{50} of the hemoglobin of newborn rabbits

It was remarked especially to those data where the depression of glycolytic rate is obtained under the influence of oxygen that the mechanism of this process could not be explained only by intracellular acidifications, which is caused by liberation of protons during oxygenation. The mechanism is of complex character, involving the key enzyme systems HK, PFK and PK [13]. The results of our experiments give the possibilities to suppose that oxygen activates the glycolysis in some cases and increases the content of DPG. The unequal reaction of glycolysis to hyperoxia depending on the species and age of animals can be a main cause of unequal changes of the affinity of hemoglobin to oxygen. So, the hyperoxygenation is followed during the first phase of the process by the retardation of glycolysis in erythrocytes of adult rabbits. The content of DPG decreases, the affinity of hemoglobin to oxygen increases. After the level of glycolytic rate and DPG content comes down to a certain minimum, the reaction to oxygen changes and the affinity begins to diminish, reverting to normal levels. In embryonic rabbit erythrocytes and in adult cats, vice versa, glycolysis increases, the content of DPG grows higher and the affinity decreases. The reaction to oxygen at different age and in different species varies because of the difference in the initial level of glycolytic enzymes and DPG. It has been shown, for example, that if the initial level of DPG is low the incubation of erythrocytes in oxygen results in the increase of this component, and if the initial level is high the result is reversed [14]. Further complex experiments are needed for the elucidation of the mechanism of the action of hyperoxia on the respiratory function of the blood.

References

[1] Чарный, А. М. Б. сб гипоксия, изд. АН УССР, Киев, 168, 1949
[2] BARKER, J. N.: Am. J. Physiol. **189**, 281 (1957)
[3] ASTRUP, P.: Försvarsmedicin **5**, 199 (1969)
[4] DUHM, J., and E. GERLACH: Pflügers Arch. ges. Physiol. **326**, 254 (1971)
[5] ROUGHTON, F. J., and J. W. SEVERINGHAUS: J. appl. Physiol. **36**, 861 (1973)
[6] WATKINS, G. M., E. J. LOVETT, M. P. ROBERTS, G. OWENS, and W. WILSON: J. Surg. Res. **16**, 504 (1974)
[7] RAPOPORT, S.: Försvarsmedicin **5**, 168 (1969)
[8] BUNN, H. F.: Science, N.Y. **172**, 1049 (1971)
[9] KUTAS, F., and M. STÜTZEL: Experientia **14**, 214 (1958)
[10] GREENWALT, T. Y., and V. E. AYERS: J. Cell. Comp. Physiol **53**, 307 (1959)
[11] ROSSI-FANELLI, A., and E. ANTONINI: Archs Biochem. Biophys. **77**, 478 (1958)
[12] WRIGHSTONE, R. N., J. B. WILSON, C. A. REINOLD, and T. H. HUISMAN: Clinica Chim. Acta **44**, 217 (1973)
[13] JACOBASCH, G., S. MINAKAMI, and S. M. RAPOPORT in: Cell. and Molecul. Biol. of Erythrocytes. H. YOSHIKAWA and S. M. RAPOPORT (Eds.). Univ. Park Press, Baltimore-London-Toronto 1974, p. 55
[14] BAKKER, J. C., E. DEKKER, and F. G. S. OFFERIJNS: Vox sang. **23**, 212 (1972)

Address of authors: Prof. Dr. L. I. IRZHAK, Assistent V. V. GLADILOV, Lehrstuhl für Physiologie der Menschen und Tiere, Universität Syktyvkar, Oktjabr. ul., 55, Syktyvkar, 167001, UdSSR

Acta biol. med. germ., Band 36, Seite 661—663 (1977)

Forschungsabteilung und Abteilung Neonatologie der Kinderklinik des Bereichs Medizin (Charité) der Humboldt-Universität zu Berlin, 104 Berlin, DDR

2,3-Diphosphoglyzerat- und Adenosintriphosphat-Konzentrationen in roten Blutzellen von Neugeborenen mit Atemnotsyndrom

B. Göldner, G. Göldner, R. Wauer und J. Gross

Zusammenfassung

Mit dem Ziel, Kriterien für die objektive Beurteilung hypoxischer Zustände des Neugeborenen zu finden, wurde die 2,3-Diphosphoglyzerat (2,3-DPG)- und die ATP-Konzentration in roten Blutzellen von normalen Neugeborenen und Neugeborenen mit Atemnotsyndrom untersucht. Es konnten keine signifikanten Unterschiede in den 2,3-DPG- und ATP-Konzentrationen innerhalb der ersten 24 Std. nach der Geburt ermittelt werden. Korrelationsanalysen zwischen der 2,3-DPG-Konzentration in roten Blutzellen und dem *p*H-Wert im Plasma, der ATP-Konzentration in roten Blutzellen und der Phosphatkonzentration im Plasma ergaben signifikant positive Beziehungen.
Zwischen der 2,3-DPG-Konzentration und dem O_2-Partialdruck im arterialisierten Kapillarblut und der Hb-Konzentration wurden negative Beziehungen gefunden. Richtung und Ausmaß der Veränderungen der 2,3-DPG-Konzentration in Abhängigkeit von ihren beeinflussenden Faktoren beim Neugeborenen entsprechen den Werten von Erwachsenen. Von dominierendem Einfluß für die in vivo-Änderungen der 2,3-DPG-Konzentration in roten Blutzellen scheint die ATP-Konzentration in roten Blutzellen und der *p*H-Wert zu sein.

Einleitung

Für die objektive Beurteilung hypoxischer Zustände des Neugeborenen fehlen bisher immer noch eindeutige und aussagekräftige Parameter. Valeri et al. [1] empfahlen als Kriterien zur Beurteilung der Hypoxie u. a. den 2,3-DPG-und ATP-Spiegel in roten Blutzellen (RBZ).
Wir untersuchten die postnatalen Veränderungen der 2,3-DPG-Konzentration und ihre Abhängigkeit vom extrazellulären *p*H-Wert, der ATP-Konzentration im Erythrozyten, des anorganischen Phosphats im Plasma (P_i), der Hämoglobin-Konzentration (Hb) sowie dem Sauerstoffpartialdruck im arterialisierten Kapillarblut (pO_2).

Material und Methoden

Die von uns untersuchte Kontrollgruppe umfaßte 9 gesunde Reifgeborene mit Geburtgewichten von 2300—3850 g. Die Gruppe der Neugeborenen mit Atemnotsyndrom bestand aus 21 Kindern mit Geburtsgewichten von 1500—4460 g. Die angewandten Methoden zur Bestimmung von 2,3-DPG, ATP, *p*H, P_i im Plasma und Hb wurden in einer vorangegangenen Mitteilung angegeben [2]. Der pO_2 wurde polarografisch mit der Meßkette „Metra" (Radebeul) bestimmt.

Ergebnisse und Diskussion

In Tab. 1 sind die innerhalb der ersten 24 Std. nach der Geburt ermittelten Parameter bei gesunden Neugeborenen und Neugeborenen mit Atemnotsyndrom

Tabelle 1

Ausgangswerte der untersuchten Parameter bei gesunden Neugeborenen im Vergleich zu Neugeborenen mit Atemnotsyndrom ($\bar{x} \pm s$)

Gruppe	Geburts-gewicht (g)	Hk (%)	Hb (g%)	P_i im Plasma (HA)	pO_2 (μ Mol/ml)	pH	ATP (μMol/ml RBZ)	2,3-DPG (μMol/ml RBZ)
gesunde Neugebo-rene ($n = 9$)	8: >2500 1: <2500	55,6 ± 7,7	18,0 ± 1,7	1,63 ± 0,16	—	—	1,01 ± 0,13	4,30 ± 0,66
Neugebo-rene mit ANS ($n = 21$)	12: <2500 g 9: >2500 g	48,0 ± 8,4	15,5 ± 2,1	1,99 ± 0,41	60,2 ± 19,6 —	7,3 ± 0,15	1,06 ± 0,23	4,08 ± 0,84

(ANS) zusammengestellt. Es sind keine signifkanten Unterschiede in den 2,3-DPG- und ATP-Konzentrationen nachweisbar.

Um die bestimmenden Faktoren für die 2,3-DPG-Konzentration zu analysieren, führten wir Regressions- und Korrelationsanalysen zwischen der 2,3-DPG-Konzentration und den anderen untersuchten Kriterien durch (Tab. 2). Die gefundenen Beziehungen erlauben gewisse quantitative Aussagen über die Veränderungen der 2,3-DPG-Konzentration in Abhängigkeit vom pH-Wert, der ATP-Konzentration, der Konzentration P_i im Plasma, pO_2 und der Hb-Konzentration.

Berücksichtigt man alle pH-Werte, die zur gleichen Zeit ermittelt wurden, wie die 2,3-DPG-Konzentration, so erhält man eine positive Korrelation zwischen 2,3-DPG-Konzentration und pH-Wert. Die Änderung der 2,3-DPG-Konzentration pro 0,1 pH-Einheiten ist mit 0,43 μMol 2,3-DPG/ml RBZ/0,1 pH-Einheiten deutlich niedriger als bei Erwachsenen (0,6—1,9 μMol/ml RBZ/0,1 pH) [3]. Ausgehend von der Tatsache, daß die Veränderungen der 2,3-DPG-Konzentration

Tabelle 2

Bestimmende Faktoren für die 2,3-DPG-Konzentration in roten Blutzellen bei Neugeborenen mit Atemnotsyndrom

		Korrelation					
x	y	n	r	$y = mx + b$	p	$\Delta y/\Delta x$ [1]	$\Delta y/\Delta x_s$
pH	2,3-DPG	66	0,44	$y = 4,33x - 27,83$	0,01	0,43	0,46
$p\mathrm{H}_a$	2,3-DPG	39	0,60	$y = 7,82x - 53,84$	0,01	0,78	0,69
ATP	2,3-DPG	90	0,57	$y = 7,14x - 3,50$	0,01	0,71	1,75
P_i	2,3-DPG	78	0,43	$y = 0,91x + 2,38$	0,01	0,91	0,45
Hb	2,3-DPG	84	−0,33	$y = 0,17x + 6,72$	0,05	0,17	0,35
pO_2	2,3-DPG	31	−0,22	$y = -0,02x + 5,70$	0,05	0,2	0,30
pO_2 (pH < 7,35)	2,3-DPG	21	−0,61	$y = 0,04x + 7,3$	0,05	0,4	0,48

$p\mathrm{H}_a$: pH-Wert, der 10—14 Std. vor der 2,3-DPG-Bestimmung gemessen wurde; x_s: Standardabweichung von x

[1] $\Delta y/\Delta x$: Änderung der 2,3-DPG-Konzentration in μMol/0,1 pH-Einheiten; 0,1 μMol ATP, 1 μMol P_i, 1 g% Hb, 10 Torr

eine gewisse Latenz gegenüber dem *p*H-Wert aufweisen, korrelierten wir den 2,3-DPG-Spiegel mit einem 10—14 Std. vorher bestimmten *p*H-Wert. Bei dieser Regression wird eine größere Änderung der 2,3-DPG-Konzentration pro 0,1 *p*H-Einheit (0,78 μMol/ml RBZ/0,1 *p*H) erreicht. Dieser Wert entspricht ungefähr dem von Erwachsenen. Dies läßt vermuten, daß in RBZ von Neugeborenen Richtung und Ausmaß der *p*H-bedingten Veränderungen der 2,3-DPG-Konzentration in RBZ von Neugeborenen wie bei Erwachsenen sind.
Zwischen der 2,3-DPG- und ATP-Konzentration wurde eine positive Korrelation festgestellt.
Nach den Modellvorstellungen von Rapoport et al. [4] verhalten sich die 2,3-DPG- und ATP-Konzentrationen in bestimmten Konzentrationsbereichen weitgehend parallel. Wahrscheinlich drückt die hier gefundene positive Korrelation zwischen 2,3-DPG- und ATP-Konzentration die Abhängigkeit der 2,3-DPG-Konzentration von der des ATP aus. Es ist bekannt, daß bei hohen ATP-Konzentrationen, d. h. niedrigen ADP-Konzentrationen vorwiegend 2,3-DPG-gebildet wird [6].
Zwischen dem pO_2 und der 2,3-DPG-Konzentration wird bei einem *p*H-Wert $\geqq 7{,}35$ eine negative Beziehung gefunden. Der Anstieg der 2,3-DPG-Konzentration um 0,4 μMol/ml RBZ bei einem Abfall des pO_2 um 10 Torr entspricht den bei Erwachsenen gefundenen Werten [7].
Die beobachteten Korrelationen lassen vermuten, daß die Regulation des 2,3-DPG-Spiegels in Erythrozyten auf ähnlichen Mechanismen beruht wie bei Erwachsenen [2].
Um eine Abschätzung der quantitativen Veränderungen der 2,3-DPG-Konzentration unter in vivo-Bedingungen bei Neugeborenen in Abhängigkeit von den biologisch vorkommenden Schwankungen der Einflußgrößen zu ermöglichen, wurden die Änderungen der 2,3-DPG-Konzentration auf die Standardabweichung (1s) der jeweiligen Faktoren bezogen. Es zeigt sich, daß die ATP-Konzentration und der *p*H-Wert in vivo von dominierendem Einfluß für die 2,3-DPG-Konzentration sind. Hingegen ist die Veränderung der 2,3-DPG-Konzentration in Abhängigkeit vom pO_2, P_i und Hb geringer.
Auf Grund dieser Zusammenhänge folgt, daß die 2,3-DPG-Konzentration allein wahrscheinlich kein Kriterium für den Grad der Hypoxie bei Kindern mit Atemnotsyndrom ist.

Literatur

[1] Valeri, C. R., C. C. Zaroulis u. N. L. Fortier in: Oxygen Affinity of Hemoglobin and Red Cell Acid Base Status. Proceedings of the Alfred Benzon Symposium IV, Copenhagen 1971. M. Rörth u. P. Astrup (Hrsg.). Academic Press, New York 1972, S. 650

[2] Gross, J., B. Göldner, G. Göldner, R. Wauer u. A. Daniel: Clinica chim. Acta (eingereicht)

[3] Gerber, G.: Bestimmende Faktoren der 2,3-DPG-Konzentration in roten Blutzellen des Menschen (in Vorbereitung)

[4] Rapoport, T. A., R. Heinrich u. S. M. Rapoport: Biochem. J. **154**, 449 (1976)

[5] Schröter, W., u. H. v. Heyden: Biochem. J. **341**, 387 (1965)

[6] Schröter, W., u. P. Winter: Klin. Wschr. **45**, 255 (1967)

[7] Rosenthal, A., W. C. Mentzer, E. B. Eisenstein, D. G. Nathan, N. M. Nelson u. A. S. Nadas: Pediatrics **47**, 537 (1971)

Acta biol. med. germ., Band 36, Seite 665—667 (1977)

[1] Chirurgische Klinik und
[2] Forschungsabteilung der Kinderklinik
des Bereichs Medizin (Charité) der Humboldt-Universität, 104 Berlin, DDR

Verhalten des 2,3-Diphosphoglyzerats in roten Blutzellen bei Kindern mit zyanotischen Herzfehlern prä- und postoperativ

D. Oldag[1], J. Gross[2], A. Michel[2], G. Evers[1] und B. Schubel[1]

Zusammenfassung

Es wird das Verhalten der 2,3-Diphosphoglyzeratkonzentration (2,3-DPG) roter Blutzellen von Säuglingen und Kindern mit zyanotischen Herzfehlern vor und nach Shunt-Operationen untersucht. Bereits bei Säuglingen im Alter bis zu 6 Monaten finden wir einen deutlichen Anstieg der 2,3-DPG-Konzentration und eine Erhöhung des Hämatokrit-(HK)-Wertes. Im späteren Lebensalter geschieht die Anpassung an die chronische Hypoxie durch weiteren Hk-Anstieg bei unverändert hohem 2,3-DPG-Spiegel. Die nach Shunt-Operationen unverändert erhöhte 2,3-DPG-Konzentration wird als effektiver Adaptationsmechanismus an die noch verbleibende Hypoxie angesehen. Die 2,3-DPG-Konzentration allein stellt kein Kriterium für die Beurteilung der Hypoxie dar.

Einleitung

Zur Beurteilung der Operationsindikation für die Totalkorrektur oder Palliativoperation von Kindern mit zyanotischen Herzfehlern gewinnen neben dem klinischen und angiologischen Befund alle Parameter an Bedeutung, die Aussagen über den Schweregrad der Hypoxie gestatten. Ziel unserer Untersuchungen war es, zu prüfen, ob der 2,3-DPG-Gehalt roter Blutzellen als weiteres Kriterium zur Beurteilung des Schweregrades der Hypoxie zu verwenden ist, da es unter Hypoxie zu einer erhöhten 2,3-DPG-Bildung mit den bekannten Auswirkungen auf die O_2-Dissoziationskurve kommt.

Material und Methoden

Die Untersuchungen wurden an gesunden Neugeborenen, Säuglingen (Lebensalter 15 Tage bis 6 Monate) und Kindern (Lebensalter 1—15 Jahre) mit zyanotischen Herzfehlern durchgeführt.

Der Einfluß von Shunt-Operationen (Waterston-Anastomose [1] oder Blalock-Taußig-Anastomose [2]) auf die Veränderung des Sauerstoffpartialdruckes im Kapillarblut (pO_2), des Hämokrit (Hk) und der 2,3-DPG-Konzentration in roten Blutzellen bei Kindern mit zyanotischen Herzfehlern im Alter von durchschnittlich 5,4 Jahren (1—15 Jahre) präoperativ und mindestens 4 Monate postoperativ wurde untersucht.

Die 2,3-DPG-Bestimmung erfolgte nach der Methode von Luisada-Opper [3]. Hb, Hk und Erythrozyten wurden nach DAB 7 [4] bestimmt. Die Messung des kapillären Sauerstoffpartialdruckes und des Blut-*p*H-Wert erfolgten potentiometrisch (BMG 32 der Fa. Radiometer, Kopenhagen).

Die Sauerstoffsättigung errechneten wir aus der Standard-O_2-Bindungskurve. Die Signifikanzprüfung erfolgte mit dem Wilcoxon-Test [5].

Ergebnisse

Tab. 1 zeigt den pO_2-Wert (Kapillarblut) und die 2,3-DPG-Konzentration der roten Blutzellen der untersuchten Patientengruppen.

Tabelle 1

pO_2, Hk und 2,3-DPG-Konzentration in roten Blutzellen bei gesunden Neugeborenen und Kindern mit zyanotischen Herzfehlern

	pO_2 (mm Hg)	Hk (%)	2,3-DPG-Konzentration (μMol/ml Zellen)
Gesunde Neugeborene ($n = 20$)	–	51 ± 8	4,4 ± 0,5
Säuglinge mit zyanotischen Herzfehlern (15 Tage–6 Monate) ($n = 19$)	30 ± 4	46 ± 13	6,0 ± 1,5
Kinder mit zyanotischen Vitien (1–15 Jahre), präoperativ ($n = 29$)	40 ± 8	71 ± 9	5,5 ± 1,3
Kinder mit zyanotischen Vitien (2–15 Jahre) mindestens 4 Monate postoperativ ($n = 16$)	56 ± 7	53 ± 9	5,3 ± 0,7

Der Normalwert der 2,3-DPG-Konzentration bei Neugeborenen beträgt 4,4 μMol pro ml Zellen und stimmt mit den in der Literatur angegebenen Werten gut überein [3,6]. Kinder mit zyanotischen Herzfehlern zeigen im Alter von 15 Tagen bis 6 Monaten bei noch normalen Hk schon eine deutliche Erhöhung der 2,3-DPG-Konzentration.

Im späteren Lebensalter (Gruppe der Kinder mit zyanotischen Herzfehlern von 1—15 Jahren) haben die Kinder eine deutliche Polyglobulie mit einem gegenüber der Norm erhöhten 2,3-DPG-Gehalt von 5,5 μMol 2,3-DPG/ml Zellen. Durch die Operation kommt es zu einem deutlichen Anstieg des Sauerstoffpartialdruckes von präoperativ 30—40 Torr auf nahezu 60 Torr postoperativ. Dieser Anstieg des pO_2 ist begleitet von einem deutlichen Abfall des Hk. Die 2,3-DPG-Konzentration zeigt keine Veränderungen. Um den Einfluß des pO_2 auf die 2,3-DPG-Konzentration zu analysieren, prüften wir die Korrelation zwischen 2,3-DPG und pO_2 (Tab. 2).

Zwischen der 2,3-DPG-Konzentration roter Blutzellen und dem pO_2 besteht keine Korrelation, wenn alle Werte zur Auswertung gelangen. Eine statistisch signifikante Beziehung wird nachweisbar, wenn für die Korrelation zwischen 2,3-DPG und pO_2 nur Blute mit einem Hk < 60 Vol.% und einem *p*H-Wert 7,35 verwendet werden.

Tabelle 2

Korrelation zwischen 2,3-DPG-Konzentration und pO_2 bei Hk < 60 Vol.% und *p*H $> 7{,}35$ sowie zwischen 2,3-DPG-Konzentration und pO_2 bei *p*H $< 7{,}35$ unabhängig vom Hk

	Geradengleichung	Korrelationskoeffizient	n
2,3-DPG/pO_2 bei Hk <60 Vol.% und *p*H $>7{,}35$	$y = -0{,}09x + 10{,}82$	$r = -0{,}59$	22 ($p < 0{,}01$)
2,3-DPG-Konzentration/pO_2 bei *p*H $>7{,}35$, unabhängig vom Hk	$y = -0{,}01 + 6{,}79$	$r = -0{,}11$	56 (nicht signifikant)

Diskussion

Die Kompensation des O_2-Mangels bei Kindern mit zyanotischen Herzfehlern erfolgt sowohl durch die Erhöhung des Hk als auch durch die Erhöhung des 2,3-DPG-Gehalts roter Blutzellen [7, 8].

Die vorliegenden Untersuchungen zeigen, daß wenige Wochen nach der Geburt die 2,3-DPG-Konzentration deutlich erhöhte Werte aufweist. Der Hk der Säuglinge mit zyanotischen Herzfehlern ist gegenüber der Norm dieser Altersgruppe deutlich erhöht. Bekanntlich zeigen Säuglinge in diesem Lebensalter eine Anämie (Trimenonanämie).

Nach den vorliegenden Befunden erfolgt die Kompensation der Hypoxie im späteren Lebensalter vor allem durch einen Hk-Anstieg und nicht durch eine weitere Erhöhung der 2,3-DPG-Konzentration.

Der postoperative Abfall des Hk läßt sich gut durch den operationsbedingten Anstieg des pO_2 erklären. Es ist bemerkenswert, daß sich die 2,3-DPG-Konzentration bei Patienten, die länger als 4 Monate mit einem funktionstüchtigen Shunt lebten, nicht ändert. Das kann als effektiver Adaptationsmechanismus an die nach Shunt-Operationen noch verbleibende Hypoxie angesehen werden.

Die Korrelation zwischen 2,3-DPG-Konzentration und pO_2 wird in unserem Patientengut deutlich, wenn der Einfluß des veränderten pH-Wertes und des Hämatokrits weitestgehend eliminiert werden. Das bedeutet, daß der pO_2 nur dann zu einem bestimmenden Faktor für die 2,3-DPG-Konzentration wird, wenn der Hk einen bestimmten Grenzwert unterschreitet. Oberhalb eines Hk von 60 Vol.% ist die 2,3-DPG-Konzentration unabhängig vom pO_2. Daraus folgt, daß die 2,3-DPG-Konzentration der roten Blutzellen allein kein Kriterium für die Beurteilung der chronischen Hypoxie darstellt.

Literatur

[1] Waterston, D. J.: Rozhl. Chir. **41**, 181 (1962)
[2] Blalock, A. u. H. G. Taussig: J. Am. med. Ass. **128**, 189 (1945)
[3] Luisada-Opper, A. V.: Clin. Chem. **19**, 118 (1973)
[4] Deutsches Arzneibuch der DDR. 7. Aufl. Akademie-Verlag, Berlin 1968
[5] Geigy Documenta, Wissenschaftliche Tabellen. 7. Aufl. J. R. Geigy, (Hrsg.). A. C. Pharma, Basel 1968, S. 193
[6] Oski, F. A., A. J. Gottlieb, W. W. Miller u. M. Delivoria-Papadopoulos: J. clin. Invest. **49**, 400 (1970)
[7] Baltzer, G., u. H. Arndt: Internist **14**, 177 (1973)
[8] Manske, H., W. Esters, K. Dörner, V. Malerczyk u. P. Lange: Z. Kardiol. **64**, 562 (1975)

Summary

D. Oldag, J. Gross, A. Michel, G. Evers, and B. Schubel: The prae- and postoperative 2,3-DPG concentration of red blood cells in children with cyanotic heart diseases

The behaviour of 2,3-diphosphoglycerate concentration (2,3-DPG) of red blood cells of babies and children with cyanotic heart diseases is studied before and after shunt operations. In babies with cyanotic heart diseases at the age of up to 6 months an increase of 2,3-DPG-level and haematocrit (HCT) is seen.

Later, the compensation of chronic hypoxia is effected by further increase of HCT at unchanged high 2,3-DPG-level. The 2,3-DPG concentration which is still increased after successfull shunt operations as compared with the normal value is considered as an effective adaptation mechanism to the residual hypoxia presenting only a small load on the circulatory system.

The 2,3-DPG concentration alone does not represent a criterion for the assessment of chronic hypoxia.

Acta biol. med. germ., Band 36, Seite 669—680 (1977)

Department of Food Science and Technology,
Fatulty of Agriculture, Kyoto University; Kyoto, 606, Japan

Multifunctionality of the enzyme in 2,3-bisphosphoglycerate metabolism of pig erythrocytes[1]

R. Sasaki, K. Ikura, H. Narita, and H. Chiba

Summary

In a continuing study of the enzymes involved in the 2,3-bisphosphoglycerate metabolism of mammalian erythrocytes, we report that 2,3-bisphosphoglycerate in pig erythrocytes is probably metabolised by one multifunctional enzyme. Bisphosphoglyceromutase, 2,3-bisphosphoglycerate phosphatase, and phosphoglyceromutase were simultaneously purified from pig erythrocytes. Three fractions (peaks I, II, and III) which had all three activities in different ratios were obtained by column chromatography. Peak I protein was extremely rich in phosphoglyceromutase, containing more than 90% of the total activity of this enzyme. In contrast, peak III protein was active in metabolising 2,3-bisphosphoglycerate, containing about 95% of both the bisphosphoglyceromutase and the 2,3-bisphosphoglycerate phosphatase activities. It seems likely that peak I protein functions as phosphoglyceromutase and that peak III protein functions in the 2,3-bisphosphoglycerate metabolism.

The homogeneity of peak III material was established by disc gel electrophoresis in the presence and the absence of sodium dodecylsulfate, as well as by ultracentrifugation. The three activities of peak III were lost at the same rate on thermal inactivation. These results indicate that the two enzyme activities for metabolising 2,3-bisphosphoglycerate, which were believed to be due to different proteins, are attributable to one protein, along with some phosphoglyceromutase activity. The bisphosphoglyceromutase activity of this protein was inhibited by the product, 2,3-bisphosphoglycerate, as well as by hydroxypyruvate phosphate, 2-phosphoglycolate, inorganic phosphate, and bisulfite. The 2,3-bisphosphoglycerate phosphatase activity was enhanced by 2-phosphoglycolate and hydroxypyruvate phosphate, and by the coexistence of chlorine ion and bisulfite, while it was inhibited by monophosphoglycerates. The native peak III protein had a molecular weight of 59,000 as determined by equilibrium centrifugation. Disc gel electrophoresis in the presence of sodium dodecyl sulfate yielded a single protein band with a molecular weight of 28,000, indicating that this protein was composed of two subunits with a similar molecular weight.

Occurrence of multifunctional proteins in pig erythrocytes is compared with that in human erythrocytes from the standpoint of the universal existence of such proteins. A new classification of 2,3-bisphosphoglycerate dependent phosphoglyceromutase is proposed, which is based on the ratios of the three enzyme activities associated with the phosphoglyceromutase proteins.

[1] A part of this paper has been submitted for publication (Sasaki, R., et al., Agric. Biol. Chem. in press).

Abbreviations: 2,3-DPG, 2,3-bisphosphoglycerate; 1,3-DPG, 1,3-bisphosphoglycerate; 3-PGA, D-3-phosphoglycerate; 2-PGA, D,L-2-phosphoglycerate; HPAP, hydroxypyruvate phosphate; SDS, sodium dodecyl sulfate; PGA mutase, phosphoglyceromutase; (EC 2.7.5.3); 2,3-DPG phosphatase, 2,3-bisphosphoglycerate phosphatase (EC 3.1.3.13); DPG mutase, bisphosphoglyceromutase (EC 2.7.5.4).

Introduction

Two enzymes, DPG mutase and 2,3-DPG phosphatase, are directly involved in the 2,3-DPG metabolism of erythrocytes. It was believed that these two enzymes were different protein molecules because of the irreversible character of the reactions catalysed. Recently, however, we have reported evidence that both DPG mutase and 2,3-DPG phosphatase activities in human erythrocytes are attributable to one protein which has the PGA mutase activity [1—3]. Subsequently, papers supporting this conclusion have appeared from other laboratories [4, 5]. Human erythrocytes have three kinds of multifunctional protein which contain the three enzyme activites in different ratios [1]. One of these kinds (peak I) is high in PGA mutase activity and probably functions in the main pathway of glycolysis, while another (peak III) has high DPG mutase and 2,3-DPG phosphatase activities, about 95% of their total activities in the erythrocytes. The ratio of the three enzyme activities of peak II is intermediate to those of I and III. It is evident from the low activities of the enzymes metabolizing 2,3-DPG in the peak II protein that this peak contributes little to the 2,3-DPG metabolism. Chemical modification experiments have shown that the three enzyme activities of peak III protein, which is chiefly responsible for the 2,3-DPG metabolism, are manifested at a common active site. Based on these results, a new regulatory mechanism has been proposed for the 2,3-DPG metabolism in human erythrocytes [2, 3].
In view of the physiological implications if such multifunctional proteins exist universally in living cells, acting as tools for metabolic regulation, we have undertaken to investigate the enzyme involved in the 2,3-DPG metabolism of other mammals. This report describes the purification of the protein metabolising 2,3-DPG in pig erythrocytes to homogeneity, and furnishes some evidence for the universality of multifunctional proteins.

Material and methods

Enzyme assays

DPG mutase, 2,3-DPG phosphatase and PGA mutase activities were measured as described previously [1]. In all instances, one unit of enzyme activity is defined as the amount of enzyme which catalyses the conversion of 1 µmole of substrate to product per min under the conditions stated. Specific activity is expressed as units per mg of protein.

Isoelectric focusing

Conditions for isoelectric focusing were as described previously [1] except that an ampholine solution (*p*H range 3.5 to 10) and a 5 mM Tris-HCl buffer, *p*H 7.5, containing 0.1 mM EDTA and 5 mM β-mercaptoethanol (buffer A) were used. Electrofocusing was carried out for 12 hrs (400 V, 0.6 mA).

Polyacrylamide gel electrophoresis

Polyacrylamide gel electrophoresis was carried out according to the procedure of Davies [6] and that of Weber and Osborn [7] in the presence of SDS. Conditions were as described previously [1] except that buffer A was used and a current of 8 mA per gel was applied for 3.5 hrs in the presence of SDS.

Ultracentrifugation

Ultracentrifugation was carried out in a Beckman model E analytical ultracentrifuge at 20 °C. Sedimentation velocity experiments were performed with a double-sector cell. The sedimentation equilibrium experiments were performed according to the meniscus depletion method of Yphantis [8].

Protein Concentration

Protein was determined by measuring the absorbance at 280 nm. Its concentration was calculated from the absorbance at 280 nm based on the assumed value $A_{280\,nm} = 10$.

Results

Purifications of enzymes

Table 1 is a summary of the procedure used in the purification of the three enzymes. All procedures were carried out at 0—4 °C.

Hemolysate: Slaughter-house blood of pigs was collected in a plastic bottle which contained 140 ml of 3.5% trisodium citrate per liter of blood. Purification of the enzyme was begun using 1750 ml of the blood. Hemolysates of the pig erythrocytes were prepared as described previously for human erythrocytes [1].

Bulk separation on DEAE-cellulose: The hemolysates were mixed with 56 g of DEAE-cellulose previously equilibrated with buffer A. After stirring the mixture for 3 hrs, the cellulose with the absorbed enzymes was filtered on a Buchner funnel and washed with 1000 ml of buffer A containing 0.05 M KCl. Then, the enzymes were eluted by washing the cellulose with 1200 ml of buffer A containing 0.35 M KCl. The protein in the eluate was precipitated by adding solid ammonium sulfate to 75% saturation.

Ammonium sulfate fractionation: The precipitate was dissolved in 510 ml of buffer A. To 570 ml of the solution was added 84 g of solid ammonium sulfate. The resulting suspension was centrifuged, and the pellet obtained was discarded. The proteins in the supernatant (600 ml) were precipitated by adding 170 g of solid ammonium sulfate. The three enzyme activities, DPG mutase, 2,3-DPG phosphatase, and PGA mutase, were recovered in this fraction with an excellent yield.

Chromatography on DEAE-Sephadex: The precipitate was dissolved in, and dialysed against, buffer A. The dialysed solution was chromatographed on a DEAE-Sephadex A-50 column (4.8 × 30 cm) equilibrated with buffer A. Protein was eluted from the column by a 4000-ml linear gradient of KCl ranging from 0 to 0.65 M KCl in buffer A. Fig. 1 shows a typical profile of the three enzyme activities and proteins. Two peaks having the three activities in different ratios appeared. The first peak material, termed peak I, contained the bulk of the PGA mutase activity and a small part of the DPG mutase and 2,3-DPG phosphatase activities. A great part of the latter two activities was found in the second peak, along with a small part of the PGA mutase activity.

Sephadex G-100 chromatography: The enzymes in the second peak were further purified because this material contained most of the enzyme activities which metabolized 2,3-DPG. Protein in the second peak was precipitated by adding solid ammonium sulfate to 75% saturation, then collected by centrifugation and dissolved in a minimum volume of buffer A. After removal of insoluble materials by centrifugation, the clear supernatant was applied to a Sephadex G-100 column (1.4 × 93 cm) equilibrated with buffer A, and the column was developed with the same buffer. As shown in Fig. 2, the three enzyme activities were found in the same elution volume. Protein in the active fractions was precipitated by adding solid ammonium sulfate to 75% saturation.

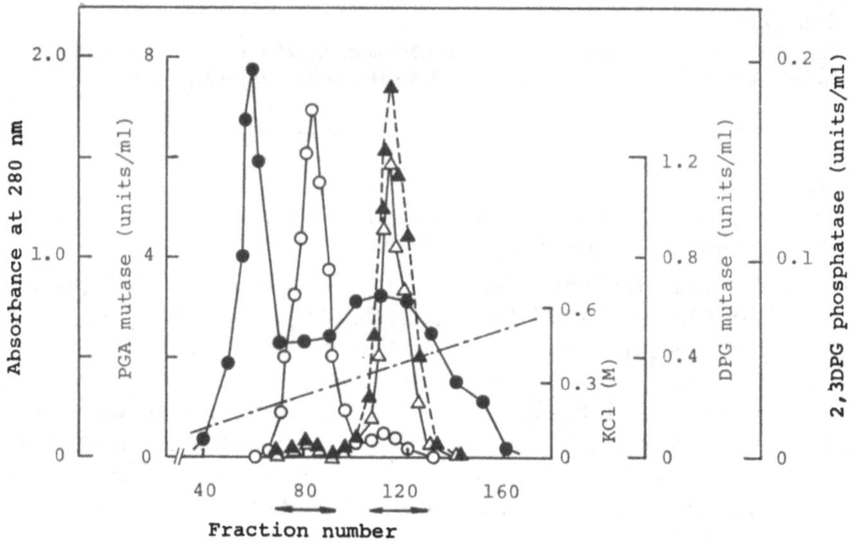

Fig. 1. Elution profile from DEAE-Sephadex.
(●——● absorbance at 280 nm; (△——△) DPG mutase; (▲------▲) 2,3-DPG phosphatase; (○——○) PGA mutase; (—·—) KCl concentration. Activities are expressed as units/ml. 1.56 g of the protein purified by ammonium sulfate fractionation was applied. Fractions were 20 ml. The flow rate was 3 ml/min. The horizontal lines with arrow heads indicate fractions pooled after elution. Active fractions eluted early were designated as peak I. DPG mutase and 2,3-DPG phosphatase activities were assayed with assay I [1]

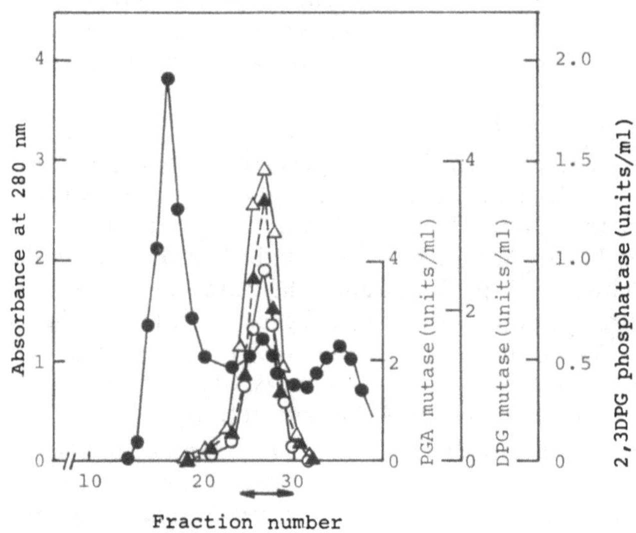

Fig. 2. Sephadex G-100 chromatogram of the second peak obtained by DEAE-Sephadex chromatography.
(●——●) absorbance at 280 nm; (△——△) DPG mutase; (▲------▲) 2,3-DPG phosphatase; (○——○) PGA mutase. Enzyme activities are expressed as units/ml. 82 mg of the second peak protein obtained by DEAE-Sephadex chromatography was applied. Fractions were 3 ml. The flow rate was 10 ml/hr. The horizontal line with arrow heads indicates fractions pooled after elution. DPG mutase and 2,3-DPG phosphatase activities were assayed with assay I [1]

Hydroxyapatite chromatography: The precipitate was dissolved in and dialysed against 10 mM potassium phosphate buffer, *p*H 6.9, containing 0.1 mM EDTA and 5 mM *β*-mercaptoethanol. The dialysed solution was added to a hydroxyapatite column (1.2 × 5.5 cm) equilibrated with the same buffer. The column was developed with a 100-ml linear gradient of potassium phosphate ranging from 10 mM to 200 mM. Two peaks carrying the three enzyme activities in differing ratios were found (Fig. 3). The first peak, termed peak II in this paper, had 33% of the total PGA mutase activity recovered, while 67% of the activity appeared in the second peak, peak III. Although a great part of the DPG mutase and 2,3-DPG phosphatase activities was recovered in peak III, peak II had these enzyme activities in detectable amounts.

CM-Sephadex chromatography of peak I material: To see if the three enzyme activities in peak I were chromatographically separable, peak I protein was further purified. To peak I was added solid ammonium sulfate to 75% saturation. The precipitate was dissolved in and dialysed against 5 mM citrate buffer, *p*H 6.0, containing 0.1 mM EDTA, 5 mM *β*-mercaptoethanol, and 1 mM 2 PGA. The dialysed solution was loaded on a CM-Sephadex C-50 (2.7 × 29 cm) equilibrated with the same buffer. The three enzyme activities were not absorbable to CM-Sephadex and appeared in the flow-through fractions (data not shown).

2,3-DPG phosphatase in the presence and the absence of 2-phosphoglycolate

As the 2,3-DPG phosphatase activity in pig erythrocytes is low, the activity has been measured throughout the purification steps in the presence of 2-phospho-

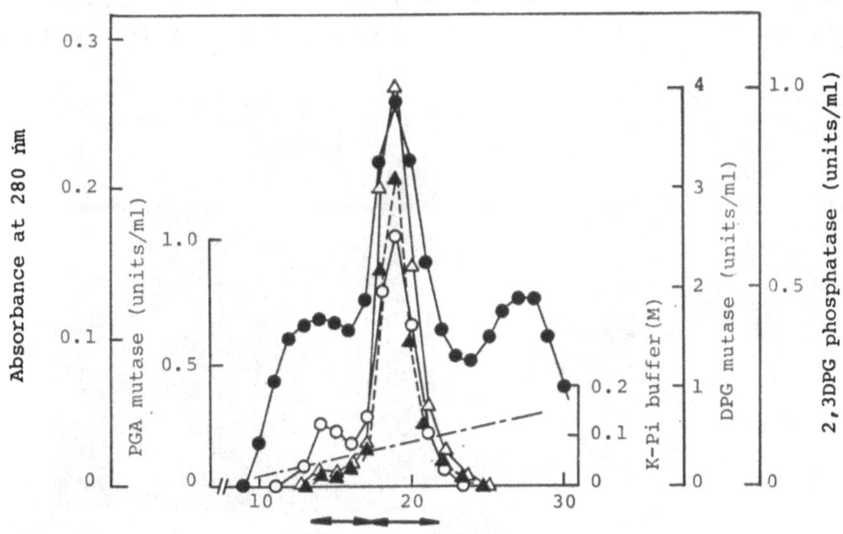

Fig. 3. Elution profile from hydroxyapatite.
(●———●) absorbance at 280 nm; (△———△) DPG mutase; (▲------▲) 2,3-DPG phosphatase; (○———○) PGA mutase; (—·—) concentration of potassium phosphate buffer. Enzyme activities are expressed as units/ml. 12 mg of the protein purified by Sephadex G-100 was applied. Fractions were 3.3 ml. The flow rate was 12 ml/hr. The horizontal lines with arrow heads indicate fractions pooled after elution. Active fractions eluted early were designated as peak II and those eluted later, as peak III. DPG mutase and 2,3-DPG phosphatase were assayed using assay I and assay II, respectively [2]

glycolate, which has been shown to be a potent activator of the 2,3-DPG phosphatase activity manifested by 2,3-DPG-dependent PGA mutase proteins from various sources [1, 9, 10]. It seemed possible that there was a 2,3-DPG phosphatase not being activated by 2-phosphoglycolate, but capable of breaking down 2,3-DPG in the pig erythrocytes. To test this possibility, the 2,3-DPG phosphatase activity in the presence and the absence of 2-phosphoglycolate was measured for samples from some purification steps: hemolysate, ammonium sulfate fractionation, Sephadex G-100 chromatography, and the final preparation (peak III). The extent of activation was almost constant (450- to 500-fold) throughout the purification, indicating that the 2,3-DPG phosphatase enhanced by 2-phosphoglycolate does indeed function in the 2,3-DPG metabolism of pig erythrocytes.

Purity and size of enzyme

Evidence that peak III is homogeneous was obtained by several methods. The enzyme preparation displayed a sharp single band on polyacrylamide gel electrophoresis in the presence and the absence of sodium dodecylsulfate (Fig. 4 and Fig. 7B). A single and symmetrical boundary with a $s_{20,w}$-value of 4.0 S was observed on ultracentrifugation (Fig. 5). The molecular weight of peak III protein was determined using the sedimentation equilibrium method. As shown in Fig. 6, a plot of the logarithm of fringe displacement versus the square of the radius gave an excellent straight line, indicating protein homogeneity. The molecular weight was calculated to be 59,000.

To investigate the subunit structure, peak III protein, was subjected to polyacrylamide gel electrophoresis in the presence of SDS (Fig. 7). The sample gave a

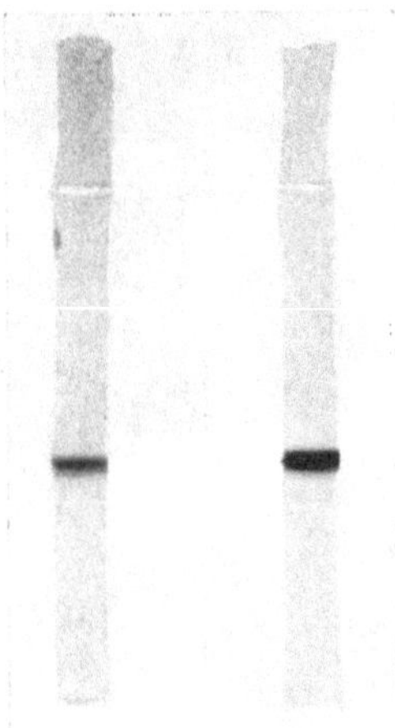

(A) (B)

Fig. 4

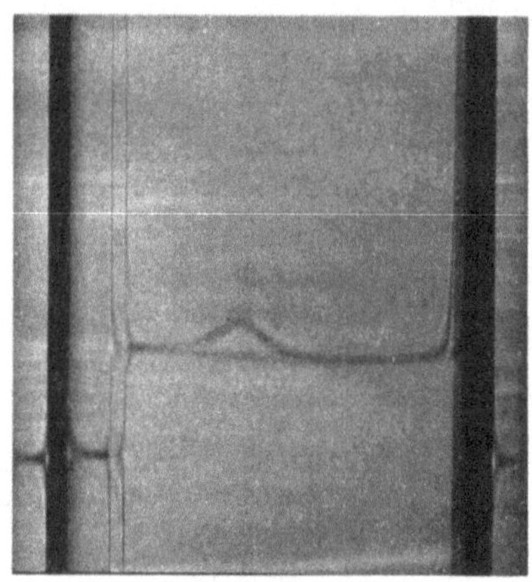

Fig. 5

Fig. 4. Polyacrylamide gel electrophoresis of peak III protein.
The proteins applied were 8 μg (A) and 16 μg (B). Migration is from top to bottom

Fig. 5. Sedimentation pattern of peak III.
The protein was 3.9 mg/ml in buffer A containing 0.1 M KCl. Migration is from left to right. The photograph was taken 48 min after reaching a speed of 60,000 rev./min. The schlieren plate angle was 55°

Fig. 6. Analytical graph of the sedimentation equilibrium pattern of peak III protein. 200 μg/ml protein in buffer A containing 0.1M KCl. The logarithm of the fringe displacement is plotted against the square of the radial position. Data were obtained from a picture taken with a Rayleigh interference optical system 9 hr after reaching a speed of 26,000 rev./min

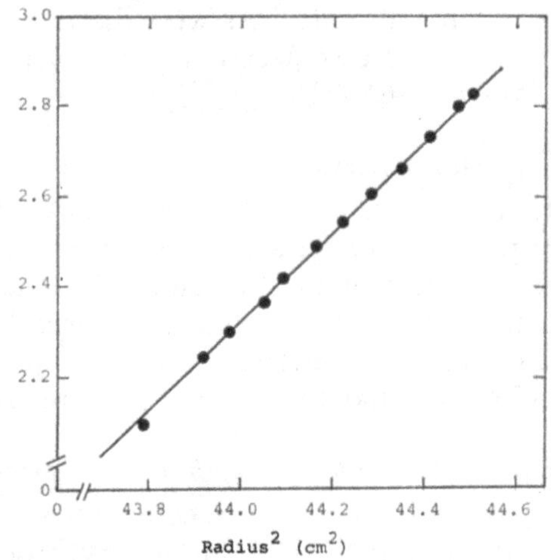

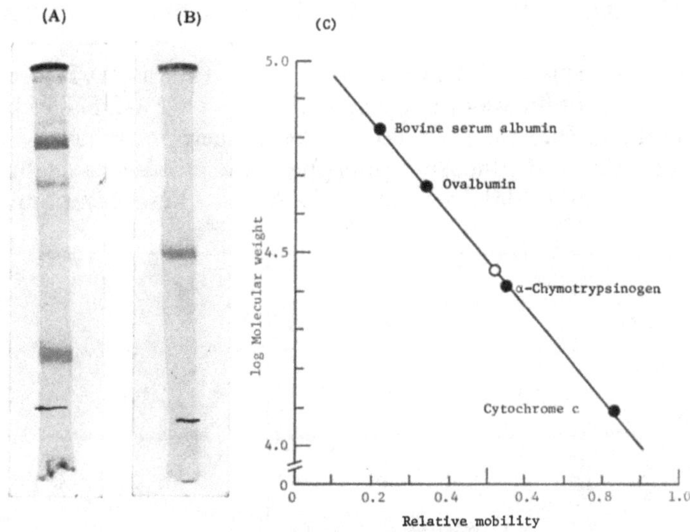

Fig. 7. Polyacrylamide gel electrophoresis of peak III protein in the presence of sodium dodecyl sulfate.
(A) Standard (top to bottom) containing bovine serum albumin (66,000), ovalbumin (47,000), α-chymotrypsinogen (26,000) and cytochrome c (12,400).
(B) Peak III protein (10 μg).
(C) Standard curve drawn from the mobilities of standard proteins. (●) Standard proteins; (○) peak III

single band, the molecular weight of which was calculated as 28,000 by use of a standard curve drawn from the electrophoretic mobilities of proteins of known molecular weight. Peak III protein, therefore, is composed of two subunits of a similar molecular weight.

Isoelectric focusing of peak III protein

Isoelectric focusing is a sensitive method for separating proteins according to their electric charges. Peak III protein was subjected to isoelectric focusing to see if the three enzyme activities were distinguishable. The three activities sharply electrofocused at an isoelectric point of 5.1; other fractions showed no enzyme activity (data not shown).

Thermal inactivation

Evidence that the DPG mutase, 2,3-DPG phosphatase, and PGA mutase activities of peaks I and III are due to one protein is shown in Fig. 8; although peak I protein purified by CM-Sephadex chromatography was heterogeneous electrophoretically, it also was subjected to thermal-inactivation experiments. In both cases the three enzyme activities were lost at the same rate during thermal inactivation, indicating that the activities are related to one protein.

Enzymatic properties of peak III protein

Specific activity: Specific activities of the DPG mutase and 2,3-DPG phosphatase of peak III from pig erythrocytes (Table 1) were very similar to those of peak III from human erythrocytes, while the PGA mutase activity of the pig enzyme was lower.

Substrate specificity of 2,3-DPG phosphatase activity of peak III protein: Peak III protein was highly specific toward 2,3-DPG in hydrolysing a phosphoester bond. 2-Phosphoglycolate, D,L-α-glycerophosphate, β-glycerophosphate, 2-PGA, 3-PGA and glucose-6-phosphate did not serve as substrates at a concentration of 0.5 mM (data not shown). All of these compounds except 2-

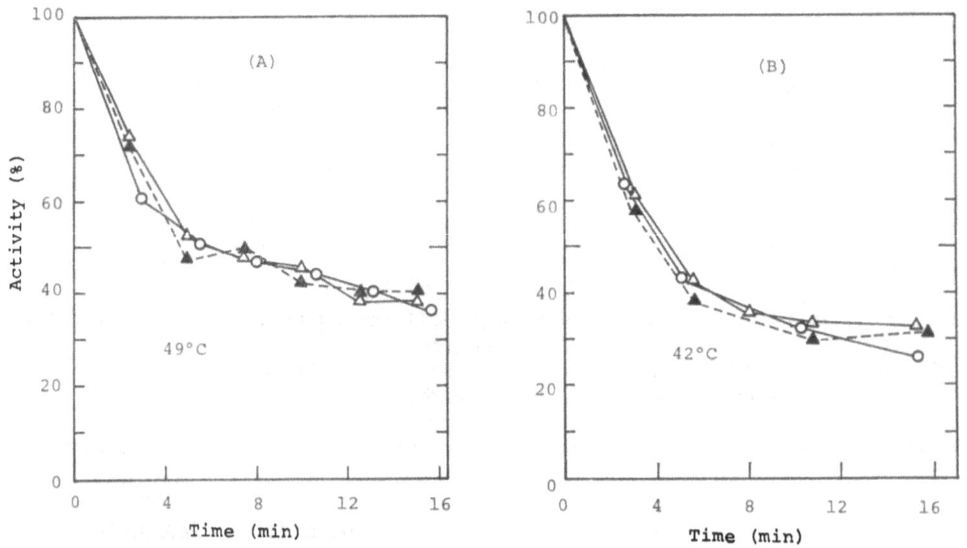

Fig. 8. Thermal inactivation curves of the three enzyme activities of peaks III and I. (A) peak III protein, (B) peak I protein. (△——△) DPG mutase; (▲– – –▲) 2,3-DPG phosphatase; (○——○) PGA mutase. The protein concentration of peak III was 216 µg/ml in buffer A and that of peak I, 1.5 mg/ml. The sample solutions (2 ml of each peak) were heated at the indicated temperature and aliquots were removed at intervals to assay the three enzyme activities. DPG mutase and 2,3-DPG phosphatase activities were assayed with assay I. [1]

Table 1
Summary of enzyme purification.
DPG mutase and 2,3-DPG phosphatase activities were assayed with assay I [1]. The activities of the hemolysates were measured after dialysing against buffer A

Fractions	Specific activity			Yield		
	DPG mutase	2,3-DPG ase	PGA mutase	DPG mutase	2,3-DPG ase	PGA mutase
Hemolysate	0.0028	0.0003	0.0165	100	100	100
DEAE-cellulose	0.094	0.0115	0.56	95.2	95.1	98.1
$(NH_4)_2SO_4$ frac.	0.200	0.025	1.22	85.7	87.4	91.1
DEAE-Sephadex						
Peak I	0.002	0.001	6.17	0.13	0.51	68.2
2nd peak	1.23	0.144	0.58	66.7	64.1	5.5
Sephadex G-100 of 2nd peak	5.87	0.856	3.4	29.4	35.0	3.0
Hydroxyapatite						
Peak II	1.05	0.562	2.63	0.43	1.9	0.19
Peak III	11.4	1.7	5.3	10.6	12.9	0.86
CM-Sephadex of peak I	0.0108	0.005	33.1	0.05	0.21	29.5

phosphoglycolate were tested in the presence and the absence of 2-phosphoglycolate, a potent activator of 2,3-DPG phosphatase activity (see Table 2).

Effects of compounds on the 2,3-DPG phosphatase and DPG mutase activity of peak III protein: The 2,3-DPG phosphatase and DPG mutase activities of peak III protein were examined for the effects of various compounds shown to affect both activities of the enzyme from human erythrocytes [1]. As shown in Table 2, monophosphoglycerates inhibited the 2,3-DPG phosphatase activity, 2-phosphoglycolate and HPAP markedly enhanced the activity. The effect of HPAP has not yet been tested on erythrocyte enzymes, but it has been shown to enhance the 2,3-DPG phosphatase activity of yeast PGA mutase [9]. Enhancement by 2-phosphoglycolate and HPAP seems to be a property of the 2,3-DPG phosphatase activity associated with 2,3-DPG-dependent PGA mutase [1, 9, 10]. Bisulfite and KCl or bisulfite and NaCl together markedly enhanced the activity, while little effect was seen from the presence of KCl alone. The chlorine ion seems to be required for display of the stimulatory effect of bisulfite. The small activation observed in the presence of bisulfite alone is probably related to chlorine ions originating from the Tris-HCl buffer. It has been shown that the inhibition of DPG mutase by its product, 2,3-DPG, is an efficient device for the regulation of 2,3-DPG level in human erythrocytes [11]. The DPG mutase activity of peak III also suffers product inhibition (Table 2). HPAP completely inhibits the activity at concentrations at which other phosphate compounds inhibited partially. It is interesting that compounds having a stimulatory effect on 2,3-DPG phosphatase activity inversely inhibit DPG mutase activity.

A new classification of multifunctional enzymes

Multifunctionality of the enzymes related to 2,3-DPG has been found not only in erythrocytes but also in various sources. 2,3-DPG-dependent PGA mutase from yeast [9, 12] and rabbit muscle [13, 14] are capable of synthesizing 2,3-DPG from 1,3-DPG and 3-PGA and hydrolysing 2,3-DPG into 3-PGA and P_i. It may be concluded that 2,3-DPG-dependent PGA mutases generally possess DPG mutase and 2,3-DPG phosphatase activities. Ratios of the enzyme activities on one protein are variable depending on the enzyme sources and probably on its physiological function. Since it is clear that there are at least two types of PGA mutase, It may be worthwhile to propose a new classification of PGA mutase based on the ratios of the three enzyme activities on one protein. 2,3-DPG-dependent PGA mutases are classified into two groups, class I and II (Table 3).

Table 2

Effects of compounds on the 2,3-DPG phosphatase and DPG mutase activities of peak III protein.

2,3-DPG phosphatase was assayed with assay I using 0.4 to 100 μg of peak III and DPG mutase was assayed with assay II using 1 μg of the enzyme [1]. The reaction mixture for 2,3-DPG phosphatase assay contained no 2-phosphoglycolate except when its effect was examined. 2,3-DPG ase = 2,3-DPG phosphatase

Enzyme	Compound	Conc. (mM)	Relative activity
2,3-DPGase	None	—	1.0
	2-PGA	1.0	0.84
	3-PGA	0.5	0.17
	HPAP	0.02	31
	phosphoglycolate	0.05	500
	KCl	100	0.94
	$NaHSO_3$	20	3.4
	KCl + $NaHSO_3$	100 20	55
	NaCl + $NaHSO_3$	100 20	50
DPG mutase	None	—	1.0
	2,3-DPG	0.1	0.4
	phosphoglycolate	0.4	0.4
	HPAP	0.4	0
	P_i	0.4	0.57
	$NaHSO_3$	6.0	0.47

Table 3

A classification of multifunctional enzymes related to 2,3-DPG

Class	Enzymes	Sources	Probable functions
I	PGA mutase Peak I (PGA mutase)	yeast, rabbit muscle human and pig erythrocytes	Glycolysis
II	Peak III DPG mutase	human and Pig erythrocytes rabbit muscle	2,3-DPG metabolism 2,3-DPG synthesis

Class I enzyme is highly active in the PGA mutase activity and functions in the main pathway of glycolysis. Class I includes peaks I protein from the erythrocytes and PGA mutase from yeast and rabbit muscle. In ruminant erythrocytes which contain 2,3-DPG at low levels, class I enzyme might be responsible for the 2,3-DPG metabolism as well as for the glycolysis. Class II enzyme is much more active in the DPG mutase and 2,3-DPG phosphatase activities as compared with class I enzyme. Peaks III protein from mammalian erythrocytes rich in 2,3-DPG and probably DPG mutase from the muscle (unpublished results) belong to this class. Both DPG mutase and 2,3-DPG phosphatase activities of class II enzyme are responsible for metabolising 2,3-DPG in the erythrocytes containing this compound at high levels. It is doubtful, however, that the 2,3-DPG phosphatase activity of class II enzyme functions to break down 2,3-DPG in the muscle, because 2,3-DPG-specific phosphatase carrying neither PGA mutase nor DPG mutase activities has been found.

Discussion

The most remarkable result described here is that 2,3-DPG synthesis and breakdown in pig erythrocytes are catalysed by one protein (peak III) carrying PGA mutase activity. This is supported by the experimental results that: (a) peak III protein had the bulk of the DPG mutase and 2,3-DPG phosphatase activities in the hemolysates, (b) peak III protein was homogeneous by various criteria of protein purity, and (c) the three enzyme activities of peak III were lost at the same rate on thermal inactivation.

Striking resemblances in physical and enzymatic properties were observed between proteins of peak III of human and pig erythrocytes [1, 2]. Inverse effects of phosphate compounds on the 2,3-DPG phosphatase and DPG mutase activities of peak III from pig erythrocytes indicate that one active site on this protein molecule is related to both activities (see Table 2). Further studies are needed, but it is likely that only these two activities but also the PGA mutase activity are manifested at one active site, as are those of the enzyme from human erythrocytes [2, 3]. Results from both erythrocytes support the suggestion that 2,3-DPG levels in mammalian erythrocytes rich in this compound may be regulated by one multifunctional protein capable of catalysing the two irreversible reactions responsible for the metabolism of 2,3-DPG. A possible significance for multifunctionality has been suggested for the 2,3-DPG metabolism of human erythrocytes [2, 3]. Results of thermal inactivations experiments of peak I protein indicate that it also is a multifunctional enzyme. However, peak I protein probably functions as PGA mutase in glycolysis, as its 2,3-DPG-metabolising activities are very low compared with those of peak III. It appears from the low activities of the three enzymes in peak II that this peak contributes neither to glycolysis nor to 2,3-DPG metabolism. Such a protein has been found in human erythrocytes [1]. It remains to be clarified whether multiple molecular forms of the enzyme related 2,3-DPG are genetic or posttranscriptional in origin.

The results described in this paper support the hypothesis of a previous paper [12] that 2,3-DPG-dependent PGA mutase generally possesses the DPG mutase and 2,3-DPG phosphatase activities.

References

[1] Sasaki, R., K. Ikura, E. Sugimoto, and H. Chiba: Eur. J. Biochem. **50**, 581 (1975)
[2] Ikura, K., R. Sasaki, H. Narita, E. Sugimoto, and H. Chiba: Eur. J. Biochem. (in press)
[3] Chiba, H., K. Ikura, H. Narita, and R. Sasaki: Acta biol. med. germ. **36**, 491 (1977)
[4] Rosa, R., I. Audit, and J. Rosa: Biochimie **57**, 1059 (1975)
[5] Hass, L. F., and K. B. Miller: Biochem. biophys. Res. Commun. **66**, 970 (1975)
[6] Davies, E. J.: Ann. N.Y. Acad. Sci. **121**, 404 (1964)
[7] Weber, K., and M. Osborn: J. biol. Chem. **244**, 4406 (1969)
[8] Yphantis, D. A.: Biochemistry **3**, 297 (1964)
[9] Sasaki, R., M. Hirose, E. Sugimoto, and H. Chiba: Biochim. biophys. Acta **227**, 595 (1971)
[10] Rose, Z. B., and J. Liebowitz: J. biol. Chem. **245**, 3232 (1970)
[11] Rose, Z. B.: J. biol. Chem. **243**, 4810 (1968)
[12] Sasaki, R., S. Utsumi, E. Sugimoto, and H. Chiba: Eur. J. Biochem. (in press)
[13] Pizer, L. I.: J. biol. Chem. **235**, 895 (1960)
[14] Laforet, M. T., J. B. Butterfield, and J. B. Alpers: Archs Biochem. Biophys. **165**, 179 (1974)

The author's address: Department of Food Science and Technology, Faculty of Agriculture, Kyoto University, Kyoto, 606, Japan

Acta biol. med. germ., Band 36, Seite 681 (1977)

Discussion

With respect to the report of MINAKAMI it was suggested to calculate the thermodynamic parameters of the degradation of glucose to 3 CO_2 and 1 glyceraldehyde-3-phosphate via the pentose pathway in the presence of methylene blue. The importance of ion transport cannot be assessed as yet. The reports of T. RAPOPORT and of GLENDE gave rise to an extensive discussion. The following points were clarified: 1) in the steady state the bypass is increased only little by greater glycolytic rates, but may be strongly influenced for short periods of time. 2) The idea of a metabolic breakdown in the absence of oscillation of the oxygenation-deoxygenation cycle does not correspond either to experimental data or the experiences of blood preservation. 3) Under conditions of blood preservation the hexokinase-phosphofructokinase system is inhibited relative to the ATP consuming processes and the DPG mutase is strongly inhibited by the low *p*H (T. RAPOPORT). FRUNDER amplified his report as follows: Addition of ruthenium red after 1 hr incubation has little effect since the ATP concentration is low. The dye has a lag phase of 10—20 min. The addition of ouabain inhibits normal and stimulated glycolysis by 20% both in intact cells and hemolysates. There is a small decrease of 2,3-DPG in the presence of ionophore and in hemolysates. Acetylcholine esterase may be solubilized to the extent of 10—15% from the membranes. The very high glycolytic rates occasionally found cannot be explained satisfactory. MARETZKI added that in 1:4 membrane-free hemolysates there is a stoichiometric relation of 1:1:1 between glucose consumption the formation of lactate and that of triose phosphates and fructose-1,6-phosphate, and that there may be observed 2,3-DPG degradation in hemolysates.

The discussion of the report of RYKSEN centered on the question whether Mg^{2+} is an allosteric activator (GERBER), or whether its effect can be explained by the inhibition by ATP^{3-}. The discussion of the report of ARESE focussed on the possible mechanism of the effect of ferricyanide which remains unclear. Furthermore it was stated (S. RAPOPORT) that in earlier work of ARESE the conditions were far from those in the cell. Addition of DPG and ATP in physiological amounts would have reduced the synthesis to 1%, which is only a tiny fraction of the glycolytic flux.

CHIBA supplemented his report by the information that beef erythrocytes do not contain the DPG mutase (peak IV) of human erythrocytes. The 2,3-DPG production of ruminant erythrocytes may be due to monophosphoglycerate mutase. In the discussion of the report of I. RAPOPORT it was clarified that no step above DPG mutase is affected by low *p*H as indicated by the unchanged 1,3-DPG concentrations. The time constants observed in vitro as well as the steady state levels correspond fully to those in vivo.

Acta biol. med. germ., Band 36, Seite 683—685 (1977)

Workshop

The following topics were discussed:

1) The properties of the multifunctional enzyme bisphosphoglycerate mutase — phosphoglycerate phosphatase;

2) the phosphofructokinase, in particular whether the cooperative effects are explainable by an allosteric transition between active and inactive states or by the association-dissociation;

3) the carbon balance in glycolysis;

4) the ATP-consuming processes in erythrocytes;

5) the correspondence between models and changes of metabolites under conditions of blood preservation.

With respect to the kinetic properties of the bisphosphoglycerate mutase S. RAPOPORT (Berlin) pointed out that at low *p*H the activity of this enzyme in intact cells is strongly inhibited. But it is not clear whether the effect has to be ascribed to a decrease of the K_i-value of 2,3-DPG or to an increase of the K_m-value of 1,3-DPG. It is very difficult do discriminate the two effects since the inhibition constant K_i is very low compared to the concentration of 2,3-DPG. The final answer has to be given by the enzymologists.
A second problem concerns the effect of phosphohydroxypyruvate. It has both a strong activating influence on the mutase activity and inhibiting influence on the phosphatase activity of the multifunctional enzyme. This effect gains some interest since it has been shown that in erythrocytes there is an operative pathway from phosphoglycerate to phosphoserine and it seems reasonable to assume that the other pathway through phosphohydroxypyruvate does also exist. In such a way the rudiment of amino acid metabolism may play a controlling role in the glycolysis at the level of the phosphoglycerates.
ARESE (Torino) pointed out that he incubated sichelcells with hydroxypyruvate, but found no change of the 2,3-DPG-concentration. He concluded from that experiment that either hydroxypyruvate does not penetrate or that it is not phosphorylated in the cell.

S. RAPOPORT (Berlin) said that hydroxypyruvate does penetrate but that the rate of phosphorylation is probably very low.

SASAKI (Kyoto) and S. RAPOPORT (Berlin) agreed that it would be worthwile to construct a kinetic model of the multifunctional enzyme. FRUNDER (Jena) suggested an analysis of the effects of the different Mg-complex species of the ligands 2,3-DPG and PG.

With respect to the phosphofructokinase HOFMANN (Leipzig) reported that the enzyme from mammalian cells may occur in different associated forms. There exists a dimer-tetramer transition and also transitions to higher polymeric forms. While the tetramer and the higher polymeric states are active, the dimer is inactive. Since F6P as an activator of the enzyme shifts the equilibrium towards the tetramer and citrate as an inhibitor and under certain conditions also ATP shift the equilibrium towards the dimer, the question arose whether the cooperative effects of the phosphofructokinase may be explained by such association-dissociation phenomena. The experiments indicate that the tetrameric form is predominant both at *p*H 7.2 and *p*H 8. It seems, therefore, that in addition to polymerization an allosteric transition between active and inactive tetramers contributes to the cooperative properties of the enzyme. At present it is not possible to give a quantitative estimate which of the mechanisms is more important within the cell. Otto (Berlin) pointed out that mathematical models which take into account only allosteric transitions explain fully the kinetic properties of the phosphofructokinase at constant enzyme concentrations. He agreed with Hofmann that a complete model of phosphofructokinase must include association-dissociation phenomena in order to describe correctly the concentration dependence of the kinetic behaviour of the enzyme. However, at the high enzyme concentrations within the cell the allosteric transition plays the predominant role. He proposed further that the phosphorylation of phosphofructokinase by protein kinase should be considered for the regulation of the enzyme in vivo.

Brandt (Tübingen) reported that he tried to calculate the bypass rate by consideration of the ATP-requirements of the cell. His estimate of 20 to 40% of the carbon flux is in agreement with results obtained by I. Rapoport who considered the decomposition rate of 2,3-DPG in the absence of substrates. Brandt proposed that the rate of the bypass increases with higher glucose consumption. S. Rapoport pointed out that a higher glucose consumption at alkaline *p*H values need not mean a higher bypass rate. One can only say that under these conditions there must be an additional ATP-wastage, which could be due also to a stimulation of the ATP-consuming reactions in the membrane. Brandt and S. Rapoport agreed that the problem may be solved only by a quantitative analysis of the ATP-consuming processes.

With respect to the balance of carbon atoms in erythrocytes Brandt reported that at low glucose uptake all carbon atoms utilized can be found in lactate and pyruvate but at high glucose concentration it is not possible to get a complete carbon balance; he never found more than 80% of the carbon atoms in lactate and pyruvate. This discrepancy remains if one takes into consideration the oxidative and nonoxidative pentose phosphate pathway and the triose phosphate pool. The carbon balance was studied not only by analytical methods but also by isotopic methods. The discrepancy exists only for glucose and not if other substrates are used.

Till (Jena) gave a short review of his work on ATPases. He was mainly interested in the question which processes in the erythrocytes consume the main part of ATP. After action of ionophores there was a drastic fall in the ATP-concentration. The same decrease of ATP and a corresponding increase of AMP and ADP was obtained if after the action of ionophore inhibitors of the different ATPase are added, ouabain for the Na^+/K^+-ATPase, ruthenium red and lanthanum chloride for the Ca^{++}-ATPase. There was no correlation between ATP-consumption and lactate formation. The only correlation which has been found was a parallelity between the decrease of ATP and the shrinkage of the cells. He suggested that the actomyosin-like system may be responsible for the shrinkage. S. Rapoport said that the ATP-consumption by Ca-ATPase is very low. The Mg/Ca-ATPase and others are not yet defined. Working with chick hemolysates Maretzki has found an ATP-consuming process with a very high K_m-value. It could well be that at different concentrations of ATP different ATPases are functional.

Maretzki (Berlin) reported that no membrane bound glyceraldehyde phosphate dehydrogenase could be found and that other authors (Schrier) could not find any binding of the phosphoglycerate kinase either. Arese replied that Tillmann and coworkers showed that in isolated membranes almost all glycolytic enzymes are present. Frunder stated that the enzymes firmly bound to the membrane amount to only 1% of the total enzymes.

The discussion turned to the problems of blood preservation. Heinrich (Berlin) summarized how the changes of the metabolite concentrations are explained by the theorists. The mathematical models of erythrocytes glycolysis show that there is only a limited range of the activity of the ATPase where the system exhibits steady states and where the concentrations of ATP and 2,3-DPG are unequal to zero. If the ATPases or the hexokinase-phosphofructokinase system are activated or inhibited too strongly, there is only one steady state with ATP and 2,3-DPG = 0. Since the degraded 2,3-DPG forms ATP at the pyruvate kinase step, ATP remains almost constant over a long period of time and falls rapidly after degradation of 2,3-DPG. The changes of the metabolites found experimentally in the ACD-medium could be fitted by the mathematical model if one assumes in addition to a relative inhibition of the hexokinase phosphofructokinase system as compared with the ATPases, also a strong inhibition of the bisphosphoglycerate mutase. The model predicts that the process is markedly affected by the degradation of the adenine nucleotides via AMP-degradation.

Aebi (Zürich) emphasized the importance of studies to find optimal conditions for blood preservation since other preservation techniques like glycerol and low temperature are not practical for general use. He pointed out that the addition of inosine, adenine and guanosine increases the time where a blood bottle can be used. In the Swiss Red Cross the conditions of blood preservation were changed from ACD to CPD-medium.

At the end of the poster session some clinical problems were discussed. Jacobasch (Berlin) said that for the clinical work results of Oski are important, namely that the degradation of 2,3-DPG in new-borns is faster than in adult erythrocytes. She asked whether this phenomenon may be derived from kinetic differences of the bypass-enzymes in newborns and adults. I. Rapoport reported that she could confirm that the degradation rate in newborns is twofold higher. She supposed that the effect is due to a higher enzyme concentration in newborns.

Section III

Enzymopathies and metabolic defects

Acta biol. med. germ., Band 36, Seite 689—701 (1977)

Department of Biochemical Genetics, City of Hope National Medical Center, Duarte, California, U.S.A.

Glucose-6-phosphate dehydrogenase abnormality and hemolysis

AKIRA YOSHIDA

Summary

Genetic abnormality of glucose-6-phosphate dehydrogenase (G6PD) is often associated with chronic and drug-induced hemolytic anemia. In order to fully understand the molecular mechanism of abnormality and hemolysis, the following studies have been carried out:

a) Structure and function of normal human G6PD:
G6PD is a peptide chain, consisting of about 500 amino acid residues, with pyroglutamic acid at the N-terminal and glycine at the C-terminal. The complete amino acid sequence has not yet been determined, but all tryptic peptides and cyanogen bromide peptides of G6PD have been isolated and analyzed. Thus, a general outline of the primary structure of the protein has become apparent. Under the optimal assay conditions in vitro, and presumably also in red cells, the enzyme consists of two subunits, i.e., dimeric form.
From the kinetic analysis, it was found that 1) binding with NADP and G6P is obligatory sequential; and 2) only one G6P and NADP, or one NADPH associates with the dimeric enzyme at a time.

b) Structural and functional abnormalities of variant G6PDs:
Due to the difficulty for obtaining sufficient amount of blood from variant subjects, structural abnormalities (i.e. exact amino acid substitutions) were elucidated in only a few variant G6PDs. Peptide maps of tryptic peptides of several variant G6PDs were examined, but exact amino acid substitutions have not yet been identified due to other technical difficulties involved.
The estimation of specific activity of variant enzymes is important for understanding the mechanism of inherited enzyme deficiency. Since direct determination of specific enzyme activity using purified homogeneous variant enzymes is often difficult, indirect immunological methods, (i.e. the quantitative neutralization method and the electro-immuno assay method) have been used for estimating the specific activity of variant enzymes. These methods are used for not only G6PD variants but also for other enzymes. However, our study disclosed that the values estimated by the indirect methods do not reconcile with the specific enzyme activity determined by the direct determination methods. The immunological methods are useful for qualitative comparison of the normal and variant enzymes, but care should be taken for deducing the quantitative specific enzyme activity from these assay methods.

c) Factors which influence G6PD activity, and physiological G6PD activity in normal and variant red cells: It is expected that physiological enzyme activity is more closely related to the clinical severity of hemolytic disorder than in vitro activity assayed under the optimal condition. Considering the concentrations of various co-enzymes and intermediate metabolites in red blood cells, together with the kinetic characteristics of the normal and variant enzymes, we estimated the physiological G6PD activity in the normal and variant red blood cells. The G6PD activity in red cells from hemolytic variant subjects are estimated to be lower than that in red cells from non-hemolytic variant subjects. The major factor that controls the physiological G6PD activity is NADPH which is a strong competitive inhibitor of the enzyme and exists at a high concentration in red blood cells. However, other undisclosed factors may also influence activity of G6PD. For example, our study demonstrated that NADPH is spontaneously converted to NADPH-X, a hydration product of NADPH, under simulated physiological conditions, and that

NADPH-X is a strong competitive inhibitor of G6PD. The concentration of NADPH-X in red cells has not yet been determined, but the study of this and other analogous materials may be important for understanding the reglutory mechanism of G6PD (and other dehydrogenases) in red cells.

Introduction

Genetic deficiency of glucose-6-phosphate dehydrogenase (G6PD) in man is often associated with chronic and drug-induced hemolytic anemia. In order to fully understand the molecular mechanism of the enzyme abnormality and hemolysis, we have been studying: a) structure (including determination of amino acid sequence and three dimensional structures) and mechanism of function of the normal human G6PD; b) structural and functional abnormalities of variant G6PDs and c) mechanism of hemolysis associated with G6PD abnormalities. Although these problems have only partly been resolved at the present time, I will report some of the progress made in my laboratory.

Results and discussion

Structure and mechanism of function of normal human G6PD

Subunit of G6PD is a single peptide chain, consisting of about 500 amino acid residues, with pyroglutamic acid at the N-terminal and glycine at the C-terminal

Table 1
Tryptic peptides from human glucose-6-phosphate dehydrogenase
(March, 1976)

Amino acid sequence was determined:
Free Lys; Free Arg; Leu-Arg; Ile-Leu-Arg; Val-Lys; Leu-Lys; Ala-Arg; Ser-Arg; Phe-Ala-Asn-Arg; Val-Val-Lys; Glu-Leu-Ser-Arg; Ala-Leu-Glu-Asn-Arg; Leu-Glu-Asp-Phe-Phe-Ala-Arg;

Amino acid sequence was partly determined:
B+: Ala-Leu-(His, Asn, *Asn*, Ser, Glu$_2$, Gly, Ala, Met, Leu)-Arg;
A+: Ala-Leu-(His, Asn, *Asp*, Ser, Glu$_2$, Gly, Ala, Met, Leu)-Arg; (Asx, Val)-Lys; (Glx, Ala, Val)-Arg; (Val$_2$, Ile)-Lys; (Glx, Val, Tyr)-Lys; (Asx, Thr$_2$, Glx, Pro, Ala, Val$_2$, Ile$_2$, Leu$_2$)-Arg; (Gly$_2$ Cys)-Lys; (Thr, Glu, Ala, Val, Tyr)-Lys; (His, Asx, Glx$_2$, Pro, Gly, Ala, Val)-Lys; (His, Asx, Gly, Ile, Leu, Cys$_2$)-Lys; Met-Met-Thr-Lys;
(His$_2$, Asx$_2$, Thr$_2$, Ser$_2$, Pro$_2$, Ala$_2$, Cys, Val$_2$, Met, Ile, Leu$_2$, Phe)-Lys; (Asx, Thr, Ser, Glx$_2$, Gly, Ala$_2$, Val$_2$, Leu, Tyr, Cys)-Arg; (Asx, Glu)-Lys; (Asx, Thr, Ser, Glx$_2$, Pro, Gly, Ala, Ile, Leu, Phe$_2$)-Lys;
(Thr, Glx$_2$, Pro, Ala, Trp)-Lys; (Glx, Gly, Phe)-Arg;
Asp-Leu-Glu-(Asp, Ser$_2$)-Arg; (Asx, Glx, Ala, Met, Leu, Trp)-Arg;
Glu-(Asx, Glx, Ile, Tyr)-Arg;
His, Asx$_2$, Thr, Ser$_2$, Glx$_3$, Pro$_2$, Ala$_2$, Ala$_2$, Val$_3$, Met, Ile, Leu$_2$)-Lys:
(Asx, Thr, Ser, Glx$_2$, Gly, Ala, Trp)-Lys; Leu-Pro-(Asx, Glx, Ala, Tyr$_2$)-Arg; (Asx$_2$, Thr, Ser, Glx$_3$, Pro$_2$, Gly$_2$, Ala, Val$_2$, Met, Leu, Tyr$_2$, Phe)-Lys; (Asx, Thr, Glx$_3$, Pro, Gly$_2$, Ala, Val$_2$, Ala, Val$_2$, Leu)-Lys; (Thr, Glx$_2$ Gly$_2$, Val$_2$, Tyr$_2$, Phe)-Lys;
(Asx, Thr, Glx$_2$, Pro, Gly, Ala, Met, Leu)-Lys;
(Asx$_2$, Thr, Glx, Pro$_2$, Gly, Val$_2$, Leu, Tyr)-Arg;
Lys-Pro-Gly-Met-Phe-Phe-(Asx$_3$, Thr, Ser, Glx$_3$, Gly, Leu$_2$, Tyr)-Arg; Ala-X-Tyr-Val-Ala-Gly-(Asx$_3$, Ser$_2$, Glx$_2$, Ala, Tyr$_2$)-Arg;
X-Gly-Tyr-Phe-X-X-Phe-(Asx, Glx, Gly, Ile)-Arg;
Glx, Pro, Gly, Val, Ile, Phe, Arg, Lys;
His, Asx$_3$, Thr$_2$, Ser$_2$, Glx$_2$, Pro, Gly, Ala, Val$_2$, Met, Ile$_2$, Leu$_2$, Tyr, Phe, Arg, Lys; His, Asx$_4$, Thr$_3$, Ser$_2$, Glx$_4$, Pro, Gly$_3$, Ala$_3$, Cys, Val$_4$, Met$_2$, Ile$_3$, Leu$_4$, Tyr$_2$, Phe$_3$, Arg$_2$, Lys
(Asx, Glx, Pro, Gly, Ala, Val$_2$, Leu, Phe)-Gly; COOH — terminal
Pyroglutamic acid; NH_2 — terminal

[1]. The complete amino acid sequence has not yet been determined, but all tryptic peptides and cyanogen bromide peptides of the enzyme have been isolated and analyzed. These peptides and several peptide portions of the enzyme including N-terminal peptides, C-terminal peptides, and the peptide involved in the structural abnormality of the common A+ variant, are listed in Tables 1—3. Comparing to other dehydrogenases whose amino acid sequences have been determined (glutamate dehydrogenase, alcohol dehydrogenase, lactate dehydrogenase, and glyceraldehyde-3-phosphate dehydrogenase) [2], human G6PD is similar to bovine glutamate dehydrogenase [3] in respect to subunit molecular size and amino acid composition. However, the N-terminal portion, which has been implicated to the co-enzyme binding site of glutamate dehydrogenase, and the C-terminal sequence are substantially different in the two enzymes.

Moreover, glutamate dehydrogenase is hexameric in the active form, while human G6PD is dimeric in the active form [4]. Human G6PD is unique in these respects, and the functional characteristic of the enzyme will be discussed below in more detail.

It has been demonstrated by the "active enzyme molecular determination method" that in the presence of both glucose-6-phosphate (G6P) and NADP, the G6PD which is functioning, is predominantly dimeric (MW = 110,000) [4].

The enzyme has tightly bound NADP which cannot be easily removed by dialysis. However, when the enzyme associated with ^{14}C-NADP was dialyzed against the buffer containing 0.1 mM NADPH, extensive release of NADP from the enzyme took place. Similarly, when the enzyme associated with ^{14}C-NADPH was dialyzed against the buffer containing 0.1 mM NADP, almost complete release of NADPH from the enzyme took place. These results indicate that NADP and NADPH compete to bind to the same binding site in the absence of G6P.

Using ^{14}C-glucose-6-phosphate, it was observed that the apoenzyme of G6PD (i.e. G6PD without bound NADP) and G6P bound with NADPH did not associate with G6P.

From these findings, it can be concluded that the association of NADP and G6P with the enzyme occur in a compulsory sequence, and the process of the enzyme reaction can be presented as shown in the following scheme:

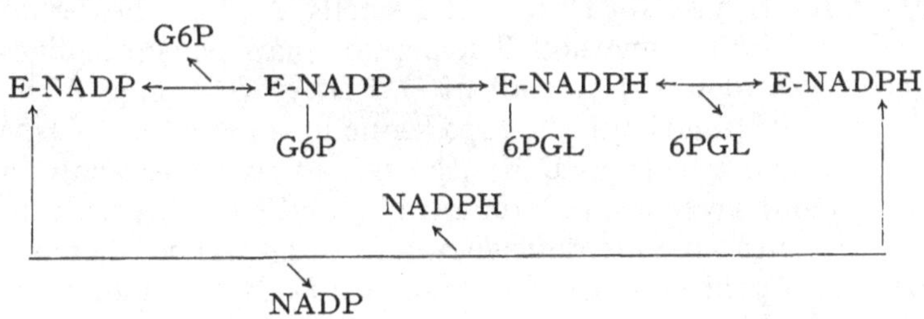

Structural and functional abnormalities of variant G6PDs

The structural abnormality, i.e. amino acid substitution of the variant enzymes, can be determined by analyzing peptide maps of the normal and variant enzymes. Using the conventional peptide mapping method on Whatman 3MM paper, thus far amino acid substitution of two G6PD variants, i.e. common A+ variant (Fig. 1)

Table 2
Amino acid composition of glucose-6-phosphate dehydrogenase CNBr peptides (March, 1976)

Peptide No.	CMGenBr A-9	CMGenBr (14-16)-1	CMGenBr (14-16)-4	CMGenBr (14-16) -5-8	CMGenBr 7,8-1cm^{c-3}	CMGenBr 4,5,6-1,2, c-1, D-1	CMGenBr A-6, CMc (2)
Asp	1		1	1	4	4	3
Thr	1	1			2	2	2
Ser	0				3	3	2
Glu	2		1	1	4	6	3
Pro	1	1			2	3	3
Gly	2	1	1		3	4	3
Ala	1				3	3	3
$^1/_2$Cys	0				1	1	0
Val	2		1		2	3	2
Met	0	1	1	1	1	1	1
Ile	1				2	3	3
Leu	1		1	2	3	4	4
Tyr	2				1	2	
Phe	1				1	2	2
Lys	2	2			3	2	2
His	1			1	1	1	1
Arg	1				3	2	3
Total (Try excluded)	19	6	6	6	39	46	37

and Hektoen variant, have been determined [5, 6]. At least 10 mg of pure enzyme preparation is required for duplicate analysis by the conventional method. The content of G6PD is about 2—3 mg per liter of normal blood, and the content of many G6PD variants associated with enzyme deficiency is lower than the normal content. It is difficult or impossible to obtain sufficient amounts of pure variant G6PD for the conventional peptide mapping. In order to overcome this problem, a micro peptide mapping method on thin layer cellulose, which required only 0.1 mg of protein, was developed [7].

Several G6PD variants, i.e. G6PD from HeLa cells, A-, Manchester, and Toronto, were examined by the micro method. The peptide map of tryptic digests of G6PD from HeLa cells was identical to that of G6PD A$^+$ (Fig. 2) [8]. Therefore, like G6PD A$^+$, the substitution should be asparagine in the normal to aspartic acid in HeLa G6PD. Peptide maps of G6PD Manchester, a variant associated with unique kinetic properties and severe hemolytic anemia, and G6PD A$^-$, a common Negro variant associated with enzyme deficiency and drug-induced hemolytic anemia, were compared with peptide map of normal G6PD. It was found that a normal peptide spot was replaced by a slower cathodal peptide spot in G6PD A$^-$, and another normal peptide spot was replaced by a faster cathodal peptide spot in G6PD Manchester. These results indicate that the two variants were induced by two separate point mutations (details will be published in a separate paper).

In the case of G6PD Toronto, a variant associated with enzyme deficiency and occasional hemolytic problem, no difference was found in the peptide maps of tryptic digests obtained from the normal and variant enzymes. Since G6PD To-

Table 2 (Continued)
Amino acid composition of CNBr peptides of glucose-6-phosphate dehydrogenase (March, 1976)

Peptide	$CMGenB_2$	$CMGenB_2$ A-3 CMC-1 A-3 CMC(*p*H)-1	$CMGenB_2$ A-5 CMC-1	$CMGenB_2$ A-5 CMC-5 A-5 CMC-7	$CMGenB_2$ A-5 CMC-8	$CMGenB_2$ 10-1 A-8 insoluble
Asp	10	10	5	4	6	2
Thr	6	5	3	2	3	1
Ser	7	5	4	3	4	1
Glu	12	10	6	4	5	2
Pro	5	5	3	2	3	2
Gly	7	7	4	2	3	2
Ala	9	7	4	3	6	1
1/2Cys	2	4	0	0	1	0
Val	5	8	4	2	2	1
Met	1	1	1	1	1	1
Ile	5	5	3	2	3	1
Leu	10	7	4	4	5	2
Tyr	6	4	3	3	2	1
Phe	6	3	3	3	4	1
Lys	6	6	3	3	4	1
His	2	3	2	2	2	1
Arg	7	5	3	3	4	2
Total (Try excluded)	106	95	55	43	57	22

Table 3
Peptides from human G6PD
(March, 1976)

N-terminal:	Pyr Glu-(Asx, Leu_2, His) Met (Asx, Thr, Ser_2, Glx_2, Pro_2, Ala_2, Val_3, Ile) Lys-
Middle part:	-Met-Met-Thr-Lys-Lys-Pro-Gly-Met-Phe-Phe- (Asx_3, Thr, Ser, Glx_3, Gly, Leu_2, Tyr)-Arg- -(Asx_3, Thr_2, Ser_2, Glx_2, Lys-Pro, Gly, Ala, Val, Ile_2, Leu_2, Tyr)-Met-His-Phe-Val-Arg-(Asx_2, Thr, Ser, Glx_2, Pro_2, Gly_2, Ala, Ile, Leu_2, Tyr, Lys, Arg)-Met--Ala-Leu-(His, Asn, Ser, Glu_2, Gly, Ala, Met, Leu)-Arg- ↓ Asp (G6PD A^+)
C-terminal:	-Arg-Pro-(Asx_2, Thr_2, Ser_2, Glx_2, Gly_2, Ala_3, Cys, Val_3 Ile_2, Leu_3, Phe_3)-Met-Val-Glu-(Asx, Gly, Leu)-Met-Arg-Glu-(Thr, Ile, Tyr_2, His)- Lys-Lys-(Asx, Glx, Pro, Gly, Ala, Val_2, Leu, Phe)-Gly-COOH

ronto has a faster anodal mobility [9], the amino acid substitution from an acidic amino acid in the normal to a neutral amino acid in Toronto, or a neutral amino acid in Toronto, or a neutral amino acid in the normal to a basic amino acid in Toronto, is expected to exist. Failure of detection of peptide substitution in the present experiment is probably due to that: a) the solvent systems used for the peptide mapping was not adequate for resolving the normal and variant peptides;

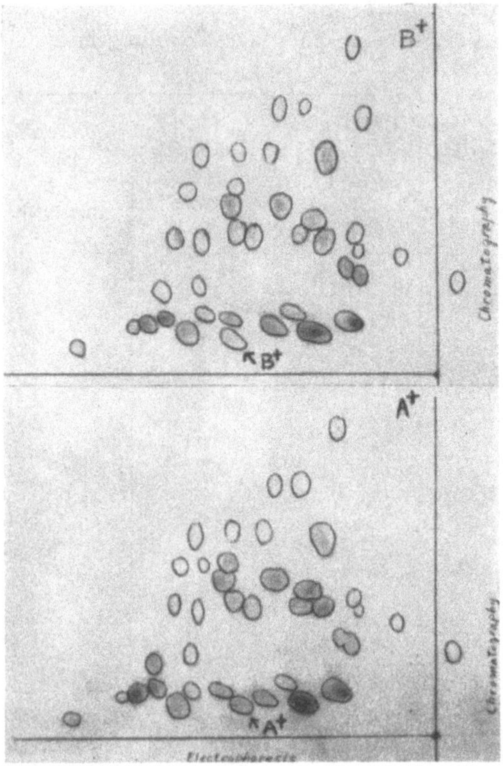

Fig. 1

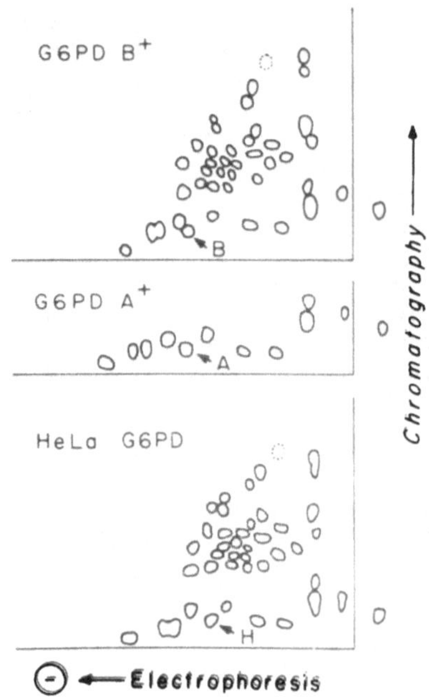

Fig. 2

Fig. 1. Peptide maps of tryptic digest of reduced, S-carboxymethylated human glucose-6-phosphate dehydrogenase on Whatman 3MM paper.
I: Normal B+ enzyme; II: Variant A+ enzyme.
First dimension, chromatography, n-butanol-acetic acid-water (4:1:5), for 18 hrs. Second dimension, electrophoresis, pyridine-acetic acid-water (1:10:289, *p*H 3.6), 2000 volts for 60 min

Fig. 2. Peptide maps of the tryptic digest of reduced, S-carboxy-methylated human glucose-6-phosphate dehydrogenase on thinlayer cellulose. Solvent systems used are same as that described in legend of Fig. 1

or b) an amino acid substitution in G6PD Toronto is located in peptides of the tryptic core (contains 15—20% of tryptic digests). A similar situation was observed in the case of G6PD Hektoen which is associated with an amino acid substitution, from histidine in the normal to tyrosine in the variant [6].

Exact amino acid substitutions of G6PD A+ and G6PD Manchester have not yet been determined due to the following technical problem: It is impossible to determine, with reasonable confidence, the amino acid composition of the peptide spots eluted from the thin layer maps, because the contamination of amino acids, particularly serine, glutamic acid, glycine and alanine, in an order of $2—3 \cdot 10^{-10}$ mole, from reagents and dust during the experimental procedures. Further technical improvement is required to overcome this problem. NOWICKI et al. [10]

reported the amino acid substitution of G6PD Frankfurt. However, I have to express serious doubt of their method and result[1].

The estimation of the specific activity of variant enzymes is important for distinguishing whether the deficiency in vivo is due to the synthesis of an abnormal enzyme with reduced specific activity or due to a decrease of the number of enzyme molecules. Direct determination of specific enzyme activity is often difficult, since it requires purification of the variant enzyme to homogeneity without causing denaturation and inactivation during the purification procedures. In case of G6PD variants, specific activity of only several variants, i.e. A^+, A^-, Hektoen, Manchester, Toronto, and G6PD from HeLa cells, have been directly determined. Except for Toronto and Manchester, whose specific activity was lower than normal (YOSHIDA, unpublished observation), these variants had normal specific enzyme activity.

Attempts were made to indirectly estimate specific activity by quantitative neutralization of the enzyme by anti-G6PD serum. From these results, it has been suggested that several G6PD variants, such as Union, Mediterranean, Toronto, etc. might have reduced specific activity, while some variants such as A^+, A^- and Hektoen, have normal specific activity [9, 11—14]. More recently, KAHN et al. [16—18] adopted the electroimmuno assay method [15] for estimating specific activity of G6PD variants. These indirect methods are valid under the assumption that the variant enzymes react identically with anti-serum to the normal enzyme, under the assay conditions. We recently examined the validity of the two indirect assay methods using the normal and variant G6PDs and three different anti-G6PD sera [19].

The values estimated by the neutralization method and that obtained by the electroimmuno assay method were substantially different. Moreover, different anti-sera gave different values for the given G6PD preparations (Table 4). The values estimated by the indirect method do not reconcile with the specific enzyme activity obtained by the direct determination.

The neutralization method and the electroimmuno assay method are useful for qualitative comparison of the normal and variant enzymes in different tissues and organisms. However, care should be taken for deducing the quantitative specific enzyme activity from the immunologic assay.

Mechanism of hemolysis associated with G6PD abnormalities

Many G6PD variants are associated with chronic nonspherocytic hemolytic anemia. More than 40 such variants have been reported at the present time. Most of the variants in this category have very severe deficiency (less than 2% of normal

[1] NOWICKI and his associates reported an amino acid substitution (lysine→glutamic acid) in G6PD Frankfurt [10]. They purified normal and variant enzymes by repetitive Sephadex gel filtration, and compared peptide maps of tryptic digests of the two enzyme preparations on thin layer silica gel plates. Analysis of peptide maps is valid only on the premise that the protein preparations are homogeneous. Human G6PD cannot be purified to homogeneity by the repetitive Sephadex gel filtration, because red cells contain many other enzymes and proteins whose molecular sizes are closely similar to G6PD. Moreover, the tryptic peptide (Asp, Ser, Glu, Gly, Ala, His, Lys) which they implicated to the amino acid substitution, does not exist in G6PD molecule (all tryptic peptides of G6PD are listed in Table 1).

Table 4
Specific activity estimated by neutralization (A) and electroimmuno assay (B)

Antiserum	No. 1		No. 2		No. 3	
	A	B	A	B	A	B
B+, partially purified	100	100	100	100	100	100
B+, purified**	106	100	75	97	75	92
B+, purified lyophilized***	60	60	29	87	65	92
Ube	88	84	91	77	88	81
Manchester****	15	*	14	*	28	*
"F"	83	82	68	93	91	88
"I"	88	103	87	87	73	95

Relative specific enzyme activity is expressed taking that of partially purified normal B+ enzyme as 100.
* Cannot be estimated, because no discernible immuno-precipitation was produced.
** Crystalline B+ enzyme was stored in 50% saturation of $(NH_4)_2SO_4$ for 18 months at 4 °C. The specific activity was 155 units/mg. The freshly prepared crystalline enzyme had specific activity of 180 units/mg, i.e., slight inactivation took place during storage.
*** Crystalline B+ enzyme was dialyzed against water and freeze-dried. The freeze-dried preparation had specific activity of 64 units/mg.
**** Partially purified preparation was stored in 50% saturation of $(NH_4)_2SO_4$ for about two years at 4 °C. The specific activity of the freshly prepared G6PD Manchester estimated by the neutralization method using anti-serum No. 1 was 55% of normal [20]. The specific activity of pure homogeneous preparation of G6PD Manchester was about 45% of normal pure (unpublished observation).

activity under optimal assay conditions) and are unstable in vitro (and presumably also in vivo) (Table 5).

Some of the variants have unusually high Michaelis constants for G6P or (and) NADP (Table 5). Therefore, it is expected that the enzyme activity is very low in the variant red cells where the concentrations of NADP and G6P are far below the Michaelis constants of the variant enzymes.

The hemolytic problem of these variants is readily correlated to the nature of their enzyme abnormalities. However, several G6PD variants associated with relatively mild enzyme deficiency and normal Michaelis constants, such as Manchester [20], Alhambra [21], and Tripler [22], cause hemolytic anemia even in the absence of exogenous agents, while some variant subjects with very severe G6PD deficiency, such as Markham [23] and Union [11], have no hemolytic problem. It is reasonable to anticipate that the physiologic enzyme activity in red cells, rather than enzyme activity measured in vitro under artificial conditions, i.e. at optimal *p*H in the presence of high concentration of NADP and G6P, is more readily related to the clinical severity of the genetic disorder. Therefore, attempts were made to estimate physiologic G6PD activity in normal and variant red cells. The red cells contain many intermediate metabolites, phosphate compounds, glutathione, and co-enzymes which may have an effect on G6PD activity. Effects of 19 such compounds existing in red cells on G6PD activity were examined [24]. Among them, NADPH, ATP, glyceraldehyde-3-phosphate, fructose-6-phosphate, fructose-1,6-diphosphate inhibited G6PD activity. Considering the physiologic concentration of these compounds, effects of glyceraldehyde-3-phosphate, fruc-

Table 5
Characteristics of hemolytic and non-hemolytic variant G6PD

Enzyme	Activity (% of normal)	K_m for G6P (μM)	K_m for NADP (μM)	Heat stability	*p*H optima
	Variants not associated with chronic hemolytic anemia				
B+ (normal)	100	50—70	2.9—4.4	Normal	8—9
A+	80	Normal	Normal	Normal	8—9
A−	8—20	Normal	Normal	Normal	8—9
Union	<3	8—12	3.6—5.2	Low	Biphasic
Markham	1.5—10	4.4—6.3		Low	Biphasic
Mediterranean	<5	19—26	1.2—1.6	Low	Biphasic
	Variants associated with chronic hemolytic anemia				
Bat-Yam	0	27		Very low	Biphasic
Ramat-Gan	0	35		Very low	Biphasic
Worcester	0	11.6	61	Very low	8.0
Oklahoma	4—10	127—200	20	Low	8.2
Ashdod	10	100		Slightly low	Biphasic
Freiburg	10—20	87—118	4		Biphasic
Albuquerque	1	115	11	Very low	8.2
Milwaukee	0.5	224			8.0
Clichy	2	178			9—10
Strasbourg	6	96	13	Low	9.0
Bangkok	5	60	5.3	Very low	8—8.5
Torrance	2.4	48—60	2.4	Very low	8—8.5
Manchester	25—30	64	6	Low	Biphasic
Alhambra	9—20	55	2.6	Low	Truncate
Tripler	35	30		Very low	Biphasic

More than 120 G6PD variants have been reported at the present time. Only typical exam-les of the variants in the two categories are listed in the Table. Enzyme activity and Michaelis constant were measured at 25°C and *p*H 8.0.

tose-6-phosphate and fructose-1,6-diphosphate are negligible, and only NADPH and ATP are expected to inhibit G6PD activity under physiologic condition.
Kinetic analysis indicated that inhibition of G6PD by NADPH was competitive with NADP, and the inhibition by ATP is competitive with G6P. It was found that the hemolytic variants associated with relatively mild enzyme deficiency (Manchester, Alhambra and Tripler) were very strongly inhibited by physiologic concentration of NADPH, while non-hemolytic variants associated with servere enzyme deficiency (Union, Markham and Mediterranean) were resistant to inhibition by NADPH (Table 6).
The physiologic concentration of NADPH in normal red cells is reported to be 40—60 μM [25], and that of NADP is lower than 2 μM [26, 27]. Under simulated physiologic conditions, i.e. in the presence of 40—60 μM NADPH, about 1 μM NADP, 40 μM G6P, 1.5 mM ATP and 1.5 mM 2,3-diphosphoglycerate, at 37 °C and at *p*H 7.3, normal G6PD activity is about 2 to 4 nmole per min per 10^{10} red blood cells. This value is comparable to the actual G6PD activity in red cells as measured by the generation of ^{14}C-CO_2 from ^{14}C-G6P.
Since the shunt pathway is the sole NADPH generation mechanism in mature red cells, the ratio of NADP to NADPH is expected to be increased in G6PD de-

Table 6

Michaelis constants (K_m) for NADP and inhibition constants (K_i) for NADPH of the normal and variant G6PD

Enzyme	K_m for NADP (μM)	K_i for NADPH (μM)	Enzyme	K_m for NADP (μM)	K_i for NADPH (μM)
B+ (normal)	12.5	9.0	Mediterranean	3.8	9.5
A+	7.2	6.7	Manchester	18.5	0.8
A−	8.0	13.0	Alhambra	19.5	3.3
Union	8.2	37.0	Tripler	80.0	2.6
Markham	6.0	16.0	Torrance	9.3	3.0

The values were determined in 0.05 M Tris-chloride, *p*H 7.3, containing 4 mM $MgCl_2$, 0.1 KCl, and 60 μM G6P at 37 °C.
Note that K_m measured under the condition is different from that determined at 25 °C and *p*H 8.0 in the presence of excess substrate (see Table 7).

ficient red cells, thus compensating for the intrinsic enzyme deficiency. Enzyme activities of normal and variant G6PDs, under various NADP/NADPH ratio, maintaining total (NADP + NAPDH) concentration as 50 μM, were determined (Fig. 3). A distinctive difference exists between the hemolytic variants and non-hemolytic variants. It is evident that a slight increase of NADP can well compensate for the severe intrinsic enzyme deficiency in red cells of the non-hemolytic variants. On the other hand, a much higher increase of NADP concentration is required for maintaining the normal level of enzyme activity in red cells of the hemolytic variants associated with relatively mild enzyme deficiency.

The problem is how high the NADP level (or how low the NADPH level) may be maintained in red cells without seriously imparing other metabolic pathways, levels of intermediate metabolites, and cell integrity.

The shunt pathway activity of red cells from a Mediterranean variant subject was reported to be about 30% of normal shunt pathway activity [28]. Since the Mediterranean variant is associated with severe enzyme deficieny and is sensitive to drug and food induced hemolytic anemia, this level (30% of normal) is presumably close to the minimum shunt pathway activity required of the unhemolyzed red cells. From the results shown in Fig. 3, the concentration of NADP required for the physiologic G6PD activity of Mediterranean variant red cells can be estimated to be 2 to 5 μM. Since the Mediterranean variant is sensitive to external reagents for hemolysis to occur, it is likely that a further increase of NADP could induce instability of red cells.

The physiologic G6PD activity in red cells from hemolytic variants (Manchester, Alhambra and Tripler) is expected to be lower than 30% of the normal level, even if NADP concentration is elevated to 10 μM (more than 10 times of normal level). (Fig. 3). This is presumably the major cause of the chronic hemolysis in these variants.

A several-fold increase of the ratio of NADP to NADPH in G6PD deficient red cells was previously reported [29, 30]. More recently KIRKMAN et al. [27] have made a determination of the total NADP + NADPH and the NADPH of A⁻ and Mediterranean variants red cells. They reported low concentrations of NADPH

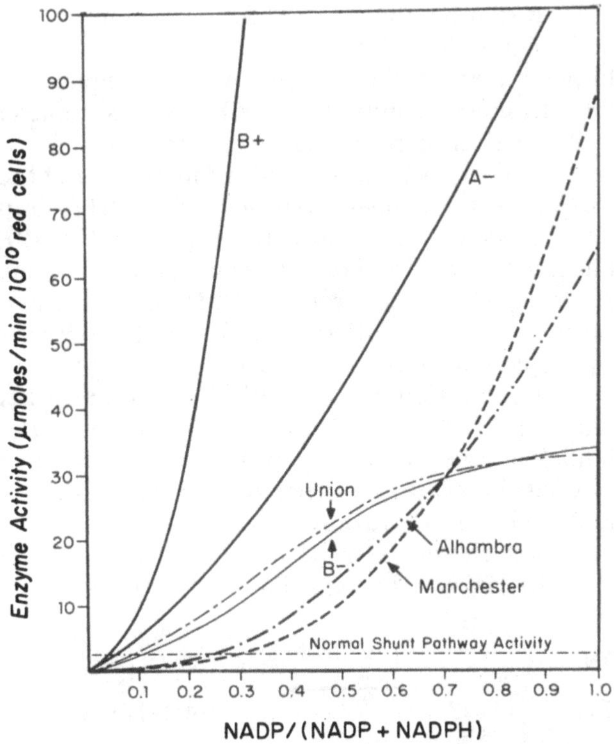

Fig. 3. G6PD activity of normal and variant red cells. G6PD activity was assayed at 37 °C under the simulated physiologic conditions, i.e. in 0.05 M Tris-chloride, *p*H 7.3, containing 4 mM $MgCl_2$, 0.1 M KCl, 1.5 mM ATP, 1.5mM 2,3-DPG, 40 μM G6P and various concentrations of NADP and NADPH, maintaining total (NADP + NADPH) concentration as 50 μM

(10—20 μM) and inversely increased concentrations of NADP (25—40 μM), i.e. more than a 50-fold increase in NADP concentration, in red cells from these G6PD deficient subjects. In the presence of 25 to 40 μM of NADP, under the simulated physiologic conditions, the physiologic G6PD activity in these variant red cells is expected to be much higher (more than 10 times) than that of normal red cells (Fig. 3). Since the shunt pathway activity of the variant red cells is below normal, one has to assume that (1) kinetic properties of the enzyme in red cells is entirely different from those determined in vitro, (2) a punative suppression mechanism (or inhibitor) for G6PD activity is operating in the variant cells, or (3) NADP and NADPH are not homogeneously distributed in red cells. The G6PD activity of the normal red cells, estimated under simulated physiologic conditions in the presence of known concentrations of NADP (about 1 μM), NADPH (40—60 μM), G6P (40 μM) and various metabolites, agrees with the physiologic G6PD activity measured by CO_2 generation. Therefore, none of the possibilities suggested above seems to exist, at least in the normal red cells.

NADPH is known to be easily oxidized to NADP, and the method used may not be adequate for protecting NADPH from the oxidization in the G6PD-deficient red cells. Development of a reliable method for determining NADP and NADPH in cells and tissue is highly desirable.

Naturally, we do not know yet all the factors which regulate G6PD activity in red cells. A possible physiologic inhibitor for G6PD was examined in our laboratory [31]. NADPH (or NADH) is spontaneously converted to a derivative designated as NADPH-X (or NADH-X), which is a hydration product of the reduced nucleotide (Fig. 4). For example, after the incubation of NADPH (sodium salt) dissolved in H_2O *p*H 6.8, for 5 days at 37 °C, more than 90% of NADPH was converted to NADPH-X. Phosphate ion stimulates the conversion, and thus, after the incubation of NADPH in 0.5 M phosphate buffer, *p*H 6.8, for 5 hrs at 37 °C, 93,6% of NADPH was converted to NADPH-X. NADPH-X strongly inhibited human G6PD, human-6-phosphogluconate dehydrogenase, and yeast G6PD. The mode of the dehydrogenase inhibition by NADPH-X was competitive with NADP (Table 7). Similiarly, NADH-X is a strong inhibitor for malatedehydrogenase (pig heart), alcoholdehydrogenase (yeast), and lactate dehydrogenase (beef heart). (Yoshida, unpublished observation). Although the rate of formation of NADPH-X and NADH-X and their concentration in vivo have not yet been determined, there is a possiblity that these materials may be produced in cells and tissues, particularly in red cells, during the log aging period. A possible regulatory role of

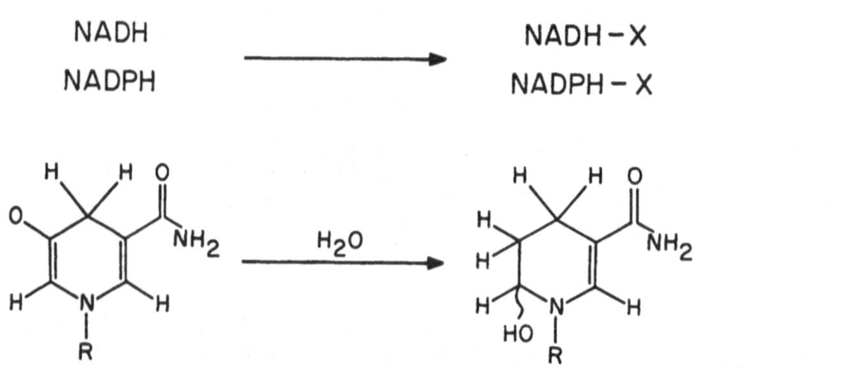

B-6-Hydroxy-1,4,5,6-tetrahydro-nicotinamide adenine nucleotide

Fig. 4. Formation of NADPH-X (or NADH-X) from NADPH (or NADH)

Table 7

Michaelis constants for NADP and inhibition constants for NADPH and NADPH-X

Enzyme	K_m (μM) CM NADP	K_i (μM) NADPH	K_i (μM) NADPH-X
Human G6PD	3.7	16.7	16.2
Human-6-phosphogluconate dehydrogenase	3.3	7.9	20
Yeast G6PD	7.0	5.5	9.1

The reaction mixtures contained 0.1 M Tris-chloride, *p*H 8.0, 0.01 M $MgCl_2$, 0.8 mM G6P, or 6-phosphogluconate, enzyme, and various concentrations of NADPH or NADPH-X.

the two compounds on dehydrogenase activities and their physiologic concentrations are worth being examined.

It is hoped that continuous effort will eventually advance our understanding of the molecular mechanism of G6PD abnormalities up to the level of our present understanding of structure and functions of hemoglobin.

Acknowledgement

I would like to thank my co-workers, Dr. I-Y. Huang, Dr. K. Nakashima, Dr. S. Watanabe, Mrs. M. Lin, Mrs. C. Ewing, and Mrs. V. Dave. This work was supported by the U.S. Public Health Service Grant HL15125.

References

[1] Yoshida, A.: Anal. Biochem. **49**, 320 (1972)
[2] Wotton, J. C.: Nature, Lond. **252**, 542 (1974)
[3] Moon, K., D. Piszkiewicz, and E. L. Smith: Proc. natn. Acad. Sci. U.S.A. **69**, 1380 (1972)
[4] Yoshida, A., and V. D. Hoagland: Biochem. biophys. Res. Commun. **40**, 1167 (1970)
[5] Yoshida, A.: Proc. natn. Acad. Sci. U.S.A. **57**, 835 (1967)
[6] Yoshida, A.: J. molec. Biol. **52**, 483 (1970)
[7] Watanabe, S., and A. Yoshida: Biochem. Genet. **5**, 541 (1971)
[8] Yoshida, A., S. Watanabe, and S. M. Gartler: Biochem. Genet. **5**, 533 (1971)
[9] Crookston, J. H., A. Yoshida, M. Lin, and O. J. Boozer: Biochem. Genet. **8**, 259 (1973)
[10] Nowicki, L., St. Strobel, and H. Martin: Klin. Wschr. **52**, 485 (1974)
[11] Yoshida, A., E. W. Baur, and A. G. Motulsky: Blood **35**, 506 (1970)
[12] Yoshida, A., G. Stamatoyannopoulos, and A. G. Motulsky: Ann. N.Y. Acad. Sci. Art 3, 868 (1968)
[13] Yoshida, A., G. Stamatoyannopoulos, and A. G. Motulsky: Science, N.Y. **155**, 97 (1967)
[14] Dern, R. J., P. R. McCurdy, and A. Yoshida: J. Lab. clin. Med. **73**, 283 (1969)
[15] Laurell, C. B.: Anal. Biochem. **15**, 45 (1966)
[16] Kahn, A., D. Cottreau, and P. Boivin: Humangenetik **25**, 101 (1974)
[17] Kahn, A., M. L. North, J. Messer, and P. Boivin: Humangenetik **27**, 247 (1975)
[18] Kahn, A., J. Hakim, D. Cottreau, and P. Boivin: Clinica chim. Acta **59**, 183 (1975)
[19] Nakashima, K., and A. Yoshida: J. Lab. clin. Med. (in press)
[20] Milner, G., I. W. Delamore, A. Yoshida, and M. Lin: Blood **43** 271 (1974)
[21] Beutler, E., and R. Rosen: Pediatrics **45**, 230 (1970)
[22] Engstrom, P. E., and E. Beutler: Blood **36**, 10 (1970)
[23] Kirkman, H. N., C. Kidson, and M. Kennedy in: Hereditary Disorders of Erythrocyte Metabolism. E. Beutler (Ed.) Grune & Stratton, New York 1968, p. 126
[24] Yoshida, A.: Science, N.Y. **179**, 532 (1973)
[25] Minakami, S., C. Suzuki, T. Saito, and H. Yoshikawa: J. Biochem., Tokyo **58**, 543 (1965)
[26] Omachi, A., C. B. Scott, and H. Hegarty: Biochim. biophys. Acta **184**, 139 (1969)
[27] Kirkman, H. N., G. D. Gaetani, E. H. Clemons, and C. Mareni: J. clin. Invest. **55**, 875 (1975)
[28] Brand, K., P. Arese, and M. Rivera: Hoppe-Seylers Z. physiol. Chem. **351**, 509 (1970)
[29] Löhr, G. W., and H. D. Waller: Klin. Wschr. **36**, 865 (1958)
[30] Schrier, S., R. Kellermayer, and A. S. Alving: Proc. Soc. exp. Biol. Med. **19**, 354 (1958)
[31] Yoshida, A., and V. Dave: Archs Biochem. Biophys. **169**, 298 (1975)

Acta biol. med. germ., Band 36, Seite 703—708 (1977)

Department of Internal Medicine, Kantonsspital CH-5000 Aarau (Switzerland)

Characterization of abnormal glucose-6-phosphate dehydrogenase variants

H. R. Marti, S. Fischer, and D. Killer

Summary

For characterizing glucose-6-phosphate dehydrogenase variants 10 functional parameters are generally used. As additional tests the determination of K_m and K_i at different *p*H values, the limiting K_m for both substrates, isoelectric focusing and electrophoresis of enzyme subunits have been recommended. Most of the variants with favourable kinetic properties do not produce chronic haemolysis. As an exception G6PD Aarau is quoted. Sporadic cases and deficiency conditions with manifest chronic nonspherocytic haemolytic anaemia should be selected for complete enzyme characterization. Individual and public health aspects are of primary importance for screening programs. Among 28,367 blood samples 424 cases with G6PD deficiency have been found in Switzerland.

A review of the international literature might provide the impression of a worldwide competition between haemoglobin and glucose-6-phosphate dehydrogenase (G6PD) investigators for detecting the higher number of abnormal mutants of the respective protein. Indeed, the haemoglobin research is favoured by the advantage that this material is usually available in large quantities allowing a complete chemical analysis. Therefore, at the present time, approximately 100 more haemoglobin variants are identified than G6PD variants are known. However, on the basis of the following three considerations, it is conceivable that the number of existing G6PD mutants might be higher than the number of haemoglobin variants. Screening studies in different parts of the world produced convincing evidence for a very high mutability at the genetic locus for G6PD [1]. Moreover by most of the screening procedures only G6PD (—) phenotypes with decreased activity are found, whereas G6PD (+) phenotypes with normal or increased activity remain undetected. Luzzatto [1] has recently pointed out that the proportion of three-quarters of G6PD (—) and one-quarter of G6PD (+) among all variants hitherto described does probably not reflect the true ratio of the two types of mutants. Finally one subunit of the G6PD molecule contains more amino acids potentially susceptible to substitutions than the α, β-dimer of haemoglobin.

Up to now approximately 150 G6PD variants have been described. In three variants, G6PD, A+, G6PD A—, and G6PD Hektoen, an abnormality of the primary structure has been detected by Yoshida [2—4]. All other variants must be characterized by functional parameters. Among the kinetic and physical properties of an enzyme mutant the catalytic activity, the substrate affinity, the product inhibition and the stability are of special importance for the red cell viability. Yoshida and Lin [5] have found that G6PD variants leading to chronic haemolysis have a low K_i for NADPH [5]. Favourable or unfavourable kinetic properties are considered to be the main determining factor for the resulting type of haemolysis [6—8]. Substrate affinity and product inhibition are likely to be decisive for the pathogenesis of chronic haemolysis, and the amount of the residual enzyme activity seems to be the crucial factor for haemolytic attacks following a redox stress situation [1]. The relationship between kinetics and haemolysis is represented in Table 1. However this is only a general rule. As an exception, the G6PD Aarau may be briefly mentioned. This new variant has recently been described by Gahr et al. [9]. Its main characteristics are given in Table 2. In spite of the favourable kinetic properties, a slight chronic nonspherocytic haemolytic anaemia is produced with persistent reticulocytosis. Furthermore the enzyme variant is characterized by a decreased thermostability and an abnormal *p*H dependence with two maxima at *p*H 7.0 and 8.5. The analogues 2-deoxyglucose-6-phosphate and galactose-6-phosphate are utilized at normal rates, the utilization of deamino-NAPD is increased. The high K_i for NADPH and the slightly decreased K_m for G6P should favour the enzyme activity within the cell. Therefore the chronic haemolysis is likely to be due to the instability of the enzyme.

Table 1

Type of haemolysis and functional properties of G6PD variants

Activity	Kinetics	Haemolysis
Low	unfavourable	chronic
Very low	normal	chronic
Low	favourable	stress-induced
Very low	favoruable	stress-induced

Table 2

Kinetic properties of G6PD Aarau [9]

	G6PD Aarau	Normal	Units
Activity	0.72	10.8 ± 2.1	IU/gHb
K_m G6P	37.9	62.3 ± 13.2	μM
K_m NADP	6.5	5.2 ± 2.1	μM
K_i NADPH	290	40 ± 4	μM

The 10 parameters generally used for characterizing a G6PD mutant are listed in Table 3. With the increasing number of G6PD variants being described, more and more parameters will be necessary in the future for distinguishing one mutant from the other. In the standardized method for G6PD assay a buffer solution of *p*H 8.0 is used [10]. However Luzzatto [1] has found that variants with the same affinity for G6P at *p*H 8.0 may have different affinities at *p*H 7.2, which is approximately the normal intracellular *p*H of the red cells. The K_m for the substrates should therefore be determined for different *p*H values. Furthermore the K_m for G6P is known to be depending on the NADP concentration and the K_m for NADP from the G6P level. The K_m for each substrate decreases with increasing concentration of the other. For this reason Balinsky et al. [11] have also determined the limiting K_m, i.e. the K_m value for each substrate at infinite concentration of the other one. This parameter was used for characterizing G6PD Kuanyama. For characterizing the Mediterranean type of G6PD NOWICKI et al. [12] have determined the electrophoretic mobility of the enzyme subunits by high voltage electrophoresis on thin layers of agarose gel at pH 8.0. Isoelectric focusing also has been used by some investigators. DREYFUS and KAHN [13] observed a progressive appearance of bands of lower isoelectric point during molecular aging of the G6PD.

Table 3

The 10 parameters generally used for the characterization of G6PD variants

Enzyme activity	Utilization Gal6P
K_m G6P	Utilization deamino NADP
K_m NADP	pH dependence
K_i NADPH	Thermostability
Utilization 2dG6P	Electrophoresis

Concluding we may say that the 10 parameters of the standard procedure for characterizing a G6PD variant can be supplemented by the determination of K_m and K_i at different pH values, by the determination of the limiting K_m for both substrates, by isoelectric focusing and eventually by electrophoresis of enzyme subunits. With the increasing number of well characterized G6PD variants new criteria will be necessary for recognizing new mutants. An adequate characterization of an enzyme will require more and more sophisticated methods and less laboratories will have the capacity to provide a complete investigation.

The laboratory capacity being limited the question therefore arises, on the basis of which criteria the cases should be selected for further studies. The problem of today is not primarily to define the parameters necessary to characterize a new mutant, but rather to find the criteria of selection of an enzyme to be fully investigated. As an illustration of this difficulty the cases with G6PD deficiency found in Switzerland will be quoted.

In our laboratory for haemoglobin analysis 28,367 blood samples have been screened for G6PD deficiency since 1962. Originating from all parts of Switzerland they were taken mainly from anaemic patients with suspected thalassaemia of haemoglobin abnormality. For screening purposes a modification of the semiquantitative methylene blue reduction test [14, 16] described by TÖNZ and BETKE [15] was used. For its easier reading this technique was found to be more convenient than the most widly used methaemoglobin reduction test [17]. Enzymes with decreased activity were determined simultaneously with the standardized method of spectrophotometrical NADPH determination [10]. The correlation of the results demonstrates the usefulness of the dye test in daily laboratory practice (Fig. 1). The same basic method has recently been used in a survey study by GANESAN et al. [18].

During 14 years 424 cases of G6PD deficiency have been found, which represents 1.5% of the blood samples examined. As a concomitant disorder an abnormal haemoglobin was present in 21 and a thalassaemia in 108 cases.

Out of the 424 patients with decreased G6PD activity 66.3% were males and 33.7% females. Among 143 female patients 44 had an enzyme activity of less than 25%

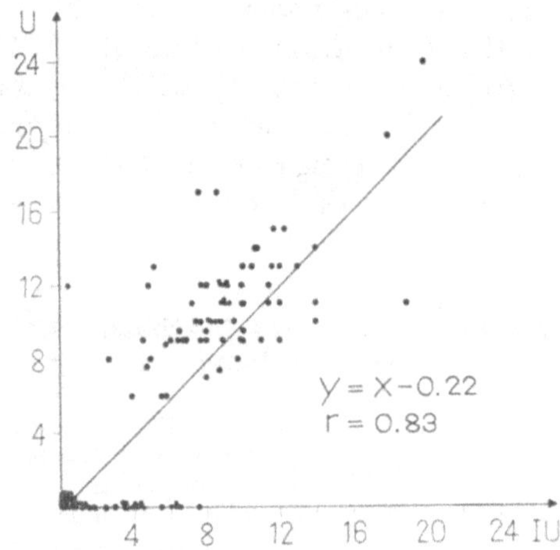

Fig. 1. Correlation between G6PD activity determined with the standard assay (abscissa) and the semiquantitative methylene blue reduction test (ordinate) in 123 female heterozygotes and homozygotes

of the lower limit of the normal range. Roughly half of the female heterozygotes can be detected by the dye test. Only in 37.3% of all cases with G6PD deficiency the enzyme screening had been requested by the treating physician. The geographical origin, which could be traced for 378 patients, is represented in Table 4. The group of 18 cases from Europe comprises 7 patients originating from Switzerland, 4 from the German Federal Republic, 1 from Austria, 5 from Poland and 1 from Roumania.

Table 4
Geographical origin of 378 patients with G6PD deficiency observed in Switzerland

Origin	Patients	
	n	%
Mediterranean	320	84.6
Europe	18	4.8
Africa	34	9.0
Middle East	6	1.6

Several new variants are likely to be present in these patients. Unfortunately the determination of the thermostability and the electrophoresis of the non-purified enzymes are of little value for selecting the cases for complete characterization. The electrophoresis on cellulose acetate plates is indeed an easy method [19]. However too many variants have a normal electrophoretic mobility [20]. By means of these routine tests the variant G6PD Aarau could not be distinguished from the Mediterranean G6PD. On the one hand isolated cases without ancestry from a region with high frequency of G6PD deficiency are more likely to be due to a new mutant. On the other hand the high frequency of G6PD deficiency in Italy and Greece is known to be heterogenous. At least 10 different variants have been described in these countries. Recently SANSONE et al. [21] found 3 rare variants among 505 newborn babies of Sardinian and south Italian origin with hyperbilirubinaemia. Large scale screening studies reported in the literature revealed a frequency of rare G6PD variants of 0.3—2.3% in Mediterranean and African countries and 6—12% in southeast Asia [1].

Since the common types G6PD A— and G6PD Mediterranean are producing no or only an extremely mild chronic haemolysis, all G6PD deficiency conditions with mainfest chronic nonspherocytic haemolytic anaemia and all sporadic cases should in principle be selected for a further characterization of the enzyme, which should be as complete as possible. The limiting factor in this recommendation will be the capacity of the few existing reference laboratories.

Fortunately the clinically important classification of G6PD variants into the two large groups with and without significant chronic haemolysis, as well as the knowledge of triggering factors inducing an acute haemolysis are based predominantly on clinical observations. The common variants without clinical manifestation of chronic haemolysis but with potentially dangerous haemolytic crisis under oxydative stress situations are a public health problem. The rare variants with significant chronic haemolysis on the other hand pose an individual health problem. At least in populations with known high incidence of G6PD deficiency large scale screening programs must be accomplished and supported by the public health authorities. They lead to the following benefit for the patients: 1) Best possible prevention of environmental triggering factors; 2) prevision of the risk and adequate treatment of neonatal jaundice; 3) best possible prevention of offending drugs given to the mother in prenatal period [1, 22].

In countries without a significant incidence of G6PD deficiency all immigrants from areas with higher frequency should be screened at the first opportunity. It is evident that for practical reasons only a limited number of G6PD variants can be fully analyzed and characterized; however considering the aspects of health as such the population screening, information and counselling may be more important than a complete biochemical investigation of all mutants.

References

[1] LUZZATTO, L.: Clinics Haemat. **4**, 83 (1975)

[2] YOSHIDA, A.: Proc. natn. Acad. Sci. U.S.A. **57**, 835 (1967)

[3] YOSHIDA, A.: Science, N.Y. **155**, 97 (1967)

[4] YOSHIDA, A.: J. molec. Biol. **52**, 483 (1970)

[5] YOSHIDA, A., and M. LIN: Blood **41**, 877 (1973)

[6] RATTAZZI, M. C., L. M. CORASH, G. E. VAN ZANEN, E. R. JAFFE, and S. PIOMELLI: Blood **38**, 205 (1971)

[7] YOSHIDA, A., E. BEUTLER, and A. G. MOTULSKY: Bull. Wld Hlth Org. **45**, 243 (1971)

[8] LUZZATTO, L.: Br. J. Haemat. **28**, 151 (1974)

[9] GAHR, M., W. SCHRÖTER, M. SUTRZENEGGER, D. BORNHALM, and H. R. MARTI: Helv. paed. Acta (im Druck)

[10] World health Organization scientific Group: Wld. Hlth Org. Techn. Rep. Ser. **366** (1967)

[11] BALINSKY, D., A. J. ROOTMAN, G. T. NURSE, E. CAYANIS, A. LANE, T. JENKINS, and I. BERSOHN: S. Afr. J. med. Sci. **39**, 1 (1974)

[12] NOWICKI, L., S. STROBEL, P. KÜHNL, H. MARTIN, G. GAETANI, and E. SALVIDIO: Proceedings of the 15th International Congress of the International Society of Hematology, Jerusalem 1974, p. 450

[13] DREYFUS, J. C., and A. KAHN: Proceedings of the 15th International Congress of the International Society of Hematology, Jerusalem 1974, p. 33
[14] MOTULSKY, A. G., J. M. KRAUT, W. T. THIEME, and D. F. MUSTO: Clin. Res. **7**, 89 (1959)
[15] TÖNZ, O., and K. BETKE: Klin. Wschr. **40**, 649 (1962)
[16] MARTI, H. R.: Experientia **24**, 416 (1968)
[17] SALVIDIO, E., and I. PANNACIULLI in: Standardization in Hematology. G. ASTALDI, C. SIRTORI, and G. VANZETTI (Eds). Franco Angeli, Milano 1970, p. 330
[18] GANESAN, J., L. E. LIE-INJO, and B. P. ONG: Hum. Hered. **25**, 258 (1975)
[19] ELLIS, N., and J. B. ALPERIN: J. clin. Path. **57**, 534 (1972)
[20] MILLER, D. R., and M. R. WOLLMAN: Blood **44**, 323 (1974)
[21] SANSONE, G., L. PERRONI, and A. YOSHIDA: Br. J. Haemat. **31**, 159 (1975)
[22] GLASS, L., B. K. RAJEGOWDA, E. BOWEN, and H. E. EVANS: J. Pediat. **82**, 734 (1973)

Acta biol. med. germ., Band 36, Seite 709—715 (1977)

Martzinovsky Institute of Medical Parasitology and Tropical Medicine, Moscow;
Kirov Institute of Medical Parasitology and Tropical Medicine, Baku;
Institute of Biological and Medical Chemistry, Moscow, USSR

Hemoglobinopathies and glucose-6-phosphate dehydrogenase deficiency in one of the regions of Azerbaijan: Mass Screening and laboratory investigations

R. B. Javadov, L. N. Grinberg, S. N. Krasnova, Sh. A. Makhmudova, and O. V. Troitskaya

Summary

Mass screening in the Geok-chai region of Azerbaijan covered 264 persons. 2 families with abnormal hemoglobin were detected. Electrophoresis in PAG and on cellulose acetate films as well as sickling test showed that in 3 out of 9 members of one of the families HbS was detected. 6 out of 8 members of the others family appeared to be HbD-carriers. In 4 members of this family abnormal hemoglobin was found out in combination with G-6-PDH deficiency. No clinical manifestations were found. A number of β-thalassemia cases were detected in screened children as well as in adults. Hemoglobin oxygen equilibrium curves were studied in patients with heterozygous β-thalassemia. In case of G-6-PDH deficiency when it is not possible to obtain active enzyme zones on the columns with PAG these zones can be detected when $CoCl_2$ (2 mM) is added into the incubation medium.

Introduction

As a result of historically established conditions there are in the Soviet Union some territories where hemoglobin abnormalities and glucose-6-phosphate dehydrogenase deficiency (G-6-PDH) are spread rather widely. As a rule these regions are former malaria foci. They include Central Asian republics [1, 2] and Transcaucasian republics [3—4]. In some regions of Azerbaijan the frequency of β-thalassemia gene carriage and G-6-PDH deficiency reaches world maximum [7, 8] The aim of this paper was to study hemoglobin abnormalities and G-6-PDH deficiency in one of the regions of Azerbaijan. The peculiarity of such investigations lies in the complex approach which included:

1) mass clinical examination;
2) study of fractional composition of hemoglobin and G-6-PDH;
3) quantitative determination of hemoglobin fractions and G-6-PDH activity;
4) study of oxygen equilibrium of hemoglobin solutions.

A total of 264 persons were examined. To determine hemoglobin fractions for 180 persons, electrophoresis (EPh) in polyacrylamide gel or on cellulose acetate films was carried out. For 6 patients suspected of thalassemia quantitative determination of minor hemoglobin fractions was made. For 15 patients of heterozygous β-thalassemia families oxygen hemoglobin equilibrium curves were registered. The G-6-PDH activity was assayed according to Bernstein in 180 persons from the Geok-chai region. Determination of NADPH reduction in a photoelectrocalorimeter was carried out in some patients.

Material and methods

Electrophoresis in polyacrylamide gel and on cellulose acetate films

Blood in a volume of 0.1 ml was collected from a fingerpuncture and was then mixed with 1 ml of distilled water. The hemolysates obtained in this way were stored at low temperature and used for EPh. Electrophoresis in polyacrylamide gel (PAG) (7.5%) was made in tubes at *p*H 8.9 of the basic gel. The hemoglobin fractions were tested by benzidine.

To obtain active zones of G-6-PDH, 0.2 ml of the hemolysate was added to each column with PAG. After EPh was finished the columns were placed in 0.5 M buffer solution of Tris-HCl, *p*H 8.5, containing NADPH (0.37 mM), G-6-P calcium salt (3.0 mM), nitroblue tetrazolium (0.12 mM) and phenacin-metosulfate (0.98 mM) [12], $CoCl_2$ (2 mM) was used as an activator in enzyme detection. This process was conducted at 37 °C during 2—3 hrs.

EPh was carried out on cellulose acetate films (made in USSR) in a special apparatus (SKB BFA, Moscow) in Tris-EDTA-borate buffer, *p*H 9.1. Hemoglobin was detected by amido black and in some cases by benzidine. To make a quantitative determination of minor hemoglobin fractions, venous blood was taken, erythrocytes were washed with physiological saline then hemolyzed by adding chloroform. The hemolysate obtained in this way contained no less than 10 g% Hb.

For quantitative determination of HbA_2 cellulose acetate films were cut into fractions and dissolved in acetic acid after EPh and after detection by amidoschwarz. Solutions were photometered at 610 nm and per cent fraction ratio was calculated by taking appropriate corrections on dilution. Fetal hemoglobin (HbF) was determined by one-minute alkaline denaturation after Singer.

Determination of G-6-PDH activity

Qualitative determination of G-6-PDH activity was carried out after BERNSTEIN in IDELSON's modification [9]. According to this method the reaction was estimated as +++ if the blue-green colour of the hemolysate with reagent did not change during 30 min. This proved a considerable enzyme deficiency.

Quantitative determination of G-6-PDH activity was made by the method introduced by KORNBERG and HORECKER [10] and also by MARK [11]. Enzyme activity was judged by the quantity of NADPH obtained in the sample, containing G-6-P (3.8 mM), NADPH (0.45 mM), $MgCl_2$ (7.7 mM). The volume of the sample was 3 ml. The reaction was started when 0.1 ml of the hemolysate was put into the cuvettes. To make evaluation of enzyme activation, the change of the solution optical density at 340 nm during 5 min was used. In this case the solution temperature and hemoglobin concentration in the hemolysate (estimated by cyanmethemoglobin method) were taken into consideration. G-6-PDH activity was indicated in international enzyme activity units on 1 g of hemoglobin (IU/g Hb).

Oxygen dissociation oxyhemoglobin curves were registered with the help of a special method based on the combination of polarographic and photometric principles. The measurement was made in hermetically closed cells, containing phosphate, *p*H 6.9, or borate, *p*H 9.0, buffer. 1 ml of hemolysate was added to 4 ml of buffer solution, and the obtained solution contained approximately 0.06 g% Hb.

The optical solution density of hemoglobin was registered in an absorption band from 600 to 630 nm. In slow titration with $Na_2S_2O_4$ the optical solution density increased due to oxyhemoglobin transformation into deoxyform. The oxygen concentration in the solution was determined polarographically using a platinum electrode. Both parameters (optical density and polarographic current) were registered on the two-coordinate selfrecorder, where the per cent hemoglobin oxygen saturation was indicated as the function of pO_2 in the solution.

Results

Two families — carriers of hemoglobin abnormalities — were detected in the Geok-chai region of Azerbaijan.

Family K: Proband K.B., 43 years old, healthy, Hb 13.2 g%. EPh on cellulose acetate films and PAG detected HbA and abnormal hemoglobin with HbS mobility (Fig. 1). Sickling test was positive. Consequently, K.B. is a heterozygote carrier of HbS. There are 8 more people in the family. A sickle-cell trait was detected in one of the proband's daughters (K.S) and in his son (K.M). Thus in the family K. 3 sickle-cell traits whithout clinical manifestations were detected (Fig. 2).

Family M: Proband M.F., 40 years old, healthy, Hb 12%. EPh on cellulose acetate films and PAG detected HbA and abnormal hemoglobin similar in mobility two HbS. Nevertheless, the sickling test was negative. 7 more members of the family M. were examined. Five of them — proband's sister, his three sons and a daughter — appeared to be carriers of abnormal hemoglobin. Thus six heterozygote carries of abnormal hemoglobin were detected in family M. Abnormal hemoglobin in the father and his three children was detected in combination with G-6-PDH deficiency (Fig. 3).

As is known HbD has an electrophoretic mobility similar to hemoglobin S [13]. In connection with the negative sickling test there are reasons to refer the detected hemoglobin to HbD. But additional investigation are required to find out the chemical structure of this hemoglobin.

Abnormal hemoglobins in Azerbaijan are not rare. Many cases of sickle-cell anemia, both pure and in combination with thalassemia [14—16], are described. There are no data on the distribution of HbD in Azerbaijan. In the Central Asian republics HbD was found out, but the structure of this hemoglobin and its relation to the already known types of HbD are not yet established [17, 18].

When mass screening in the Agdash region was carried out, it turned out that there was anemia in nearly 10% of the population (Hb lower than 10 g%). Several families with the detected anemia were examined more carefully and as a result of this examination a number of β-thalassemia cases were found out (Table 1).

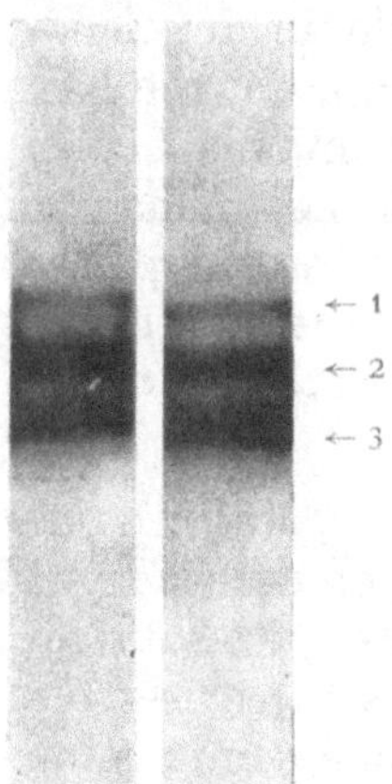

◄Fig. 1

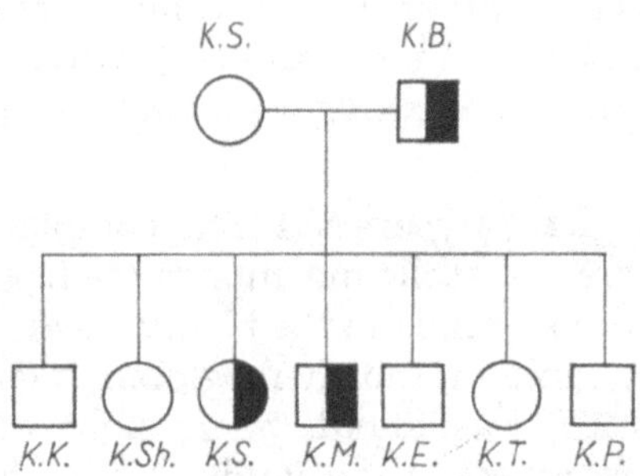

Fig. 2

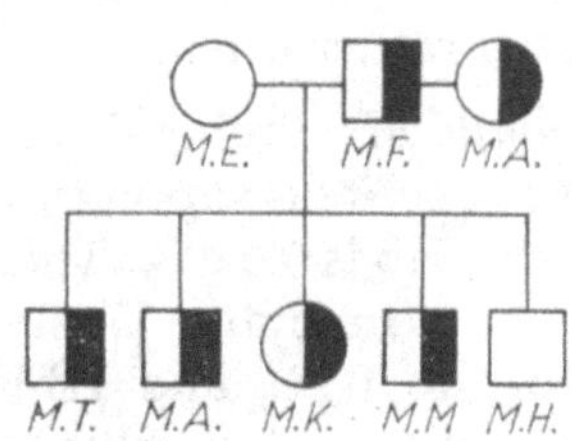

Fig. 3

Fig. 1. Polyacrylamide gel electrophoresis.
A) Hb of proband K. B.; B) HbAS: 1 = HbA_2; 2 = HbS; 3 = HbA

Fig. 2. Distribution of hemoglobin S in the family K.
○□ HbAA; ◑◨ HbAS

Fig. 3. Distribution of hemoglobin D and G-6-PDH deficiency in the family M.
□○ HbAA; ◑◨ HbAD; ◑◨ G-6-PDH deficiency

Table 1
β-thalassemia in the Agdash district of Azerbaijan

N	Initials	Sex	Age	Hb (g%)	HbA_2* (%)	HbF** (%)	Target cells	Diagnosis
1.	A. H.	f	2.5	6.8	—	72.8	+	Thalassemia major (anemia Cooley)
2.	M. A.	f	13	9.5	4.2	3.7	+	Thalassemia minor with increased HbA_2 and HbF
3.	B. E.	m	12	11.8	5.7	1.2	+	Thalassemia minor with increased HbA_2
4.	Yu. I.	m	11	10.0	2.3	3.6	+	Thalassemia minor with increased HbF
5.	[18]	m	adult	13–16	2.4	1.6	—	Normal values
		f	adult	12–14				

* Electrophoresis on acetate cellulose; ** Singer method

Family A: Proband A. Kh., a girl, 2.5 years old. The child is pale, weak, stands up with difficulty; she has a large head, a saddle nose, hepatosplenomegaly. Blood count: Hb = 6.8 g%, erythrocytes = 2×10^6, color index = 0.8; leukocytes = 25×10^3. There was hypochromia in the blood smear, marked anisopoikilocytosis, and target-like erythrocytes.

EPh on cellulose acetate films showed the main fraction hemoglobin deficiency. This fraction consists of HbF, its content is very high (= 72.8%). On the basis of these data thalassemia major (monozygous β-thalassemia, or anemia Cooley) was diagnosed in the woman-patient A. Kh. Nevertheless we did not manage to find out even signs of thalassemia minor in this woman's relative on the part of her mother (4 persons) (her father was not examined). Obviously the members of the family A. have a silent form of heterozygous β-thalassemia, which is encountered rather often.

Family M: Proband M. A., a girl, 13 years old. She complains of pain in bones, her appetite is suppressed. There is systolic murmur in the heart, sometimes hematuria; Hb is 9.6 g%. There is hypochromia in the blood smear, anisocytosis, target-like erythrocytes. While determining minor hemoglobin fractions the increase of HbA_2 = 3.7% and HbF = 4.2% was found. Some signs of thalassemia minor were detected in three other members of the family M: in the father (target-like erythrocytes), in the sister (anemia) and in the brother (target-like erythrocytes).
Family B: Proband B. E., a boy, 12 years old. "Tower skull", a saddle nose, target-like erythrocytes in the blood smear are marked, Hb is 11.8 g%. EPh on acetate cellulose films detected the increase of HbA_2 (= 5.7%), hemoglobin F is normal (= 1.2%). Diagnosis of the patient B. E. is heterozygous β-thalassemia. Signs of thalassemia minor (target-like erythrocytes and anemia) were detected in one of the proband's sisters and in one of his brothers.

Family Yu: Proband Yu. I., a boy, 11 years old. Typical changes in the skull bones (the head is narrowed in the front); Hb is 10.0 g%. There are hypochromia and target-like erythrocytes. The increase of fetal hemoglobin (HbF) of 3.6% was recorded, at the same time HbA_2 is normal (= 2.3%). Other members of the family (4 persons) have these or those signs of thalassemia: target-like erythrocytes, anemia, typical changes of the skull bones. Three children died of anemia at the age of 7—10.

Thus in the Agdash region of Azerbaijan several families suffering from β-thalassemia of different types were detected (Table 1). This affirms the genetic and clinical heterogeneity of this disease [2, 19]; β-thalassemia occurs more often in Azerbaijan than in other Transcaucasian republics [4]. The majority of the papers deals with the description of the thalassemia major [3, 20]. Thalassemia minor is less investigated in view of the poorly expressed clinical manifestations and the difficulties of quantitative determination of HbA_2. Taking into consideration our results, quantitative determination of HbA_2 using EPh on cellulose acetate films can be successfully applied in detection and investigation of thalassemia minor.

It is now well-known that different variants of human hemoglobin differ from HbA in their oxygen-linking properties [21]. Nevertheless there are no literature data concerning this funtion of hemoglobin in thalassemia. In the course of the mass screening oxygen equilibrium curves of hemoglobin were studied in patients

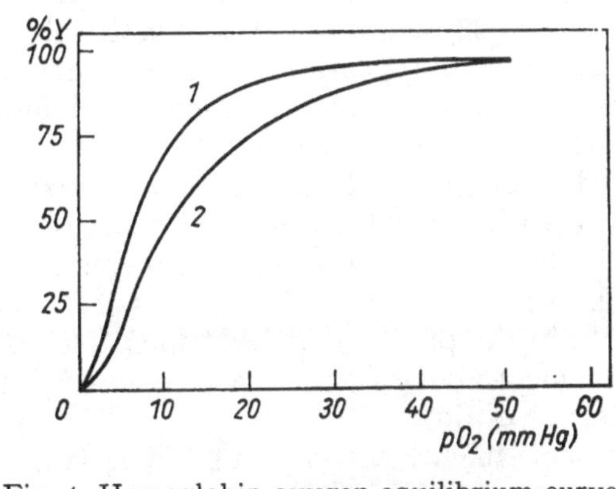

Fig. 4. Hemoglobin oxygen equilibrium curves.
1 = β-thalassemia heterozygous; 2 = normal

Table 2
Oxygen affinity and Bohr effect in hemoglobin solutions in cases of heterozygous β-thalassemia

°C	*p*H	P_{50} (mm Hg)		log P_{50}*		Δ log P_{50}/Δ *p*H	
		N	Th	N	Th	N	Th
30	7.0	24.5	20.8	1.39	1.32	0.18	0.33
30	9.0	10.7	4.61	1.03	0.66		

* Calculated from data at *p*H 7 and 9.
N: The average data for 8 healthy persons; Th: those for 15 persons with heterozygous β-thalassemia

with heterozygous β-thalassemia. Typical curves are shown on Fig. 4. Quantitative results are given in Table 2. It is clearly seen that at *p*H 9.0 the value of P_{50} in hemoglobin carriers of β-thalassemia is lower (i.e. the affinity to oxygen is higher) than in healthy people. This results in the fact that the Bohr's effect in the cases of β-thalassemia is manifested much stronger than normally (Table 2).

To explain this phenomenon, additional investigation are needed but it is possible to point out that the fraction composition of hemoglobin in the persons examined is noticeably changed. At the same time each of the minor components of hemoglobin has its own typical oxygen affinity and, it is possible, that in thalassemia HbA_2 and HbF have more influence on the measured value of P_{50} than normally.

As is known, G-6-PDH defiency is widely spread in a number of regions of Azerbaijan [5, 7]. With the introduction into practice in our country of the express-test after BERNSTEIN (cit. in [9]) there appeared a possibility of carring out a large scale mass screening on mutant gene G-6-PDH carriage.

Determination of G-6-PDH activity by optical method showed that some persons with deficiency according to BERNSTEIN (see "Methods") retain up to 40—50% of enzyme activity (1.85—2.70 IU/g Hb).

EPh of hemolysate of blood with normal G-6-PDH (5—9 IU/g Hb) activity gave the possibility to obtain two coloured enzyme zones on the column with PAG. As is known the G-6-PDH human erythrocyte molecule can exist in two active forms — tetramer and dimer — the correlation of which depends on divalent ions of metal, particularly on Co^{++} [22]. Adding $CoCl_2$ (2 mM) solution into the incubation mixture reduces the time of appearance of active enzyme zones 2—3 fold.

We did not manage to get coloured enzyme zones on columns with PAG in 40 examined patients with G-6-PDH deficiency (activity < 1.5 IU/g Hb). When CoCl (2 mM) was added there gradually appeared two enzyme zones though less coloured than in control with normal activity. EPh of hemolysate of blood from patients with G-6-PDH deficiency detected four enzyme zones with low activity and low mobility (Fig. 5) in some cases. It was possible to detect them on the columns with PAG only in the presence of Co^{++} ions. The exception was only in the cases when the enzyme activity was nil: active G-6-PDH zones were not detected even in the presence of Co^{++}.

One case of the increased enzyme activity (13.0 IU/g Hb) was found out. EPh on PAG showed that the enzyme was subdivided into 3 zones with higher mobility and activity than in control with normal activity (Fig. 5).

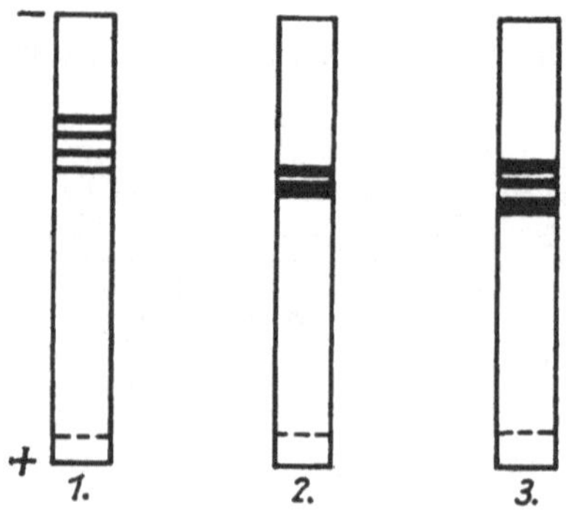

Fig. 5. Polyacrylamide gel electrophoresis of G-6-PDH.
1. Low activity (1.05 IU/g Hb) (developed with $CoCl_2$, 2 mM); 2. normal activity (5.70 IU/g Hb); 3. high activity (13.04 IU/g Hb)

On the whole, we consider the methods with colour indicators, such as the Bernstein method, to be undoubtedly justified in mass screening. Nevertheless the results obtained by these methods should be considered as preliminary. To estimate enzyme activity in patients suspected of G-6-PDH deficiency, it is necessary to make quantitative determination of enzyme activity.

As to the influence of Co^{++} ions on G-6-PDH activity in screened population it should be mentioned that activation was not detected in all the patients with enzyme deficiency, for example, there was no case of complete absence of enzyme activity. Vitamin B_{12}, as well as divalent cations (Ca^{++}) did not change the activity of the enzyme. It is known that Co increases G-6-PDH normal activity most effectively [22]. Until now it is difficult for us to explain the peculiarities of the Co^{++} effect on the G-6-PDH activity detected in Azerbaijan in patients deficient in this enzyme. This problem is under further investigation.

References

[1] Марычев, А. Г.: Вестн. Акад. Мед. Наук СССР Т. 11, 85, (1965)

[2] Гарькавцева, Р. Ф.: Генетика № 10, 137 (1972)

[3] Ахундова, А. М.: Талассемия, Баку (1972)

[4] Воронов, А. А.: Пробл. гематол. № 11, 32 (1973)

[5] Джавадов, Р. Б. Фавизм, Баку (1975)

[6] Местиашвили, И. Г., И. Ш. Зедгенидзе, М. Г. Абдушелишвили, Л. М. Фридман и Е. Н. Цверева: Пробл. гематол. № 11, 37 (1973)

[7] Токарев, Ю. Н: Пробл. гематол. № 10, 3, (1974)

[8] Лысенко, А. Я. и др.: Пробл. гематол. № 11, 16. (1973)

[9] Идельсон, Л. И. и Э. Р. Котоян: Лабор. дело, № 7, 428 (1970)

[10] Kornberg, A., and B. L. Horecker: Meth. Enzym. 1, 323 (1955)

[11] Mark, P. A.: Meth. Enzym. 9, 131, (1966)

[12] Сафонов, В. И. и М. П. Сафонова: Электрофорез в полиакриламидном геле и его применение в биологии, сельском хозяйстве, медицине и пищевой промышленности. Москва, с. 7 (1971)

[13] Доклад научной группы Воз. Лечение гемоглобинопатий и родственных заболеваний. Серия технич. докладов. № 509 (1974)

[14] Дикова, В. Ю.: Пробл. гематол. № 4, 55 (1968)

[15] Назарли, А. Г., и А. Р. Абдулаев: Пробл. гематол. № 10, 22 (1974)

[16] Назарли, А. Г., и Х. Н. Абраханова: Пробл. гематол. № 1, 61 (1976)

[17] Гарькавцева, Р. Ф., Е. Н. Мусницкая и А. Ю. Асанов: Пробл. гематол. № 11, 49 (1973)

[18] Гарькавцева, Р. Ф., Е. Н. Мусницкая и Б. А. Кауров: Пробл. гематол. № 1, 37 (1976)

[19] Гарькавцева, Р. Ф., Е. Н. Мусницкая, Н. П. Соловьева и Д. А. Ахмедов: Клинич. медицина № 10, 109 (1974)

[20] Ахундова, А. М.: Пробл. гематол. № 10, 18 (1974)

[21] Lehmann, H., and R. G. Huntsman: Man's Hemoglobin. Academic Press, New York 1974

[22] Bonsignore, A., I. Lorenzoni, R. Cancedda, M. F. Cosulick, and A. De Flora: Biochem. biophys. Res. Commun. 42, 159 (1971)

Acta biol. med. germ., Band 36, Seite 717—730 (1977)

Institut für Physiologische und Biologische Chemie der Humboldt-Universität zu Berlin
104 Berlin, Hessische Str. 3—4, DDR

Energiestoffwechsel roter Blutzellen bei Pyruvatkinase-Enzymopathien

G. Jacobasch, M. Grieger, Ch. Gerth und K. Bier

Zusammenfassung

Pyruvatkinase-Enzymopathien sind bisher nur für rote Blutzellen nachgewiesen worden. Daraus leiten sich strukturelle Besonderheiten für das Erythrozytenenzym, die noch nicht völlig geklärt sind, ab. Pyruvatkinase (PK)-Defekte können die Ursache nichtsphärozytärer hämolytischer Anämien sein. Die Enzymopathie ist durch einen Polymorphismus charakterisiert. Er findet seinen Ausdruck in mehr als 5 verschiedenen PK-Varianten. In der DDR wurden bisher 16 Gendefektträger für diese angeborene Stoffwechselanomalie nachgewiesen.
Folgende Parameter wurden zur Enzymcharakterisierung angewendet: Phosphoenolpyruvat (PEP)-Abhängigkeit, ATP- und Alaninhemmbarkeit, Nukleotidspezifität, Temperatur- und Harnstoffstabilität sowie die Reifungsabhängigkeit. 2 PK-Varianten mit stark herabgesetzter PEP-Affinität werden beschrieben. Die Besonderheiten des Energiestoffwechsels der roten Blutzellen unter aeroben und anaeroben Bedingungen werden herausgestellt.

Die am häufigsten auftretende Anomalie eines Glykolyseenzyms ist ein Defekt der Pyruvatkinase (PK). Da diese Enzymopathie ausschließlich auf die rote Blutzelle beschränkt ist, sind Besonderheiten des Pyruvatkinase-Isoenzyms roter Blutzellen zu postulieren. Trotz intensiver Untersuchungen in mehreren Arbeitsgruppen [1—3] ist die Struktur des für den Erythrozyten typischen Isoenzyms der PK bisher nicht völlig aufgeklärt. Die kinetischen Eigenschaften der Erythrozyten-Puyruvatkinase zeigen eine weitgehende Übereinstimmung mit denen des Leberenzyms [4, 5]. Unterschiede in der elektrophoretischen Wanderungsgeschwindigkeit sind dagegen Hinweise für eine Nichtidentität beider Isoenzyme [1, 2]. Ergebnisse der SDS-Polyakrylgelelektrophorese sind in Übereinstimmung mit der Annahme, daß die Erythrozyten-Pyruvatkinase des Menschen aus zwei verschiedenen Untereinheiten zusammengesetzt ist, die ein annäherndes Molekulargewicht von je 63000 aufweisen, was einem Molekulargewicht von 241000 für das aktive Tetramer entspricht [6]. Die Existenz von zwei verschiedenen Untereinheiten konnte durch Peptidanalyse bestätigt werden [6]. Darüber hinaus wurde in vergleichenden Untersuchungen festgestellt, daß eine Untereinheit der Erythrozyten-PK homologe Sequenzen mit dem Muskeltyp (M) der PK des Menschen aufweist. Es wird vermutet, daß die Pyruvatkinase des Erythrozyten ein Hybrid darstellt, das aus PK-L- und PK-M_2-Untereinheiten aufgebaut ist. Nicht völlig auszuschließen ist jedoch, daß das Erythrozytenenzym ein Isomer der L-PK ist, da die PK-L dieselben antigenen Eigenschaften wie die Erythrozyten-PK besitzt [1, 3].

Vor kurzem wurde bekannt, daß der Pyruvatkinasemangel nicht nur in roten Blutzellen des Menschen auftritt, sondern daß diese Anomalie wahrscheinlich bei mehreren Säugetieren vorkommt, wie z. B. beim Hund [7—9]. Diese Tatsache

bietet für das experimentelle Arbeiten gewisse Vorteile. Allerdings muß in diesem Zusammenhang darauf hingewiesen werden, daß sich die PK der Erythrozyten zwischen den einzelnen Säugetieren sowohl in ihren immmunologischen als auch kinetischen und elektrophoretischen Verhalten unterscheidet [10, 11].
Die z. T. beobachteten Kreuzreaktionen bei vergleichenden immunologischen Untersuchungen an Pyruvatkinasen von Erythrozyten verschiedener Säugetiere mit Kaninchenantiseren gegen Erythrozyten-PK des Menschen stimmen mit der Vermutung überein, daß diese Enzyme weitgehend homologe Proteinsequenzen aufweisen [12].
Der Pyruvatkinasedefekt der Erythrozyten des Menschen geht ebenso wie der des Hundes mit einer nichtsphärozytären hämolytischen Anämie einher. Eine Grobeinteilung in 2 Gruppen von nichtsphärozytären hämolytischen Anämien läßt sich anhand des Autohämolysetestes vornehmen [13]. Der PK-Defekt zählt danach zu den Typ II-Anämien, da die Autohämolyse in vitro durch Glukosezusatz nicht verringert wird. In vielen PK-Mangelfällen kann dagegen eine deutliche Abnahme der Autohämolyse durch Zusätze von Adenosin oder ATP erreicht werden. Der Grad der Hämolyse, der unter den Bedingungen des Autohämolysetestes auftritt, scheint mit der klinischen Schwere der Erkrankung zu korrelieren.
Neben relativ selten auftretenden milden Verlaufsformen der Erkrankung kennen wir auf der anderen Seite hämolytische Anämien mit einer extremen Verkürzung der Lebenserwartung der Zellen bis auf 1—2 Tage. Viele der Patienten zeigen stark gesteigerte Erythropoeseraten mit auffallend hohen Retikulozytenwerten. Nicht selten sind Retikulozytenwerte von mehr als 50% festzustellen. Diese Patienten sind zeitweise transfusionsbedürftig, insbesondere dann, wenn im Verlaufe von Infektionserkrankungen oder bei der Einwirkung anderer exogener Noxen die Erythropoese gehemmt wird.
Die Retikulozytose wirft die Frage nach der Reifungsabhängigkeit der Pyruvatkinase auf. Die Pyruvatkinase zählt zu den reifungsabhängigen Enzymen der roten Blutzelle [14, 15]. Retikulozyten, die bis zu 10fach höhere PK-Aktivitäten als Erythrozyten aufweisen können, zeigen im Elektropherogramm eine zusätzliche PK-Aktivitätsbande (Abb. 1). Der strukturelle Aufbau dieses für Retikulozyten

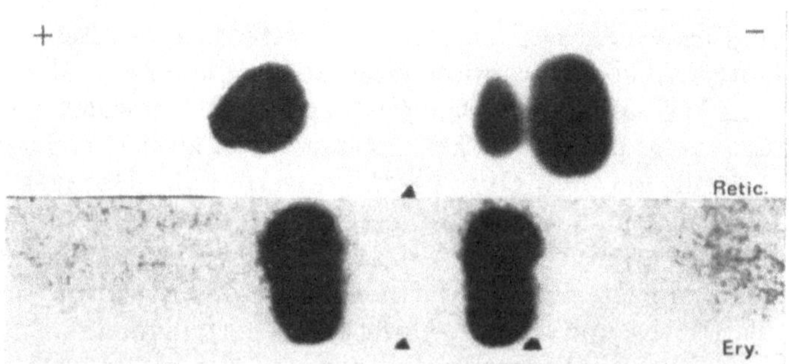

Abb. 1. Elektropherogramm der Pyruvatkinase von Kanincheneryth rozyten (unten) und -retikulozyten (oben).
Die Elektrophorese wurde im SFH auf Cellogel in 30 mM Kaliumphosphatpuffer, *p*H 6.9, bei 225 V/cm und −2 °C durchgeführt. Die Aktivitätsdarstellung erfolgte anschließend durch Inkubation der Cellogelstreifen im Reaktionsgemisch, wodurch es zur Fluoreszenzlöschung am Ort der PK kommt

typischen PK-Isoenzyms, das wesentlich instabiler als die Erythrozyten-PK ist, ist bisher noch ungeklärt [16].

Die Reifungsabhängigkeit der PK ist bei PK-Mangelpatienten kaum noch nachweisbar. Abb. 2 zeigt das Verhalten der PK-Aktivität von drei PK-Defektträgern, einem G6PD-Mangelpatienten in einem Kontrollblut jeweils in Abhängigkeit vom Reifegrad der Zellen. Im Gegensatz zu den Bluten von PK-Mangelpatienten lassen die roten Blutzellen des Kontrollblutes ebenso wie das an Retikulozyten reichere Blut eines G6PD-Mangelpatienten eine deutliche Abnahme der PK-Aktivität mit zunehmendem Reifegrad der Zellen erkennen. Abb. 2 veranschaulicht weiterhin, daß der Grad der Enzymaktivitätsverminderung zwischen den einzelnen PK-Defektträgern über einen größeren Aktivitätsbereich variiert. Das gleiche trifft sowohl für eine Reihe kinetischer Enzymparameter als auch physikochemische und immunchemische Charakteristika zu. Die bisher vorliegenden Ergebnisse der Enzymcharakterisierungen unterstützen die Auffassung, daß die PK-Enzymopathie einen Enzympolymorphismus aufweist, etwa in gleicher Weise wie einige andere Proteine mit genetisch aberranten Varianten, wie z. B. die Hämoglobine und die Glukose-6-phosphat-Dehydrogenase. Einige atypische pathologische PK-Varianten roter Blutzellen sollen nachfolgend ausführlicher vorgestellt werden.

16 PK-Mangelfälle aus 12 Familien unseres Landes wurden bisher anhand kinetischer und physikochemischer Parameter charakterisiert. Tab. 1 gibt eine Über-

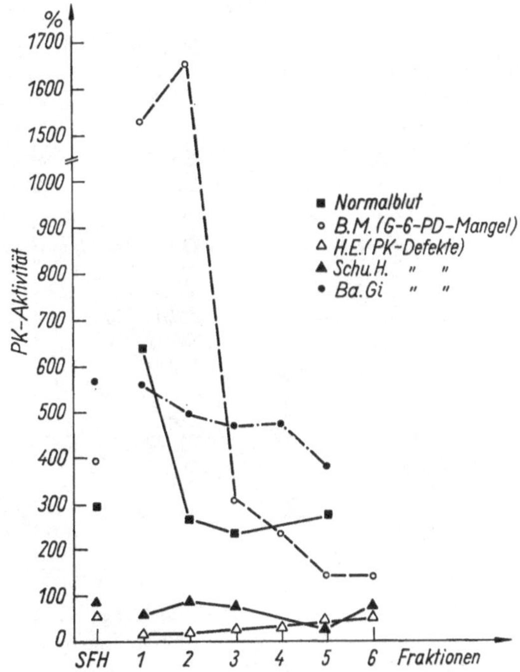

Abb. 2. Reifungsabhängigkeit der Pyruvatkinase roter Blutzellen.

Die retikulozytenhaltigen Zellsuspensionen wurden mit Hilfe der Dextrandichtegradientenzentrifugation isopyknisch über einen Dichtebereich von 1,063 bis 1,084 in 6 Fraktionen aufgetrennt [27]. Die jüngsten Zellen mit der geringsten Dichte sind in Fraktion 1 lokalisiert

sicht über die PK-Restaktivitäten roter Blutzellen der einzelnen Patienten unter optimalen Meßbedingungen. Die Ergebnisse erlauben eine Grobeinteilung in 4 Gruppen. Am häufigsten traten mit 11 Fällen PK-Enzymopathien mit geringen Restaktivitäten auf, die etwa 10% der Norm entsprachen. 3 Patienten wiesen etwa doppelt so hohe Restaktivitäten auf. In einem Fall (Mö. A.) betrug die Aktivitätsverminderung nur nahezu 40% und bei dem Patienten Ba. G. wurden unter sättigenden Substratkonzentrationen Maximalaktivitäten bestimmt, die im Vergleich zu Kontrollwerten nahezu einer Verdoppelung entsprachen.
Die Ergebnisse der Familienuntersuchungen bestätigten in allen Fällen einen autosomal rezessiven Erbgang [17].

Die 4 Patientengruppen, die sich anhand der PK-Restaktivitäten einteilen ließen, erwiesen sich auch relativ einheitlich in bezug auf ihr kinetisches Verhalten. Die Erythrozyten-PK von Kontrollpersonen ist durch ein kooperatives Verhalten mit einem Hill-n von etwa 1,7 gekennzeichnet (Abb. 3) Der $S_{0,5}$-Wert für PEP liegt bei 563 μM. Fru-1,6-P hebt die Kooperativität auf. Parallel dazu verringert sich der $S_{0,5}$-PEP-Wert auf 114 μM.

Die Gruppe des PK-Mangels mit den geringsten Enzymaktivitäten („quantitativer" PK-Mangel Typ 1) zeigt eine im Vergleich zur Norm höhere PEP-Affinität, wie sie nur in Gegenwart von Fru-1,6-P normalerweise erreicht wird. Im Gegensatz zu Kontrollbluten zeigt diese Gruppe von PK-Defektträgern keine Beein-

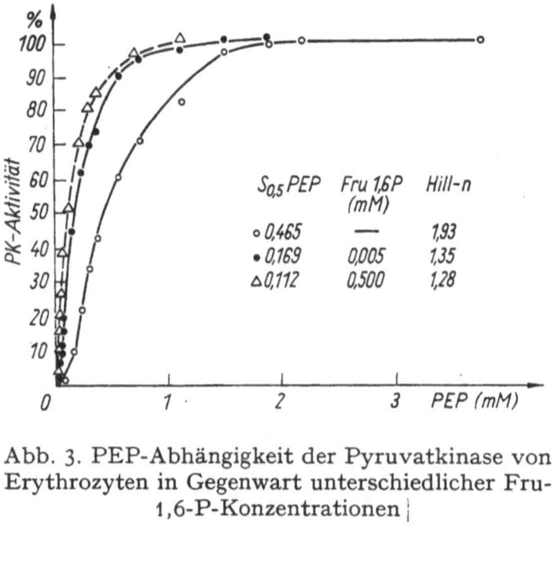

Abb. 3. PEP-Abhängigkeit der Pyruvatkinase von Erythrozyten in Gegenwart unterschiedlicher Fru-1,6-P-Konzentrationen

Tabelle 1
Pyruvatkinasedefektträger in der DDR

Proband	PK	Geschlecht
Be. J.	18	♂
Be. D.	35	♂
Schi. C.	26	♀
Schi. St.	33	♀
Wa. M.	24	♂
Wa. W.	39	♂
Schu. M.	28	♀
Schu. T.	36	♂
Ha. E.	35	♀
Gr. H.	35	♂
Ster. R.	36	♂
Ko. G.	55	♂
Sim. J.	59	♂
Sch. H.	65	♂
Mö. A.	104	♂
Ba. Gi.	505	♂
Kontrollen	274 ± 42	

Die PK-Aktivitäten, jeweils bestimmt unter sättigenden Substratkonzentrationen, sind in μMol PEP-Umsatz/ml Zellen · Std. bei 37 °C angegeben. Testansatz siehe [17]

Tabelle 2
PEP-Abhängigkeit pathologischer Pyruvatkinasevarianten roter Blutzellen

	Anzahl	PK-Aktivität	ohne Fru-1,6-P-Zusatz		+ Fru-1,6-P 50 mM	
			$S_{0,5}$ PEP (mM)	Hill − n	$S_{0,5}$ PEP (mM)	Hill − n
		$\bar{x}$ s	$\bar{x}$ s		$\bar{x}$ s	
Kontrollen	4	312 ± 28,7	0,563 ± 0,075	1,78	0,114 ± 0,020	1,16
PK-Defekte	1	505	4,848	1,65	0,670	1,29
	1	96	2,006	1,07	1,321	0,69
	7	46 ± 15,0	0,307 ± 0,057	1,06	0,093 ± 0,032	1,00
	7	20 ± 3,5	0,124 ± 0,042	0,98	0,077 ± 0,230	0,85

Angabe der PK-Aktivität in µMol PEP-Umsatz/ml · Std. bei 37 °C, pH 7,6

flussung des kinetischen Verhaltens durch Fru-1,6-P-Zusatz und keine positive Kooperativität (Tab. 2). Je höher dagegen die PK-Restaktivitäten sind (siehe Gruppe 2), desto höher sind auch die $S_{0,5}$-Werte für PEP und die Ansprechbarkeit des Enzyms auf positive Effektoren, wie Fru-1,6-P, oder auf negative Effektoren, wie ATP oder Alanin [5].

Die 2 PK-Enzymopathien mit relativ hohen bzw. über der Norm liegenden Maximalaktivitäten sind im Vergleich zu den bisher beschriebenen PK-Anomalien durch eine stark verminderte PEP-Affinität charakterisiert. Sie entsprechen dem „qualitativen" PK-Mangel Typ 2. Der Patient Ba. Gi. wurde 1963 geboren und am 4. Lebenstag wegen einer Hyperbilirubinämie (16,9 mg%, direkt reagierend 5,4 mg% Bilirubin) stationär aufgenommen. Die osmotische Resistenz lag im Normbereich (0,42—0,32), Antikörper waren nicht nachweisbar. Die Lebenserwartung der roten Blutzellen war auf $T/2 = 3{,}5$ Tage verkürzt.

Es bestand eine beträchtliche Retikulozytose und Erythroblastämie sowie eine erhebliche Hepatosplenomegalie. Im Alter von 8 Monaten wurde die Splenektomie durchgeführt. Der histologische Befund ergab eine Hämosiderose und gesteigerte Erythrophagozytose. Das Blutbild zeigte eine deutliche Anisozytose und Polychromasie. Die Price-Jones-Kurve war zweigipflig mit einer deutlichen Rechtsverschiebung. Die Retikulozytenwerte lagen seit der Splenektomie zwischen 20 und 60%. Bluttransfusionen sind in größeren Abständen, insbesondere beim Auftreten von Infektionskrankheiten, erforderlich. Infektionskrankheiten führten mehrfach zu lebensbedrohlichen anämischen Zuständen, bei denen im Extremfall die Hb-Konzentration auf 2 g% und die Zahl der Erythrozyten auf $0{,}7 \times 10^6$ abfielen, bei einer gleichzeitigen Abnahme der Retikulozytenwerte auf 1 ‰. Erst einige Tage nach der durchgeführten Bluttransfusion und dem Abklingen der Infektion war wieder eine Erythropoese zu beobachten. Dieses Blutbild unterstreicht die Auffassung, daß der Reifungsstop während der Infektionskrankheit auf der Stufe der Proerythroblasten erfolgt; denn als Ausdruck der wieder einsetzenden Hämatopoese war eine starke Erythroblastämie (33 500/mm³) bei noch tiefen Retikulozyten nachweisbar.

Abb. 4 zeigt den Erbgang des pathologischen PK-Merkmals in der Familie. Die Mutter des Patienten Ba. M. ist heterozygot für den PK-Defekt Typ 1, zu erkennen an der maximalen PK-Aktivität von 120 und einer der Norm entsprechenden

PEP-Abhängigkeit mit einer Konstanten für $S_{0,5}$PEP von 563 μM. Der Vater, (Ba. G.) repräsentiert das heterozygote Merkmal für den „qualitativen" PK-Mangel Typ 2. Die maximale Enzymaktivität von 317 entspricht Kontrollbluten, der $S_{0,5}$-Wert für PEP ist dagegen mit 2359 μM auf das 4—5fache gegenüber der Norm erhöht. Die Erythrozyten-Pyruvatkinase der Schwester des Patienten (Ba. G.) weist die gleichen Enzymcharakteristika wie die der Mutter auf, d. h., sie ist heterozygot für den „quantitativen" PK-Defekt Typ 1. Der Patient Ba. Gi. ist auf Grund der geringen PEP-Affinität der pathologischen Enzymvariante als homozygot für den PK-Defekt Typ 2 einzustufen. Der $S_{0,5}$PEP liegt mit 4848 μM um nehezu eine Größenordnung über der Norm. Das bedeutet, daß in der intakten roten Blutzelle unter nicht sättigenden PEP-Konzentrationen trotz Berücksichtigung des erhöhten Substratspiegels und der hohen PK-Kapazität ein Defizit in der aktuellen Enzymaktivität resultiert; denn die errechenbare PK-Aktivität entspricht etwa nur 10% der Norm.

Die kooperativen Eigenschaften sind bei dieser PK-Anomalie mit einem Hill-n von 1,6 voll erhalten (Abb. 5). Fru-1,6-P-Zusatz erhöht die PEP-Affinität, d. h. der $S_{0,5}$PEP sinkt auf 670 μM ab.

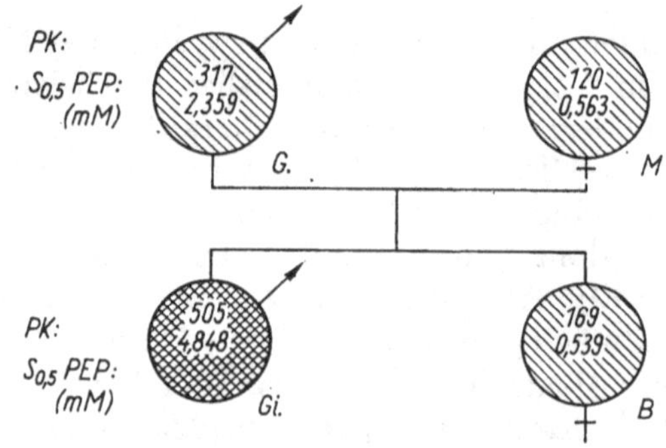

Abb. 4. Verhalten der Pyruvatkinaseaktivität roter Blutzellen und der jeweiligen PEP-Abhängigkeit in der Familie Ba.

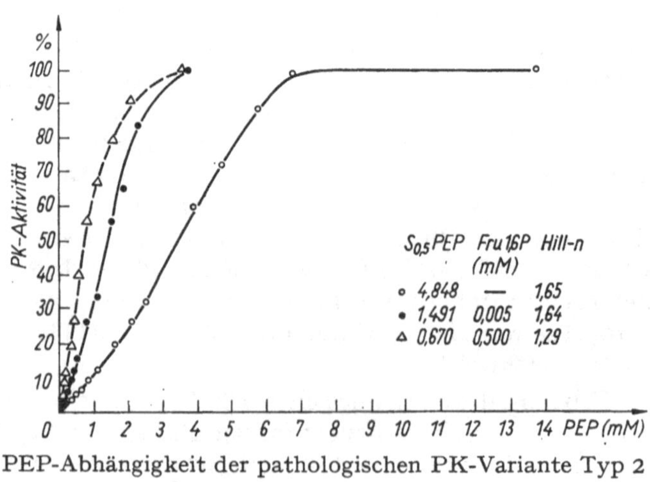

Abb. 5. PEP-Abhängigkeit der pathologischen PK-Variante Typ 2 (Ba. Gi.)

Abb. 6 und 7 demonstrieren die Empfindlichkeit dieser PK-Mutanten gegenüber ATP und Alanin, sowie die nahezu vollständige Aufhebung dieser Hemmung durch Fru-1,6-P-Zugabe. Die experimentellen Bedingungen in den 2 dargestellten Versuchen unterscheiden sich durch die Höhe der eingesetzten PEP-Konzentrationen.

Ein weiterer Patient mit einer atypischen PK-Enzymopathie ist der 1956 geborene Mö. A. (Abb. 8). Er ist das 2. Kind von 4 Geschwistern. Aus der Familienanamnese geht hervor, daß die ältere Schwester (Mö. R.), bei der ebenfalls eine angeborene hämolytische Anämie bestand, mit 3 Jahren verstarb. In den ersten

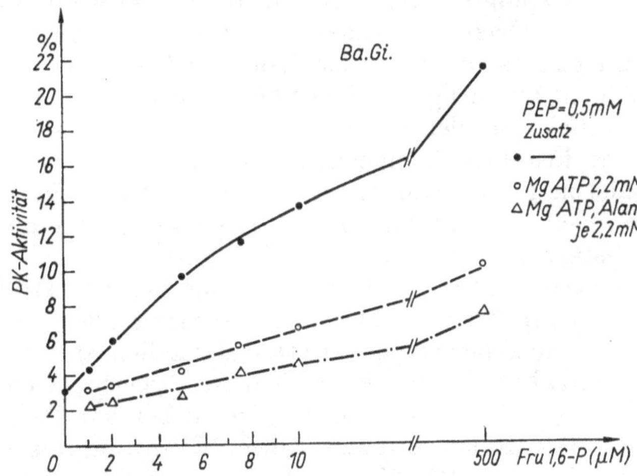

Abb. 6. ATP- und Alaninhemmung der pathologischen Pyruvatkinasevariante Typ 2 (Ba. Gi.) und Reaktivierbarkeit durch Fru-1,6-P

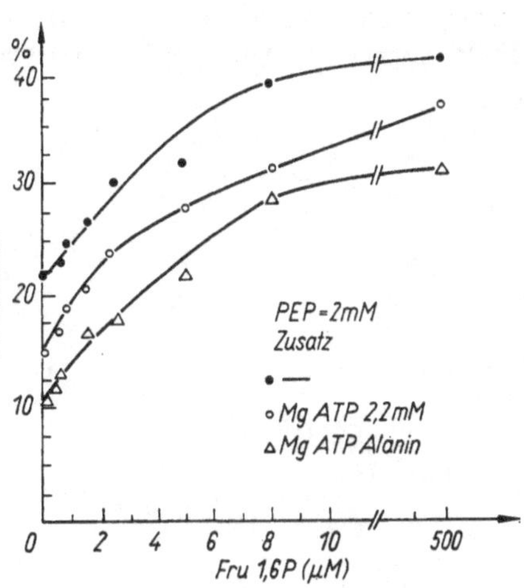

◂ Abb. 7

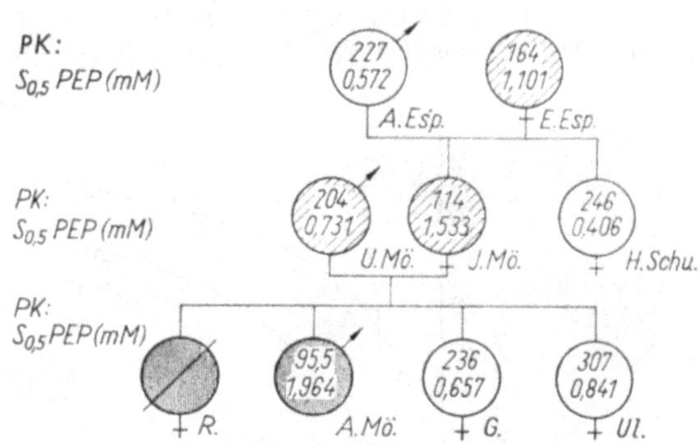

Abb. 8

Abb. 7. Hemmung und Reaktivierung der PK Typ 2 (Ba. Gi.) bei erhöhtem PEP-Spiegel

Abb. 8. Erbgang des pathologischen Pyruvatkinasemerkmals roter Blutzellen in der Familie Mö.

Lebensjahren traten bei dem Patienten Mö. A. ca. alle 2 Monate, auch unter Prednisongabe, hämolytische Krisen auf. Das klinische Bild war durch eine nichtsphärozytäre, hämolytische Anämie mit Hepatosplenomegalie bei bestehendem Ikterus charakterisiert. 1962 wurde die Splenektomie durchgeführt, da die Sequestration der roten Blutzellen, insbesondere der Retikulozyten, in der Milz erfolgte. Zu dieser Zeit betrug der Bilirubinspiegel 2,03 mg% (davon 0,45 mg% direkt reagierendes Bilirubin). Die osmotische Resistenz lag im Normbereich (0,46—0,26% NaCl), die mechanische Resistenz wurde mit 16,8% bestimmt. Das Blutbild zeigte eine deutliche Anisozytose und Polychromasie; die Price-Jones-Kurve war verbreitert mit einer ausgeprägten Rechtsverschiebung. Die Retikulozytenwerte lagen zwischen 10—20%. Nach der Splenektomie traten sehr viel seltener hämolytische Krisen auf, meist nur im Zusammenhang mit Infektionskrankheiten. Wiederholte Gallenkoliken bei nachgewiesener Cholelithiasis machten 1973 die Cholecytektomie erforderlich. Bei der zu diesem Zeitpunkt ebenfalls durchgeführten Leberbiopsie wurde histologisch eine massive Siderose mit vorwiegender Ansammlung des Siderins in den Leberzellen und nur geringer Speicherung in den Kupffer'schen Sternzellen nachgewiesen. Außerdem wurde eine ausgeprägte periportale Fibrose festgestellt.

Die Retikulozytenwerte sind seit der Splenektomie deutlich erhöht, wobei Extremwerte bis zu 86% auftraten. Darüber hinaus sind in großer Zahl Howell-Jolly-Körper nachweisbar. Die Lebensdauer der roten Blutzellen ist mit T/2 = 3,6 Tage (Normwert 26,5) deutlich verkürzt. Der Patient ist körperlich wenig belastbar.

Die maximale PK-Aktivität der roten Blutzellen ist bei dem Patienten auf etwa 40% im Vergleich zu Kontrollen vermindert. Die PEP-Affinität ist deutlich herabgesetzt; der $S_{0,5}$PEP ist mit 2000 μM auf das 4fache der Norm erhöht (Abb. 8 und 9). Das kooperative Verhalten ist bei dieser pathologischen PK-Variante nahezu aufgehoben. Der Erbgang des pathologischen PK-Merkmals über 3 Generationen (Abb. 8) läßt die Schlußfolgerung zu, daß es sich möglicherweise in diesem Fall nicht um einen homozygoten Defektträger einer bisher unbekannten pathologischen PK-Variante, sondern um einen doppelt heterozygoten Merkmalsträger

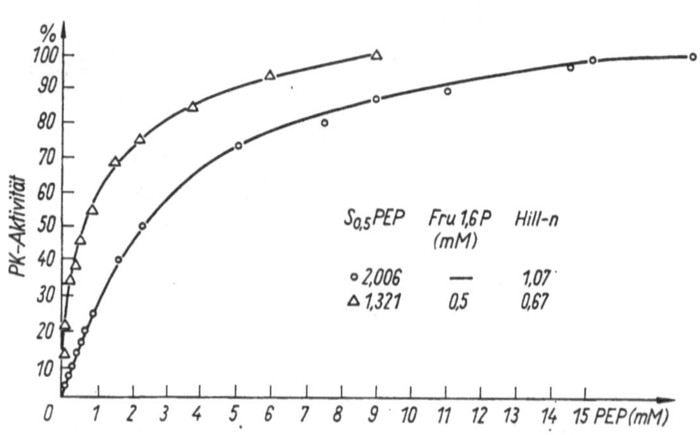

Abb. 9. PEP-Abhängigkeit der pathologischen PK-Variante roter Blutzellen des Patienten Mö. A.

handelt. Eine sichere Unterscheidung zwischen homozygoten und doppelt heterozygoten Defektträgern ist jedoch anhand der bisher bestimmbaren Parameter nicht möglich.

Beide pathologischen PK-Varianten mit hohen $S_{0,5}$PEP-Konstanten (Ba. Gi. und Mö. A.) sind relativ stabil im Temperaturstabilitätstest. Abb. 10 zeigt das Verhalten der PK-Aktivität im stromafreien Hämolysat beider Patienten und von Kontrollpersonen in Abhängigkeit von der Zeit bei 53 °C (der schraffierte Bereich entspricht dem von Kontrollbluten).

Die Stabilität gegenüber 2 M Harnstoff ist aus der Abb. 11 ersichtlich. Während sich die PK des Patienten Mö. A. unter dem Einfluß von Harnstoff als sehr stabil erweist, ist die Harnstoffstabilität der pathologischen PK-Variante des Patienten Ba. Gi. im Vergleich zu Kontrollen deutlich herabgesetzt. Ein Vergleich der Harnstoffstabilität mit den in Japan beschriebenen PK-Varianten „Tokyo" und „Sapporo", die ebenfalls durch hohe $S_{0,5}$PEP-Konstanten charakterisiert sind, jedoch sehr Harnstoff-empfindlich sind [18], macht u. a. wahrscheinlich, daß die Gruppe der pathologischen PK-Varianten Typ 2 heterogen ist. Ähnliche Schlußfolgerungen lassen sich auch aus der Nukleotidspezifität ableiten [18]. Tab. 3 gibt eine Übersicht über die relative Aktivität der PK roter Blutzellen mit UDP, GDP und CDP als Reaktionspartner im Vergleich zu ADP. Aus der Zusammenstellung ist ersichtlich, daß alle 4 aufgeführten pathologischen PK-Varianten vom Typ 2 im Vergleich zu Kontrollbluten relativ hohe Umsatzgeschwindigkeiten mit UDP und

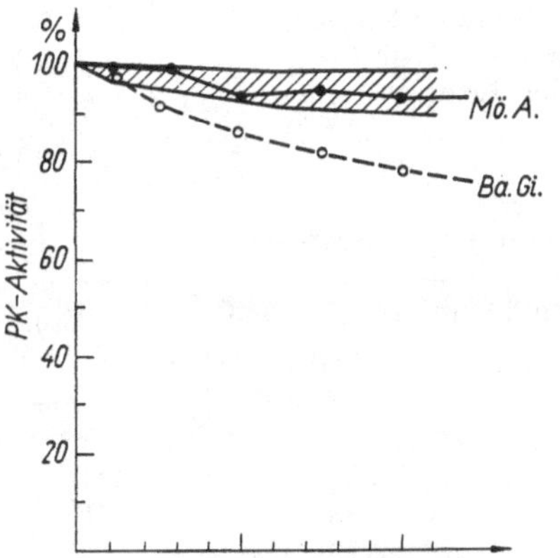

Abb. 10

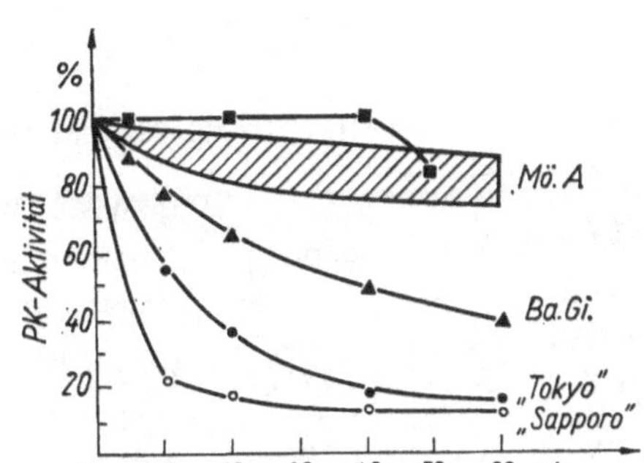

Abb. 11

Abb. 10. Temperaturstabilität der Pyruvatkinase im Hämolysat der PK-Defektträger Ba. Gi. und Mö. A.

Die Inkubationstemperatur für das SFH betrug 53 °C, die PK-Aktivitätsbestimmung wurde bei 37 °C durchgeführt. Der schraffierte Bereich entspricht dem Verhalten von Kontrollbluten

Abb. 11. Harnstoffstabilität der Pyruvatkinase roter Blutzellen von Kontrollpersonen und PK-Defektträgern Typ 2.

Die Harnstoffkonzentration betrug 2 M.

Die experimentellen Daten für die PK-Varianten „Tokyo" und „Sapporo" wurden der Arbeit von Miwa et al. entnommen [18]

GDP aufweisen, obwohl zwischen den einzelnen Defektträgern beträchtliche Differenzen bestehen.

Lange Zeit war die Frage, ob die Glykolyserate im PK-Mangel erniedrigt ist, umstritten. Einige Autoren kamen zu falschen Einschätzungen, da sie als Bezugsparameter den reifen Erythrozyten wählten. Das ist jedoch auf Grund der meist starken Retikulozytose nicht korrekt. Ein besseres Maß ist die Beurteilung der Glykolyserate anhand der bestehenden Korrelation zur Glutamat-Oxalazetat-Transaminase (GOT) (Tab. 4). Entsprechende Untersuchungen wurden von der Arbeitsgruppe um OSKI [19] durchgeführt. Während für die sog. Kontrollversuche — rote Blutzellen von Patienten mit Sichelzellanämien und Sphärozytosen — die experimentell ermittelte Glykolyserate sehr gut mit der auf der Grundlage der

Tabelle 3

Nukleotidspezifität der Pyruvatkinase roter Blutzellen % PK-Aktivität verglichen mit ADP als Reaktionspartner

Proband	UDP	GDP	CDP
Kontrollen	68	67	13
Ba. Gi.	152	118	34
Mö. A.	74	82	18
PK „Sapporo"	183	136	23
PK „Tokyo"	98	94	15

Die jeweils eingesetzte Endkonzentration der Nukleotide betrug 2 mM. Die Angaben für die PK-Defekte „Sapporo" und „Tokyo" wurden der Literatur entnommen [18].

Tabelle 4

Verhalten der Glykolyserate retikulozytenhaltiger Zellsuspensionen

Probanden	Zellsusp.	Hämolysat	berechnete Werte	Hämolysat + PK	GOT (Einh./10^{10} Zellen)
	(μMol Glukoseverbrauch/ml Zellen · Std.)				
PK-Defekte					
A. Y.	2,77	2,84	3,60	3,88	4117
J. Y.	1,91	2,14	3,20	3,35	3682
L. Y.	1,72	1,82	2,80	2,75	2725
C. M.	1,29	1,94	3,00	6,49	2939
D. S.	1,70	1,64	3,55	3,95	4090
M. R.	1,29	1,42	3,05	3,83	2974
Hereditäre Sphärozytose	2,64	2,77	2,80	2,70	2727
Sichelzellanämie	3,04	2,92	3,00	2,87	3001
Normale Kontrolle	1,76	1,82	1,64	1,72	1146

Die Glykolyseraten wurden in Krebs-Ringer-Bikarbonat-Puffer bei pH 7,4 nach OSKI und BOWMAN [19] bestimmt. Die Vorhersage der Glykolyserate erfolgte anhand der bestehenden Korrelation zur GOT-Aktivität. Der Muskel-PK-Zusatz zum Hämolysat betrug 1 IE/ml.

GOT-Aktivität berechneten übereinstimmt, liegt die Glykolyserate bei allen untersuchten PK-Mangelfällen deutlich unter den Erwartungswerten. Dieses Defizit ist in Versuchen mit Hämolysaten von PK-Defektträgern durch Zusatz gereinigter PK ausgleichbar. In allen übrigen Fällen erfolgt dagegen im Hämolysatsystem bei PK-Zusatz keine Erhöhung der Glykolyse, da dieses Enzym nicht geschwindigkeitsbestimmend für die Glykolyse ist [20, 21].

Besteht auf Grund der gesteigerten Erythropoese eine starke Retikulozytose, wie bei den in dieser Arbeit vorgestellten Patienten, so ist neben der ATP-Bildung bei der Glykolyse die noch bestehende Atmungskettenphosphorylierung zu berücksichtigen. Die Funktionstüchtigkeit der Retikulozytenmitochondrien kann dabei in Abhängigkeit vom Reifegrad der Zellen beträchtlich differieren und der Pasteur-Effekt mehr oder weniger abgeschwächt sein. Bei den PK-Defektträgern mit verminderter PEP-Affinität ist der Pasteureffekt mit 21,9% für Ba. Gi. und 15,7% für Mö. A. auf Grund der Unreife der roten Blutzellen sehr deutlich ausgeprägt. Die Höhe der Atmung sowie das Verhalten der Glykolyse und glykolytischen Metabolitspiegel unter anaeroben und aeroben Bedingungen ist für Zellsuspensionen beider Patienten bei *p*H 7,2 in den Tab. 5 und 6 zusammengefaßt. Die Metabolitkonzentrationen zeigen die für den PK-Mangel typischen Veränderungen; der PEP- sowie die Mono- und Diphosphoglyzeratwerte sind erhöht und der ATP-Gehalt im Vergleich zur Norm beträchtlich vermindert. Die Abnahme des ATP wird besonders deutlich, sobald die Atmungskettenphosphorylierung gehemmt wird. Daraus läßt sich die Schlußfolgerung ableiten, daß die defekten roten Blutzellen mit der Reifung sehr schnell einem ATP-Mangel unterliegen und der Energiestoffwechsel nicht mehr aufrecht gehalten werden kann [22].

Die Glykolyserate sowie die Konzentrationsveränderungen von Glykolysemetaboliten roter Blutzellen, die bei Inkubationsexperimenten in vitro unter Bedin-

Tabelle 5
Energiestoffwechsel der roten Blutzellen des Patienten Ba. Gi.

Metabolit	Aerob	Anaerob
	(μMol/ml Zellen)	
ATP	815	337
ADP	160	236
AMP	77	226
$\sum$ Adeninnukleotide	1052	799
Glu-6-P	220	44
Fru-1,6-P	17	13
DOAP + GAP	100	73
1,3-DPG	16	–
2,3-PDG	9505	8350
3-PG	217	134
PEP	80	94
Pyruvat	20	35
Laktatbildung (μMol/ml Zellen · Std)	1,52	6,94
Atmung (mm³ O_2/ml Zellen · Std.)	79	

Die Inkubation gewaschener Zellsuspensionen wurde 60 min bei *p*H 7.2 und 37 °C durchgeführt.
Der Retikulozytengehalt betrug 50%.

Tabelle 6
Energiestoffwechsel der roten Blutzellen des Patienten Mö. A.

Metabolit	Aerob	Anaerob
	µMol/ml Zellen	
ATP	933	350
ADP	121	235
AMP	51	151
$\sum$ Adeninnukleotide	1105	736
Glu-6-P	191	25
Fru-1,6-P	14	18
DOAP + GAP	67	53
2,3-DPG	9290	8540
3-PG	109	148
PEP	44	80
Pyruvat	33	44
Laktatbildung (µMol/ml Zellen · Std.)	0,67	4,27

Die Inkubationszeit der gewaschenen Zellsuspensionen betrug 60 min. Der *p*H-Wert war 7.2, die Temperatur 37 °C. Die Suspension enthielt 68,5% Retikulozyten.

gungen des quasi steady state meßbar sind, lassen sich für den PK-Mangel auf der Grundlage des theoretischen Glykolysemodells mit guter Genauigkeit vorausberechnen [23]. Für die in Tab. 7 zusammengestellten Ergebnisse wurde ein PK-Defektträger vom Typ 1 (Schi. C.) ausgewählt, bei dem nur eine mäßige Retikulozytose bestand, so daß das für Erythrozyten ausgearbeitete Glykolysemodell noch ohne wesentliche Einschränkungen anwendbar war [17]. Die maximale PK-Aktivität der roten Blutzellen dieser Patientin war mit 26 µMol/l Zellen · Std. auf etwa 10% der Norm herabgesetzt; alle übrigen Enzymkapazitäten der Glykolyse waren unverändert. Die in Tab. 7 angegebenen Werte belegen, daß auch aus der Anwendung des theoretischen Glykolysemodells bei Herabsetzung der PK-Kapazität auf 10% der Norm sich ein Abfall der Glykolyserate und des ATP-Spiegels sowie ein Anstieg der PEP- und 2,3-DPG-Konzentrationen ableitet, und die errechneten Daten in guter Übereinstimmung mit den experimentell ermittelten Ergebnissen sind.

Tabelle 7
Vorhersage einiger Metabolitenkonzentrationen und der Glykolyserate roter Blutzellen bei einem PK-Defekt Typ 1

Metabolit	theoretisch berechneter Wert (µM)	experimentell ermittelter Wert (µM)
ATP	870	940
PEP	106	144
2,3-DPG	6300	6100
Glykolyse (µMol Glukoseumsatz/ml Zellen · Std.)	1200	1600

Aus Ergebnissen verschiedener Studien läßt sich die Schlußfolgerung ableiten, daß der ATP-Spiegel in roten Blutzellen von PK-Defektträgern durch 2 Faktoren negativ beeinflußt wird 1. durch die verminderte Glykolyserate und Bevorzugung des 2,3-DPG-Weges und 2. durch einen gesteigerten ATP-Verbrauch, dessen Ursache noch nicht in allen Einzelheiten geklärt ist. Als ein Beispiel ATP-verbrauchender Prozesse sei der energieabhängige Kaliumtransport angeführt. Ergebnisse der Arbeitsgruppe von Oski belegen, daß im PK-Mangel sowohl der Kalium-Einstrom als auch -Ausstrom im Vergleich zur Norm erhöht ist, wobei der Kaliumverlust überwiegt, so daß eine negative Bilanz resultiert [19].

Schröter und Tillmann [24] vermuten in diesem Zusammenhang, daß die PK direkt durch ATP-Bereitstellung eine wichtige Rolle für energieabhängige Prozesse der Zellmembran spielt, ähnlich wie es für die Phosphoglyzeratkinase angenommen wird. Sie konnten zeigen, daß ein Anteil der PK, der z. T. maskiert ist, fest an die Zellmembran gebunden ist. Für 3 PK-Defektträger vom Typ 1 wiesen sie nach, daß der PK-Mangel auch den membrangebundenen Anteil betrifft. Da dieser Aktivitätsverlust in der Membran nicht mit der PK-Verminderung im Zytosol korrelierte, diskutierten sie eine direkte Beziehung zwischen der Schwere der Anämie und dem PK-Aktivitätsverlust in der Zellmembran.

Berkel et al. [25, 26] postulierten, daß in einigen Fällen abweichende PK-Eigenschaften auch die Folge sekundärer Defekte sein können, die auf oxydative Einflüsse z. B. durch den Anstieg des oxydierten Glutathionspiegels zurückzuführen sind und somit Zusammenhänge zu Stoffwechselprozessen der Zellmembran wahrscheinlich machen.

Die bisher vorliegenden Befunde reichen jedoch insgesamt noch nicht für eine zufriedenstellende Deutung der verschiedenen pathologischen PK-Varianten aus. Auch die kausalen Zusammenhänge zwischen der PK-Enzymopathie und dem Zustandekommen der nichtsphärozytären hämolytischen Anämie lassen sich noch nicht in allen Einzelheiten erklären.

Literatur

[1] Imamura, K., T. Tanaka, T. Nishina, K. Nakashima u. S. Miwa: J. Biochem., Tokyo **74**, 1165 (1973)

[2] Whittel, N. M., D. O. K. Ng., K. Prabhakbararao u. R. S. Holmes: Comp. Biochem. Physiol. **46** B, 71 (1973)

[3] Kahn, A., J. Marie u. P. Boivin: Hum. Genet. **33**, 35 (1976)

[4] Rosengurt, E., L. Jimenez de Asua u. H. Carminatti: J. biol. Chem. **244**, 3124 (1969)

[5] Tanaka, K. R., u. D. E. Paglia: Semin. Hemat. **8**, 367 (1971)

[6] Peterson, J. S., C. J. Chern, R. N. Harkins u. J. A. Black: FEBS Lett. **49**, 73 (1974)

[7] Searcy, G. P., D. R. Miller u. J. B. Tasker: Cand. J. comp. Med. **35**, 67 (1971)

[8] Standerfer, R. J., M. B. Rittenberg, C. J. Chern, J. W. Templeton u. J. A. Black: Biochem. Genet. **13**, 341 (1975)

[9] Dhindsa, D. S., J. B. Black, R. D. Koler, D. A. Rigas, J. W. Templeton u. J. Metcalfe: Respiration Physiol. **26**, 65 (1976)

[10] Bashan, N., S. Moses, Y. Gross u. A. Liune: FEBS Lett. **54**, 323 (1975)

[11] Black, J. A., C. J. Chern u. M. B. Rittenberg: Biochem. Genet. **13**, 331 (1975)

[12] Rittenberg, M. B., C. J. Chern, D. Lincoln u. J. A. Black: Immunochemistry **12**, 491 (1975)

[13] Selwyn, J. G., u. J. V. Dacie: Blood **9**, 414 (1954)

[14] Sass, M. D., F. Vorsanger u. P. W. Spear: Clinica chim. Acta **10**, 21 (1964)

[15] GRIEGER, M., G. JACOBASCH u. CH. GERTH: VII. Internationales Symposium über Struktur und Funktion der Erythrozyten, Berlin 1973. (= Abhandlungen der Akademie der Wissenschaften der DDR 1973). S. RAPOPORT u. F. JUNG (Hrsg.). Akademie-Verlag, Berlin 1975, S. 561

[16] PAGLIA, D. E., u. W. N. VALENTINE: J. Lab. clin. Med. **76**, 202 (1970)

[17] JACOBASCH, G., W. HELBIG, I. SYLLM-RAPOPORT, H. PESTER, K.-M. HEINE, CH. BOESE, F. M. G. OTTO u. H. J. BLAU: Dte. GesundhWes. **24**, 1441 (1969)

[18] MIWA, S., K. NAKASHIMA, K. ARIYOSHI, K. SHINOHARA, K. ODA u. S. TANAKA: Br. J. Haemat. **29**, 157 (1975)

[19] OSKI, F. A. u. H. BOWMAN: Br. J. Haemet. **17**, 289 (1969)

[20] JACOBASCH, G., S. MINAKAMI u. S. M. RAPOPORT in: Cellular and Molecular Biology of Erythrocytes. H. YOSHIKAWA u. S. M. RAPOPORT (Hrsg.). University of Tokyo Press, Tokyo 1974, S. 55

[21] RAPOPORT, T. A., R. HEINRICH, G. JACOBASCH u. S. M. RAPOPORT: Eur. J. Biochem. **42**, 107 (1974)

[22] KEITT, A. S.: Am. J. Med. **41**, 762 (1966)

[23] RAPOPORT, T. A., R. HEINRICH u. S. M. RAPOPORT: Biochem. J. **154**, 449 (1976)

[24] SCHRÖTER, W., u. W. TILLMANN: Klin. Wschr. **53**, 1101 (1975)

[25] BERKEL, T. J. C., J. F. VAN KOSTER u. G. E. J. STAAL: Biochim. biophys. Acta **321**, 496 (1973)

[26] BERKEL, T. J. C., G. E. J. STAAL, J. F. VAN KOSTER u. J. G. NYESSEN: Biochim. biophys. Acta **334**, 361 (1974)

[27] GROSS, J., S. ROSENTHAL u. I. SYLLM-RAPOPORT: Acta biol. med. germ. **26**, 643 (1971)

Summary

G. JACOBASCH, M. GRIEGER, CH. GERTH, and K. BIER: Energy metabolism of red cells in pyruvate kinase enzymopathies

Until now pyruvate kinase enzymopathies have been described only for red blood cells. On the basis of these results special structural properties of the erythrocyte PK was assumed, which are not yet totally established. PK defects may cause a nonspherocytic hemolytic anemia. This enzymopathy is characterized by a polymorphism, which is expressed in more than 5 different pathological variants. Up to now 16 cases of PK deficiency have been diagnozed in the GDR. The following parameters are used for the characterization of the PK: the PEP-dependence, the inhibition by ATP and alanine, the specificity to nucleotides, the stability to temperature and urea and the maturation dependence. Two pathological variants of the PK with a decreased PEP-affinity are described. Furthermore the differences in the energy metabolism of the red blood cells of these two patients under aerobic and anaerobic conditions are discussed.

Acta biol. med. germ., Band 36, Seite 731—734 (1977)

Deparment of Pharmacology, New York University School of Medicine New York, New York, U.S.A

Superoxide anion and drug-induced hemolysis

B. GOLDBERG and A. STERN

Summary

Superoxide anion, either generated during the autooxidation of dihydroxyfumaric acid or by the interaction of 1,4-naphthoquinone-2-sulfonate and intracellular hemoglobin in red cells pretreated with the intracellular superoxide dismutase inhibitor, diethyldithiocarbamate, produces structural changes in red cells hemoglobin and hypotonic lysis. No evidence for lipid peroxidation was found in red cells exposed to either 1,4 naphthoquinone-2-sulfonate in the presence of diethyldithiocarbamate or to dihydroxyfumaric acid, although the membranes of these cells retained a green pigment. These results suggest that superoxide anion reacts with cellular hemoglobin to form hemoglobin breakdown products which bind to the red cell membrane and thereby increase the osmotic fragility of the cell.

Much interest is currently being focused on the role of the copper-zinc enzyme superoxide dismutase (SOD) in the red cell. This enzyme catalyses the dismutation of the highly reactive superoxide anion (O_2^-), to the less reactive H_2O_2 and O_2 [1]. It has been postulated that the SOD present in the red cell protects these cells against O_2^- released during the autooxidation of oxyhemoglobin [2]. In the past few years, we have been interested in the possibility of O_2^- acting as a mediator of the toxic effects of hemolytic drugs on the red cell. We have demonstrated that the hemolytic drugs, phenylhydrazine [3, 4] and menadione [5], interact with hemoglobin in a manner that results in the production of relatively large amounts of O_2^-.

In the present studies we have attempted to demonstrate a toxic role for O_2^- in the red cell. Two basic procedures were utilized for this investigation. Red cells were exposed to a flux of O_2^- from the outside of the cell and the toxic effects observed were ascribed to O_2^- by noting inhibition in the presence of externally added SOD. The source of O_2^- in this case was derived from the autooxidation of dihydroxyfumaric acid (DHF). O_2^- was also generated inside red cells by the use of 1,4 naphthoquinone-2-sulfonate (NQ), a water soluble derivative of the hemolytic drug, menadione, which reacts with hemoglobin resulting in O_2^- production. In this procedure, the role of O_2^- as a mediator of hemolysis was investigated by manipulating the intracellular levels of SOD with the SOD inhibitor diethyldithiocarbamate (DDC).

When red cells are exposed to DHF, their cellular hemoglobin undergoes rapid breakdown to methemoglobin and a green pigment. This hemoglobin breakdown (Fig. 1) is inhibited by SOD and catalase (CAT) and accelerated by lactoperoxidase (LP), an enzyme which catalyses the autooxidation of DHF and increases the production of O_2^- at the expense of H_2O_2 [6].

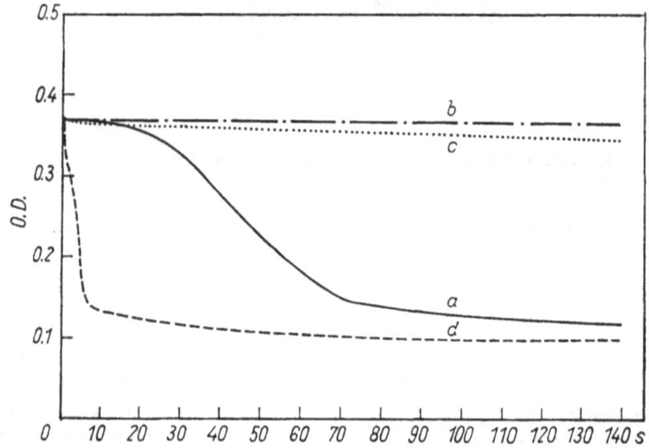

Fig. 1. Kinetics of hemoglobin breakdown in red cells exposed to DHF, recorded with intact cells as a decrease in absorbance at 576 nm on an Amino-Chance spectrophotometer in the dual wavelength mode.
Concentrations: red cells (0.2.% in 10 mM sodium phosphate buffer, *p*H 7.4, at 28 °C containing 125 mM NaCl and 10 mM glucose), DHF (4.5 mM), SOD (8.3 μg/ml), CAT (2920 μg/ml),LP (34.2 μg/ml).
a) red cells treated with DHF; b) effect of SOD on red cells treated with DHF; c) effect of CAT on red cells treated with DHF, d) effect of LP cells treated with DHF

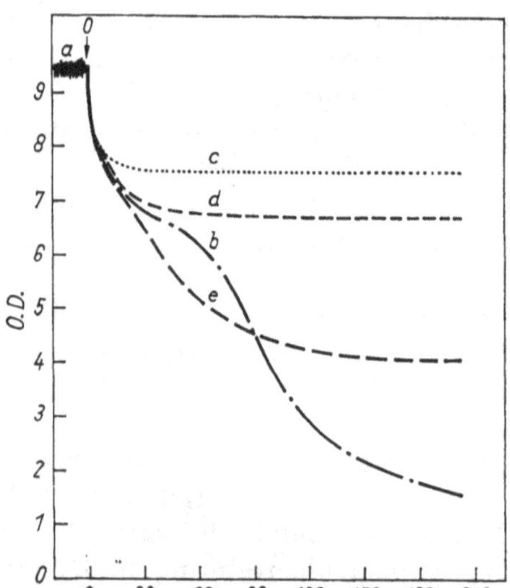

Fig. 2. Kinetics of hypotonic lysis of red cells exposed to DHF measured as a decrease in light scattering at 740 nm on a Cary 14 spectrophotometer.
Concentrations: red cells (0.2% in 10 mM sodium phosphate buffer, *p*H 7.4, at 28 °C containing 68.8 mM NaCl and 10 mM glucose), DHF (4.5 mM), SOD (8.3 μg/ml), CAT (2920 μg/ml), LP (34.2 μg/ml).
a) light scattering at 740 nm of the red cell suspension prior to the addition of DHF; at the arrow, DHF was added to the suspension, resulting in the hemolysis shown in curve b; c) effect of preincubation of the red cell suspension with SOD; d) effect of preincubation of the red cell suspension with CAT; e) effect of preincubation of the red cell suspension with LP

Addition of DHF to a hypotonic suspension of red cells causes a rapid lysis of the cells. This hypotonic lysis is inhibited by SOD and CAT, and initially accelerated but eventually inhibited by LP (Fig. 2). The lysis induced by DHF is preceded by a lag phase of approximately one minute during which the bulk of the hemoglobin breakdown occurs in red cells, suggesting that the hemoglobin breakdwon precedes hemolysis. To substantiate this idea, we exposed an aerated suspension of carbonmonoxyhemoglobin-containing red cells to an aqueous solution of DHF. No hemoglobin breakdown or hemolysis was observed in the carbonmonoxyhemoglobin-containing red cells even though O_2^- production was demonstrated in the reaction medium by the SOD-inhibitable reduction of nitroblue tetrazolium. The membranes of red cells exposed to DHF, were found to retain a green pigment, which was not seen when the preparations contained externally added SOD, CAT or when the cellular hemoglobin was converted to carbonmonoxyhemoglobin. Both membrane sulfhydryl oxidation and lipid peroxidation were not observed in red cells exposed to DHF.

A suspension of red cells exposed to NQ exhibited a gradual disappearance of their oxyhemoglobin with formation of methemoglobin as the major product. If the red cells were pretreated with DDC, which caused an 83% decrease in the SOD activity, and then exposed to NQ, they showed a significant increase in the rate of oxyhemoglobin disappearance when compared to red cells not treated with DDC, but exposed to NQ (Fig. 3). Addition of NQ to a hypotonic suspension of red cells results in a gradual lysis of the cells which is also seen when red cells are treated with DDC and not exposed to NQ. By comparison, the hemolytic process is dramatically increased when NQ is added to a hypotonic suspension of red cells pretreated with DDC (Table 1). The membranes of red cells exposed to either NQ or DDC retained small amounts of hemoglobin compared to untreated controls, while the membranes of red cells pretreated with DDC and then exposed to NQ retained large amounts of a green pigment. Lipid peroxidation did not occur in red cells pretreated with DDC and exposed to NQ, even though significant hemolysis was observed.

The preceding experiments clearly show that cellular damage to the erythrocyte in the presence of either DHF or NQ is manifest as hemoglobin breakdown and

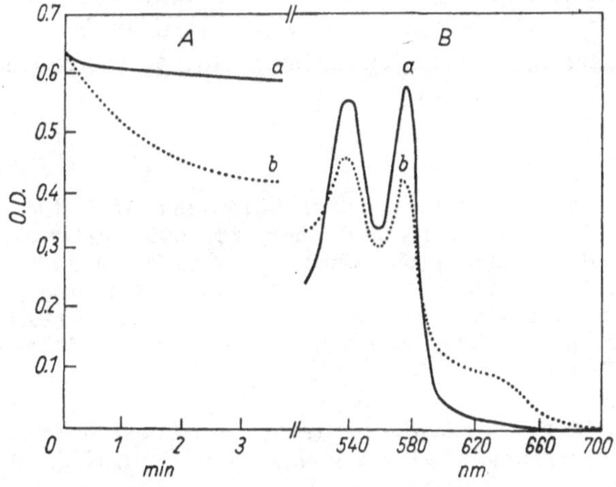

Fig. 3. Kinetics (A) and spectra (B) of oxyhemoglobin disappearance caused by the action of NQ in lysates prepared from cells in the presence or absence of DDC, recorded on a Cary 14 spectrophotometer. The kinetic curves were recorded at 576 nm. Spectra were run 6 m after mixing of the reactants. Concentrations: red cells (0.5%), NQ (0.4 mM), DDC (2.4 mM). Buffer: 7 mM sodium phosphate, *p*H at 7.4 at 28 °C containing 0.1 mM EDTA.
a) NQ added to lysates; b) lysates pretreated with DDC followed by the addition of NQ

Table 1
Effect of inhibition of red cell superoxide dismutase on 1.4-naphthoquinone 2-sulfonate induced hemolysis*

SOD activity (% of control)	Duration of exposure to NQ or an equal amount of buffer (hrs)	% Lysis + NQ	% Lysis − NQ
100	2	5.0	3.2
17	2	79.4	6.6
100	5	82.5	3.2
17	5	88.2	34.7

* Percent hemolysis was expressed as $100 \times \frac{O.D._{super}}{100\ O.D._{100\%}}$ lysis, where $O.D._{super}$ is the optical density at the peak of the Soret^{super} band of hemoglobin present in the supernatant of the red cell suspensions after centrifugation. $O.D._{100\%}$ lysis is the optical density at the peak of the Soret band of hemoglobin present in the entire red cell suspension after 10× dilution with distilled water to effect 100% lysis.

hypotonic lysis. Inhibition of both hemoglobin breakdown and hypotonic lysis by catalytic amounts of SOD added to the external medium of red cells exposed to DHF, suggests that O_2^- serves to mediate red cell damage. In the case of NQ, inhibition of red cell SOD by DDC fully exposes the red cell to the large amounts of O_2^- generated by the NQ-hemoglobin interaction, thereby resulting in hemoglobin breakdown and hypotonic lysis.

The absence of lipid peroxidation in red cell exposed of DHF or NQ suggests that lipid peroxidation plays a minor role in O_2^--induced hemolysis. The O_2^- dependent formation of green hemoglobin-breakdown pigment which attaches to the red cell membrane in erythrocytes exposed to either DHF or NQ may prove to be important in initiating membrane changes leading to increased osmotic fragility and lysis.

References

[1] McCORD, J. M., and I. FRIDOVICH: J. biol. Chem. **243**, 5753 (1968)
[2] MISRA, H. P., and I. FRIDOVICH: J. biol. Chem. **247**, 6960 (1972)
[3] GOLDBERG, B., and A. STERN: J. biol. Chem. **250**, 2401 (1975)
[4] GOLDBERG, B., A. STERN, and J. PEISACH: J. biol Chem. **251**, 3045 (1976)
[5] GOLDBERG, B., and A. STERN: Biochim. biophys. Acta **437**, 628 (1976)
[6] YAMAZAKI, I., and L. H. PIETTE: Biochim. biophys. Acta, **77**, 47 (1963)

The author's address: B. GOLDBERG and A. STERN, Department of Pharmacology, New York University School of Medicine, 550 First Avenue, New York, N.Y. 10016, U.S.A.
Supported by a grant-in-aid from the New York Heart Association

Acta biol. med. germ., Band 36, Seite 735—741 (1977)

Medizinisch-chemisches Institut der Universität Bern, Bern, Switzerland

Unstable mutants and molecular hybrids in enzyme deficiency conditions

H. Aebi, S. R. Wyss, and B. Scherz

Summary

Multiple molecular forms contribute to various types of enzyme heterogeneity: "Iso(en)-zymes" and "Allozymes" (enzyme variants) are of genetic origin whereas "Metazymes" represent secondary modifications of epigenetic nature. The concept of variability originates from the discovery of a large number of enzyme variants. Structural gene mutations can lead to enzyme variants of low specific activity or reduced stability and can cause enzyme deficiencies.

In heterozygous carriers of oligomeric enzyme defects, the formation of hybrid molecules is possible by random assembly of simultaneously synthetized normal and mutant subunits.

The study of normal processes as well as enzyme anomalies contributes essentially to better understanding of biochemical individuality and evolutionary events at a molecular level.

Introduction

The better the possibilities to detect minor differences analytically, the more evidence is obtained for a multitude of variations previously unknown in enzymes. As a consequence, the concept of uniformity of the species specific molecular architecture had to be replaced by the concept of variability representing the existence of multiple molecular forms of the the same species. Presently the study of enzyme heterogeneities observed also in humans has developed into a specialized field of medically oriented enzyme research. For practial reasons the investigation of erythrocyte and leukocyte enzymes is being emphasized [1, 2].

Multiple molecular forms can be the manifestation of various types of enzyme heterogeneity: (1) Isoenzymes („Isozymes") represent the biochemically distinguishable forms of one enzyme present in the same organism. (2) Enzyme variants ("Allozymes") are mutated forms which differ from normally occurring enzymes with respect to structure and function. The multitude of enzyme variants observed when different individuals are compared is responsible for the concept of biochemical individuality. (3) Secondary enzyme modifications ("Metazymes") which can contribute to enzyme heterogeneity are of epigenetic origin. No definite distinction can be made between molecular alterations in regulation processes, aging phenomena and preparatory artefacts [3—5].

This discussion deals with the importance and the problems involved in the detection of enzyme variants and with the special circumstances encountered in oligomer enzymes. Since our own investigations were concerned with human erythrocyte and leukocyte catalase, this tetramer enzyme will be used as an example in the ensuing presentation.

Enzyme variants

The existence of multiple alleles for a single structural gene is much more common than originally assumed. According to HARRIS [1], heterozygosity for an enzyme anomaly has to be anticipated in more than one fifth of the population if the gene frequencies calculated from 20 known enzyme polymorphisms are taken as a basis. Fig. 1 serves as an illustration of the relationship between the frequencies of heterozygote and homozygote individuals: The rarer the gene for a mutant enzyme, the higher the ratio of heterozygotes versus homozygotes.

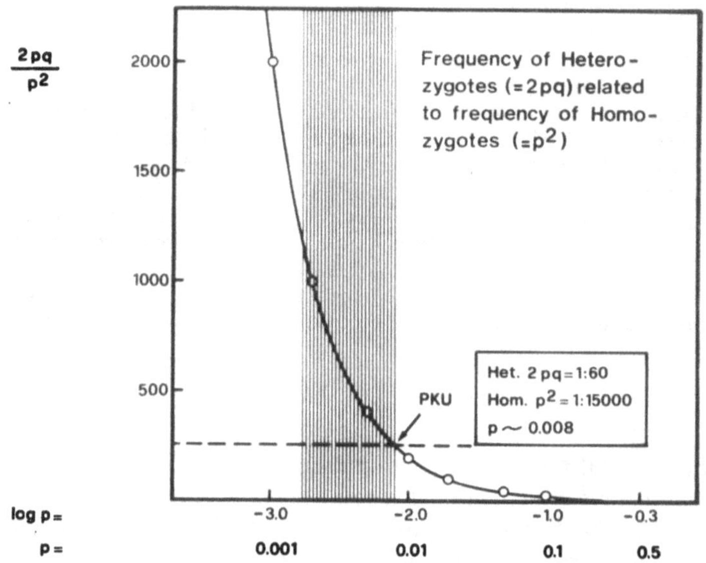

Fig. 1. Relationship between the frequency of individuals heterozygous and homozygous for a rare structural gene mutation.
Ordinate: Ratio of $2pq:p^2$; abscissa: frequency of the mutant gene (p) plotted as log p. The data were calculated using Hardy-Weinberg's formula. The shaded area designates the range valid for most inborn errors observed in humans

Depending upon the activity of an enzyme variant, the condition will manifest itself as a silent anomaly or as an enzyme defect. It defects — especially in erythrocyte enzymes — are due to an extensive lack of enzyme activity, two types of structural gene mutations have to be distinguished, as shown in Fig. 2: (a) Enzyme variants of low specific activity, probably due to a mutational event close to the active site, affecting the catalytic activity of the enzyme rather than the stability. (b) Enzyme variants of normal specific activity but decreased stability, possibly resulting from a mutation near the contact interphase and thus affecting the proper binding to the adjacent partner subunit.

This discrimination is both of theroretical and practical importance. Whereas in the former type all organs are similarly affected by the deficiency and thus the enzyme activity found in blood reflects the situation in other tissues, this is not the case in the latter type. Deficiency conditions due to the synthesis of an unstable enzyme variant become predominantly manifest in cells with a slow

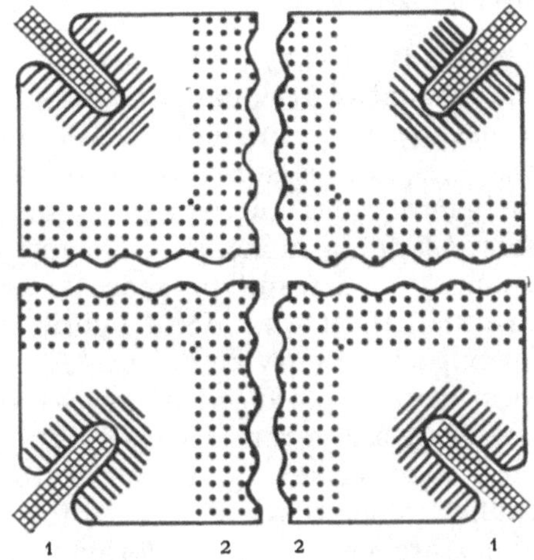

Fig. 2. Tentative model of a tetramer molecule illustrating the assumed localization of structural gene mutations.
1 = Mutations close to the active site probably affecting the specific activity of the enzyme rather than the stability. 2 = Mutations at the contact interphase leading to an unstable enzyme variant of approximately normal specific activity

turnover rate — like erythrocytes — therefore, the activity found in blood cannot be taken as an indicator for the overall situation. For example, acatalasemia (extensive lack of erythrocyte catalase) discovered in 3 Swiss kinships is due to the synthesis of an unstable catalase variant which easily dissociates into inactive dimer or monomer subunits [6]. The properties of this mutant enzyme as compared with normal catalase are summarized in Fig. 3. In contrast, the enzyme variant present in most Japanese acatalasemia individuals has low specific activity. This concurs with the observation that residual catalase activity in various tissues (notably cultured fibroblasts) of Japanese patients was found to be lower than in Swiss acatalasemics and that clinical symptoms (e.g ulceration of the oral cavity) have occurred in Japanese cases only.

	Normal enzyme	Unstable variant (AB)
molecular distribution pattern	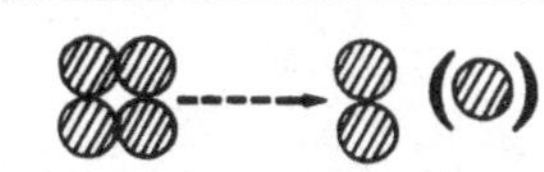 Tetramer >> Dimer ~96 % ~4 %	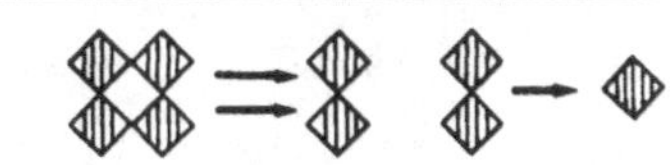 Tetramer << Dimer ~2 % ~98%
activity in red cells	2.2 mg Catalase / g Hb (= 100 %)	20 μg / g Hb ~1-2 % of normal
Signs in vivo	t/2 > 120 d equal distribution in red cell population	t/2 ~ 24h pseudomosaicism, residual act. in reticulocytes
inactivation in vitro	50% after 10 min at 63 °C	50 % after 10 min at 55 C

Fig. 3. Forms of human erythrocyte catalase: Quantitative and qualitative differences between the normal enzyme and the catalase variant found in Swiss Type Acatalasemia

Hybrid enzymes

The study of enzyme variants in humans is rather restricted primarily because the amount of residual activity still detectable in blood of homozygote patients is usually very small. However, more insight into the nature of such mutant enzymes can be gained by detailed investigation of the enzyme species present in heterozygous carriers of these defects. Whereas cells of individuals heterozygous for a monomeric enzyme defect contain the normal enzyme in half quantity as a result of gene dosage (e.g. uridyl-gal-l-ph-transferase in galactosemia), the formation of hybrid molecules can occur in heterozygous carries for oligomeric enzyme defects. As shown in Fig. 4, such intermediate molecular form can be produced if normal and mutant subunits are synthetized simultaneously and a random assembly to the oligomeric enzyme form takes place in the second phase only. Recently studies have shown that this situation obviously applies to heterozygous carries for Swiss type acatalasemia [7, 8]. The presence of erythrocyte catalase hybrids in these heterozygotes would be a plausible explanation for the previously observed absence of a linear gene dosage [9]. Fig. 5 demonstrates that blood catalase activity values obtained form Swiss heterozygotes fall within the normal range. One must exercise caution, however, in choosing an assay method which reflects true activity by measuring the initial reaction rate rather than enzyme stability. The properties of this heterozygote catalase species were found to be intermediate with regard to heat stability, electrophoretic mobility and stability in urea when compared with the normal enzyme and the catalase variant present in acatalasemic homozygotes. These findings were strongly suggestive of hybrid enzyme formation in Swiss acatalasemic heterozygotes. Consequently, steps were taken to imitate such an hybridization process in vitro by using dimer subunits obtained from purified normal and heterozygote erythrocyte catalases.

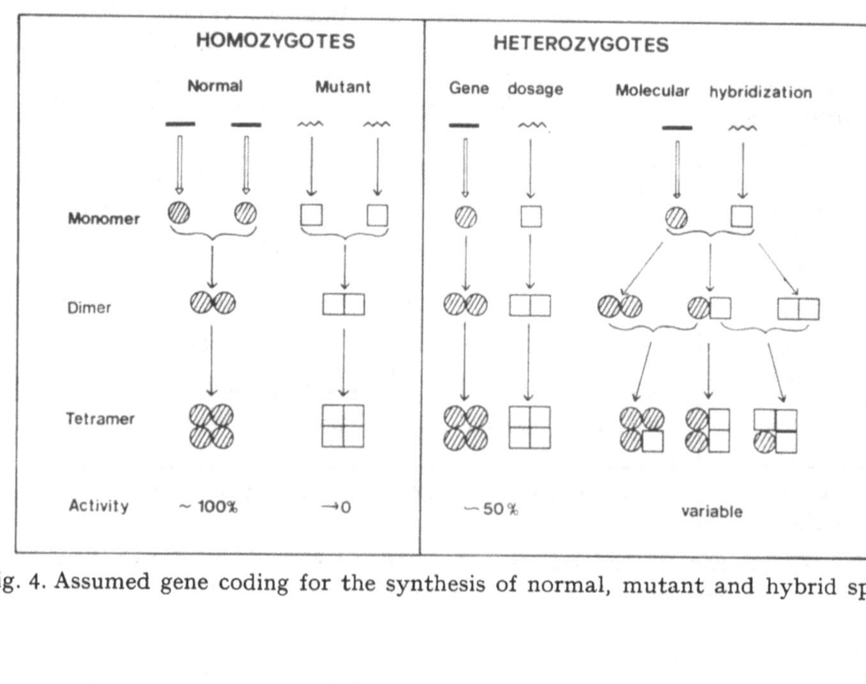

Fig. 4. Assumed gene coding for the synthesis of normal, mutant and hybrid species

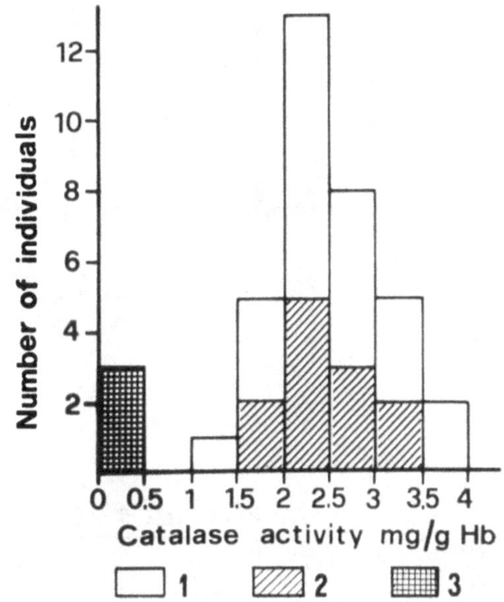

Fig. 5. Catalase activity in hemolysates of normal individuals as compared to heterozygotes and homozygotes for Swiss type acatalasemia.
Distribution is shown in ranges of 0.5 mg catalase/g hemoglobin. 1 = normal ($n = 34$; 54 samples); 2 = heterozygotes ($n = 12$; 19 samples); 3 = homozygotes ($n = 3$; 17 samples)

Dissociation of the active catalase tetramer into inactive dimer subunits can readily be accomplished by treatment with 8 M urea [10, 11]. The loss of catalatic activity is accompanied by a complementary generation of peroxidatic activity and an alteration of antigenic properties. Moreover, ultracentrifugal analysis of the dissociated material reveals a molecular weight of 120,000, which is consistent with dimer. However, this process is reversible if the perturbant (urea) is eliminated by subsequent dialysis. Restoration of catalatic and disappearance of peroxidatic activity as well as a molecular weight of 240,000 corresponding to catalase tetramer provide evidence for the recombination of the dimer subunits. A typical experiment demonstrating this shift in the peroxidase/catalase activity pattern upon dissociation of the catalase tetramer after urea treatment and recombination of the dimer subunits during dialysis is shown in Fig. 6. The sensitivity towards urea is more pronounced in the heterozygote acatalasemic catalase species, dissociation occurring at a 6 M urea concentration as compared with 8 M urea for normal catalase. Moreover, a small amount of residual peroxidase activity is seen to persist after dialysis of the heterozygote enzyme. If equal amounts of inactivated normal and heterozygote catalases are mixed and subjected to dialysis, a "new" (hybrid) active catalase species is formed. The electrophoretic pattern shown in Fig. 7 reveals intermediate mobility of the hybrid enzyme when compared to its native constituents and intermediate properties are also encountered in comparative heat stability tests [8, 11].

Nevertheless, some questions remain to be elucidated. Theoretically an hybridization of the tetramer enzyme catalase should result in 3 hybrid enzyme forms in a relationship of 4:6:4. So far there is no evidence of heterogeneity in the catalase species found in Swiss acatalasemic heterozygotes nor in the hybrid enzyme obtained in vitro. However, it is conceivable that monomer subunits may have

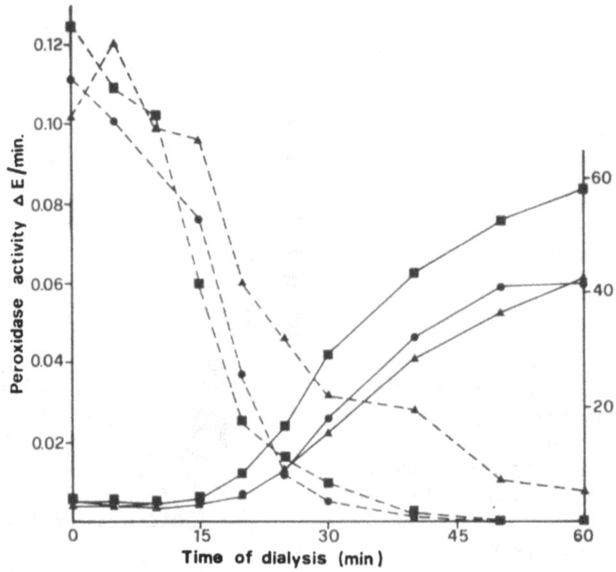

Fig. 6. Dissociation, reactivation and hybridization of human erythrocyte catalases. Samples were inactivated in 8 M urea for 3 min prior to dialysis at 20 °C. Peroxidase activity and relative catalase acticity were measured immediately after inactivation and at various time intervals during 60 min dialysis.
(●———●) normal erythrocyte catalase; (▲———▲) heterozygote acatalasemic erythrocyte catalase; (■———■) mixture of normal and heterozygote erythrocyte catalases; (———) catalase activity; (-------------) peroxidase activity

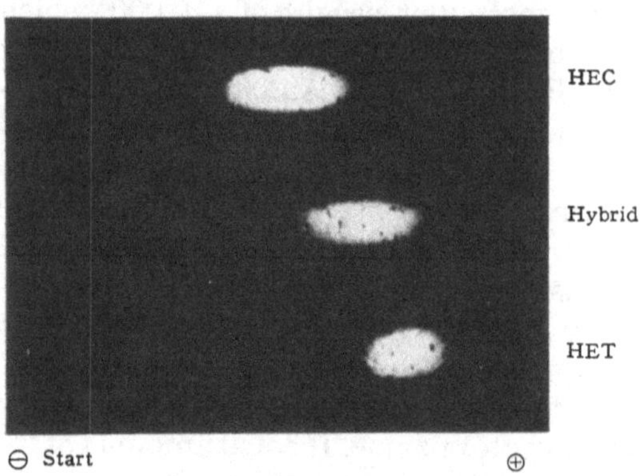

Fig. 7. Starch-gel electrophoresis of purified erythrocyte catalases after treatment with 8 M urea and reactivation by dialysis for 90 min at 20 °C.
HEC = normal human erythrocyte catalase; HET = heterozygote acatalasemic erythrocyte catalase, Hybrid = dialyzed mixture of inactivated HEC and HET.

a tendency to combine preferentially with the partner monomer rather than with the same type of subunit, thus forming one molecular enzyme species only.

Concluding remarks

If the results and information obtained from these experiments are viewed in a general context, it becomes evident that enzyme research, much like biology on the whole, is characterized by a dualism in its aims. On one hand the strive for better knowledge and understanding of the laws and principles of enzyme action by searching for the ordinary and the generally valid, while on the other hand, due to the systematic search for the extraordinary and the study of anomalies, essential new insight is gained into the variability of enzyme proteins. The discovery of increasing numbers of enzyme variants has led to a better understanding of biochemical individuality and the evolutionary processes taking place at the molecular level. The concept of isozymes, previously an exception but presently almost a rule, has actually caused maximum specialization and adaptation to changing environmental conditions.

Acknowledgement

The catalase studies were part of project No. 3.384.74 subsidized by the Swiss National Science Foundation.

References

[1] HARRIS, H.: The Principles of Human Biochemical Genetics. 2nd ed. North-Holland Publishing Company, Amsterdam/Oxford 1975
[2] YUNIS, J. (Ed.): Biochemical Methods in Red Cell Genetics. Academic Press, New York/London 1969
[3] MASTERS, C. J., and R. S. HOLMES: Adv. comp. Physiol. Biochem. **5**, 109 (1974)
[4] MARKERT, C. I. (Ed.): Isozymes. Vol. **1**: Molecular Structure. Academic Press, New York 1975
[5] AEBI, H.: Chimia **30**, 91 (1976)
[6] AEBI, H., S. R. WYSS, B. SCHERZ, and F. SKVARIL: Eur. J. Biochem. **48**, 137 (1974)
[7] WYSS, S. R., and H. AEBI: Enzyme **20**, 257 (1975)
[8] AEBI, H. S. R., WYSS, B. SCHERZ, and J. GROSS: Biochem. Genet. **14**, (1976)
[9] AEBI, H. in: Hereditary Disorders of Erythrocyte Metabolism, Duarte, Calif. 1967. Vol. **1**, 41. E. BEUTLER (ed.) Grune & Stratton, New York/London 1968
[10] AEBI, H., B. SCHERZ, Y. BEN-YOSEPH, and S. R. WYSS: Experientia **31**, 397 (1975)
[11] SCHERZ, B., E. KUCHINSKAS, S. R. WYSS, and H. AEBI: Eur. J. Biochem. (1976) submitted for publication

The author's address: H. AEBI, S. R. WYSS, and B. SCHERZ, Medizinisch-chemisches Institut der Universität,
3000 Bern, 9, Brühlstraße 28, Switzerland

Acta biol. med. germ., Band 36, Seite 743—748 (1977)

Central Institute of Hematology and Blood Transfusion, Moscow, USSR

Valency hybrids of hemoglobin in red cells of patients with hereditary enzymopenic methemoglobinemia

A. P. Andreyeva, M. G. Dmitriyeva, A. A. Levina, L. M. Tsibulskaya, Ye. G. Kazanetz, I. I. Ilyinskaya, I. V. Derviz, and Yu. N. Tokarev

Summary

Twenty eight patients with hereditary methemoglobinemia have been found and examined.

Using the techniques of analytic isoelectrofucossing in polyacrylamide gel and the preparative isoelectrofocussing on a column in sucrose gradient we have managed to disclose an abnormal fraction in hemolysate of the above mentioned patients. It has been proved that the obtained fraction contains hemoglobin valent hybrids, possesses high affinity to oxygen and low cooperativity.

Due to the presence of valency hybrids about one-half of the quantity of oxy-Hb fails to participate in the normal function of oxygen delivery, which evidently is the molecular cause of disturbance of oxygen delivery in patients with hereditary enzymopenic methemoglobinemia. This also explains a high degree of cyanosis in them.

Methemoglobinemia is characterized by increased methemoglobin contents of blood as compared with the physiological norm. It may be brought about by various causes: 1) presence in blood of abnormal hemoglobins referring to M-hemoglobinopathy and containing heme oxydized iron either in α- or in β-chains; 2) a decrease in the activity of enzyme systems participating in met-Hb reduction in the red cells; 3) action of a number of toxic substances.

During recent years in the USSR a number of patients with hereditary enzymopenic methemoglobinemia has been described [1—3].

The given disorder characterized by $NADH_2$-methemoglobin-reductase deficiency refers to a rather rare hereditary disease. According to WHO data the incidence of distribution of the gene of this enzyme deficiency in the heterozygous state is less than 1% among most of the populations.

Materials and methods

Determination of met-Hb concentration in blood and in hemolysates was performed following the method described in [4], met-Hb reductase activity was determined using the technique of Haggesh et al. [5].

Isoelectrofocussing in polyacrylamide gel was done using 7% gel concentration and ampholines at *p*H 6—9. Gels were stained by 0.1% Coomassi solution and subjected to densitometry (apparatus DMU-2) at 560 nm.

Hemolysates of patients were fractionated on DEAE-cellulose in 0.1 M Tris-buffer using *p*H gradient from 8.3 to 6.8. Preparative isoelectrofocussing was done with a LKB apparatus on 100 ml-columns in sucrose gradients.

Oxygenation cûrves were studied spectrophotometrically with a Perkin-Elmer apparatus in 0.1 M phosphate buffer, *p*H 7.15, 20 °C.

Results and discussion

We have detected and studied 28 patients from 22 families aging 10—67 years. It should be noted that 16 families have been found among Yakuts during an expedition to the Eastern Siberia Autonomous Soviet Republic of Yakutia. Consequently a new area of this disease has been discovered. A similar area was described earlier by SCOTT and MCGRAW [6] among Eskimos and Indians, the resident population of Alaska. In all patients a considerable cyanosis was observed. While performing differential diagnosis we excluded heart failure and presence in blood of group M-hemoglobins on the basis of spectrophotometry of hemolysate at 630, 600 and 500 nm. It was found that values of Betke's coefficients in the patients under investigation were higher than 1.25 and 2.80 that excluded HbM. Diaphorase I activity was markedly decreased (0.00—0.37 μmoles/mg Hb/min compared to 2.5—3.5 in donors). This demonstrated the homozygous state of the given patients. In patients-homozygotes total Hb was within 13—21 g% and the met-Hb quantity varied from 6 to 42% as presented in Table 1.

Table 1
Total Hb, met-Hb and oxy-Hb contents in blood of the patients

Patients	Hb		
	Total Hb (g%)	Met-Hb (% of total)	Hb O_2 (g%)
1	15.7	23	12.1
2	21.0	33	14.1
3	16.8	19	15.8
4	24.0	41	14.2
5	13.4	11	12.0
6	19.8	38	12.9

The table shows that functionally active oxy-Hb content was rather high in the given patients (12—15 g%) while cyanosis was marked in them.
Oxygenation curves of hemolysates of the patients are presented in Table 2.

Table 2
Functional characteristics of hemolysates of the patients

Patients hemolysate	Met-Hb (%)	Functional properties	
		Δ lg P_{50} (%)	Hill's constants ("*n*")
1	11	0	2.73
2	12	0	2.42
3	19	0	2.43
4	23	0	2.13
5	26	20	2.00
6	33	13	2.20
7	38	20	2.12
8	41	43	1.65
Artificial mixture	20	0	2.80—3.00

Table 2 shows that met-Hb presence in hemolysates in quantities up to 20% does not influence Hb affinity to oxygen, but lowers Hill's coefficient to a certain degree, i.e. disturbs cooperativity on oxygen binding. Met-Hb in hemolysates higher than 20% raises Hb affinity to oxygen within 13—40% and considerably disturbs the form of the oxygenation curve; a curve twist on Hill's plot was observed (Fig. 1). It should be noted that oxygen affinity values and Hill's coefficient did not differ form native Hb in artificial mixtures containing 80% oxy-Hb and 20% met-Hb. Consequently alterations in oxygen supply induced by hemolysates of the given patients is not determined by the mere fact of the presence of these or other met-Hb quantities.

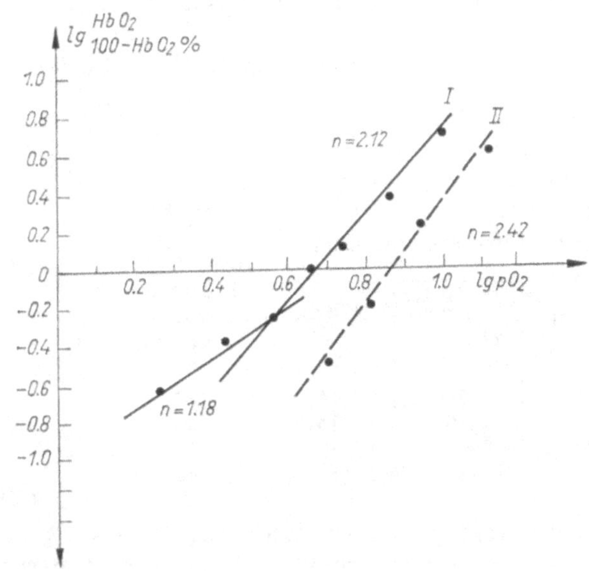

Fig. 1

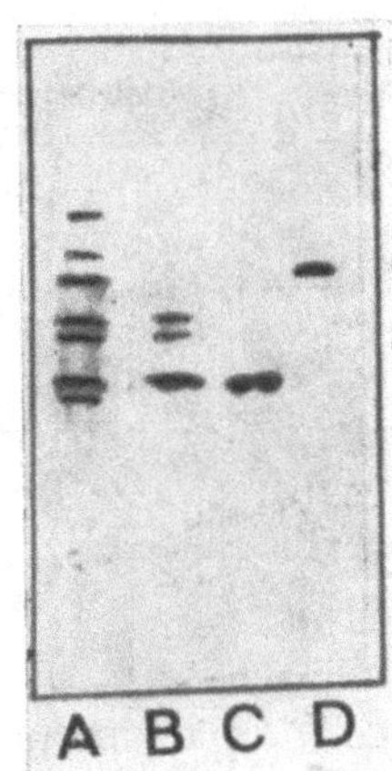

Fig. 2

Fig. 1. Oxygenation curves with 38% met-Hb content I: before, II: after treatment

Fig. 2. Isoelectrofocussing in polyacrylamide gel.
A) hemolysate of the patients; B) the patient hemolysate after dithionite treatment; C) oxy-Hb; D) met-Hb;

A detailed investigation of patient hemolysate content was performed by a specially developed original technique of analytic isoelectrofocussing in polyacrylamide gel. Strict division of met- and oxy-forms of Hb was achieved in such conditions. A divaricating fraction, which is intermediate between oxy- and met-Hb possessing the isopoint value of 7.45 (Fig. 2), was found in patients hemolysates besides met- and oxy-Hb. Oxy-Hb scarlet fraction the brownish intermediate fraction and met-Hb dark brown fraction could be observed visually in pherograms. It was suggested that the identified abnormal fraction represented Hb valency hybrids having one, two or three oxidized hemes in the Hb molecule.

It should be kept in mind that hemes may be oxidized either in α-or in β-chains. Due to it, the number of possible variants of valent hybrids may be great. The suggestion about partial oxidation of the abnormal fraction hemes was supported by the results of densitometry of nonstained pherograms at 560 nm (near maximum of oxy-Hb) and 630 nm (maximum of met-Hb).

While recording the densitogram at 560 nm the following per cent ratios of fractions of oxy-Hb, valency hybrids and met-Hb were obtained: 54:37:8. At 630 nm (maximum for the oxidized Hb form) these ratios were 25:58:14, i.e., an increase in fractions of valency hybrids and met-Hb was observed. Partial oxidation of hemes in the given fraction is supported by the results of experiments on reduction of hemolysates by sodium dithionite. During this process disappearance of abnormal fraction with a simultaneous increase in oxy-Hb fraction was observed (Fig. 2B).

The quantitative evaluation of fractions of oxy-Hb, met-forms of Hb and valency hybrids is given in Table 3.

Table 3

Results of isoelectrofocussing in polyacrylamide gel

Patients	Met-Hb in hemolysate (%)	Results of isoelectrofocussing			% of oxidized hemes in fraction of valency hybrids
		Met-Hb	Valency hybrids	Hb O_2	
1	19	5	51	44	27
2	33	23	39	38	25
3	38	23	42	35	35
4	41	24	36	40	47

Met-Hb content in hemolysates of different patients varied from 19 to 41% of total Hb. The quantity of real met-Hb amounted to 5—24% according to results of isoelectrofocussing ingel. The quantity of the fraction of valency hybrids was comparatively constant 36—50%, i.e. practically equal to the oxy-Hb fraction, irrespective of total met-Hb in hemolysate. But a rise in met-Hb quantitiy in hemolysate was accompanied by an increase in percentage of oxidized hemes in the fraction of valency hybrids. When the met-Hb content was about 40%, almost 50% hemes in the fraction of valency hybrids was in the oxidized form. A considerable rise in Hb affinity to oxygen in hemolysate and a bend of the curve in Hill's plot was observed in these two particular patients.

Consequently only 35—50% of total Hb are fully functional, i.e. oxy-Hb in patients with enzymopenic methemoglobinemia. This is why high values of oxy-Hb obtained in them with the deduction of met-Hb content are fictitious. The real contents of oxy-Hb functionally capable of oxygen delivery constitutes only one-third or one-half of this quantity in different patients. This is the molecular cause of oxygen delivery disturbance by hemoglobin in such patients that determines a high degree of cyanosis in them.

A preparative isolation of the fraction of valency hybrids by the method of isoelectrofocussing on a column in sucrose gradient and by chromatography on

DEAE-cellulose or ion exchange Sephadex (Fig. 3) was performed which yielded three fractions. The first fraction corresponded to met-Hb according to the isopoint value (met-Hb constituted 96% of total Hb in it). The second fraction — fraction of valency hybrids — contained 50—54% of oxidized hemes and the third fraction did not practically have met-Hb and represented oxy-Hb.
Absorption spectra of the isolated fractions in the visible part of spectrum are given in Fig. 4. This figure shows that ratio D_{578}/D_{541} in the fraction of valent hybrids is decreased due to the presence of the oxidized form of Hb.

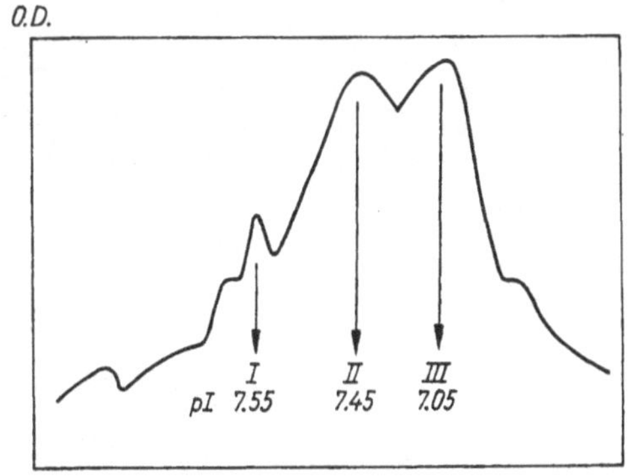

Fig. 3

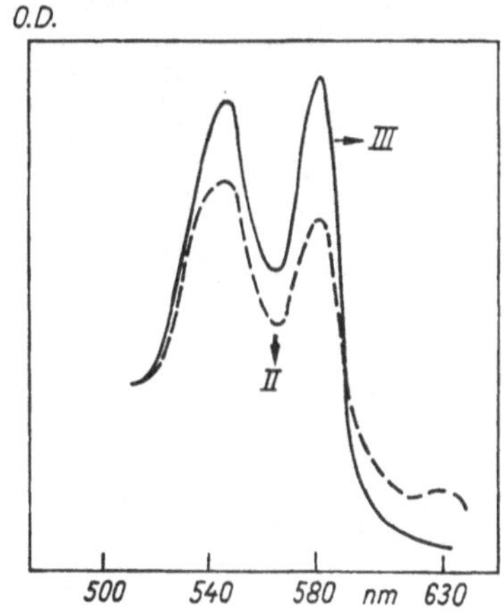

Fig. 4

Fig. 3. Fraction of the patient hemolysate by isoelectrofocussing in a column in sucrose gradient:

Fraction I contains 96% of met-Hb, fraction II 50—54% of met-Hb, fraction III 2—4% of met-Hb

Fig. 4. Spectral characteristics of the fraction of valency hybrids (II) and oxy-Hb fraction (III) from the patient hemolysate after fractionation by the technique of isoelectrofocussing in sucrose gradient

It was found that the fraction of valency hybrids has very high affinity to oxygen and is practically devoid of cooperativity in binding and releasing oxygen (lg $P_{50} = 0.1$ mm Hg, $n = 1.1$). Changes in quantity of valency hybrids in the process of reduction of hemolysate in vitro by sodium dithionite and the process of treatment of the patients by ascorbic acid or methylene blue in the glucose solution were studied. It was found that a decrease in fractions of met-Hb and valent hybrids with a simultaneous increase of oxy-Hb fraction occurs in the process of reduction in vivo and in vitro. A normalization of Hb affinity to oxygen and a rise in Hill's coefficient is observed (Fig. 1).
The above mentioned data allow to make the following conclusions: the presence of Hb valency hybrids constituting 35—50% of total Hb has been proved in the blood of patients with methemoglobinemia of enzyme origin.
Due to the presence of valency hybrids, about 50% of oxy-Hb quantity fail to perform the function of oxygen delivery because of low cooperativity and high affinity of valency hybrids to oxygen.

References

[1] KUSHAKOVSKIY, M. S.: Clinical forms of hemoglobin injury. Publishing House Meditsina, Moscow 1968, p. 157
[2] DERVIZ, G. V., and R. E. KIMMERAL: Problemy Gemat. Pereliv. Krovi **12**, No. 12, p. 37 (1967)
[3] TOKAREV, YU. N., F. E. FEINSTEIN, I. A. ALEXEYEV, and I. V. DERVIZ: Problemy Gemat. Transfus. **11**, 112 (1976)
[4] DERVIZ, G. V.: Lab. Delo **9**, 527 (1966)
[5] HAGGESH, F. T., M. SALMANOVISY, and M. J. AVROW: Lab. clin. Med. **72**, 339 (1968)
[6] SCOTT, E. M., and I. C. MCGRAW: J. biol. Chem. **237**, 267 (1962)

Acta biol. med. germ., Band 36, Seite 749—751 (1977)

Institut des sciences médicales, departement de biochimie, Université' d'Oran, Laboratoire de Biochimie Chuoran

Protection de la G-6-PDH erythrocytaire pendant l'hemolyse

H. Lachachi, S. Benharrat, M. Rusu et L. Ababei

Ces derniéres années, des études approfondies ont été faites sur l'activité de la G6PH dans les érythrocytes humains, en corrélation avec le déficit de cet enzyme à l'origine des principales anémies hémolytiques enzymoprivés héréditaires [1]. L'activité de la G-6-PDH est déterminante dans le métabolisme intermédiaire des érythrocytes [2, 3]; son déficit entraine une diminution du taux de NADPH, par conséquent, il limite la formation de glutathion réduit dans la réaction catalysée par la glutathion réductase [3]. Or le taux de glutathion réduit est essentiel pour l'activité d'un nombre important d'enzymes sulfhydrylés. Des facteurs génétiques ont été incriminés dans le déficit partiel ou total en G-6-PDH [4, 5]. Dans notre travail, nous avons étudié les différents mécanismes impliqués dans l'inactivation ou destruction de la molécule de l'enzyme, en particulier, une éventuelle hydrolyse peptidique intracellulaire (in vivo) ou in vitro au cours de la préparation de l'hémolysat. Morrison et Neurath [6] ont déterminé trois protéinases (I, II, III) dans le membrane érythrocytaire. Heller [7] a montré l'éxistence dans la membrane érythrocytaire humaine de deux protéases: l'une à *p*H optimum de 7,4 et l'autre à *p*H 8,0. Tökes et Chambers [8] a isolé par ultracentrifugation et chromatographie sur Sephadex trois protéinases de la membrane érythrocytaire: la première à *p*H optimum de 7,4, la seconde à *p*H 8,0, et l'autre à *p*H 3,0.

Nous avons postulé l'intervention de protéases intraérythrocytaires dans la destruction de la molécule protéinique de l'enzyme durant l'hémolyse; pour cela, nous avons introduit un facteur antiprotéasique: la trypsine inhibitrice. Pour étudier d'éventuelles liaisons peptidiques structurelles essentielles pour l'activité de la G-6-PDH, nous avons également utilisé la phospholipase D; dans un travail antérieur [9], nous avons obsérve que la phospholipase A_2, la phospholipase C, ainsi que la lysozyme n'ont aucune influence sur l'activité de la G-6-PDH érythrocytaire.

Matériel et méthodes

Le sang est prélevé sur héparine, puis les érythrocytes ont été séparés par centrifugation du sang á 3000 tours/min pendant trois minutes, lavés á trois reprises avec de l'eau physiologique froide (+ 4 °C). Pour le dosage de l'activité, nous avons utilisé des hémolysats dans un rapport de 1/10 préparés avec de l'eau distillée froide et une solution de digitonine à 0,02%. Nous avons ajouté successivement de la trypsine inhibitrice et de la phospholiphase D. Pour les dosages, nous avons utilisé le surnageant de l'hémolysat, obtenu par centrifugation à 6000 tours/min pendant dix minutes. Les déterminations ont été effectuées sur un spectrophotomètre Perkin Elmer 124 à 340 nm [10, 11]. La phospholipase D provient du chou (Boehringer), la trypsine inhibitrice du blanc d'oeuf de poulet (Boehringer).

Resultats et discution

Dans le tableau 1: Nous remarquons que dans l'hémolysat osmotique à l'eau distillée aussi bien que celui avec digitonine, la trypsine inhibitrice augmente l'activité de la G-6-PDH érythrocytaire.

Dans le tableau 2: La phospholipase D inactive complètement l'enzyme. Dans ces conditions la trypsine inhibitrice ne diminue pas le degré d'inhibition de la G-6-PDH par la phospholipase D.

Dans le tableau 3: Nous constatons que l'activité de la G-6-PDH dans l'hémolysat à l'eau distillée est fortement diminuée après conservation du sang à 4 °C pendant 24 h. Cette diminution est partiellement bloqueé par la trypsine inhibitrice; donc dans ce cas, nous pouvons considérer que même après conservation à 4 °C pendant 24 h., il existe dans l'hémolysat des protéases actives qui peuvent hydrolyser une certaine quantité de protéines endogènes parmi lesquelles figure la G-6-PDH. Une étude systématique est en cours dans cette direction, afin d'établir les différents

Tableau 1

Effets de la trypsine inhibitrice sur l'activité de la G-6-PDH dans un hémolysat érythrocytaire humain

Nature de l'hémolysat	Activité en mol/ml cellules/min
Eau distillée	0,679
Eau distillée + trypsine inhibitrice (1 mg/ml)	1,540
Solution de digitonine 0,02%	1,050
Solution digitonine 0,02% + trypsine inhibitrice (1 mg/ml)	2,260

Tableau 2

Effets de la phospholipase D sur l'activité de la G-6-PDH dans l'hémolysat érythrocytaire

Nature de l'hémolysat	Activité en µmol/ml cellule/min
Témoin (eau distillée)	0,679
+ Phospholipase D (1 mg/ml)	0,00
+ Phospholipase D (1 mg/ml) + trypsine inhibitrice (1 mg/ml)	0,00

Tableau 3

Effets de la trypsine inhibitrice sur l'activité de la G-6-PD en fonction du temps de conservation de l'hémolysat

Nature de l'hémolysat	Temps de conservation de l'hémolysat (à 4 °C) (heures)	Activité en µmol/ml/min
Eau distillée	1	0,706
	24	0,291
Eau distillée + trypsine inhibitrice (0,3 mg/ml)	1	0,752
	24	0,529

Tableau 4

Effets de la phospholipase D et de la trypsine inhibitrice sur la G-6-PD enzyme purifié de la levure, dilué àu 1/100

	Activité en µmol/ml/min
Enzyme	4,505
Enzyme + phospholipase D (0,5 mg/ml)	4,501
Enzyme + trypsine inhibitrice (1 mg/ml)	4,500

paramètres cinétiques de ces protéases (optimum de température, *p*H, vitesse, K_m), et dépendance du temps et de la concentration en trypsine inhibitrice. Par ailleurs, nous remarquons que la trypsine inhibitrice est capable d'arrêter à 4 °C l'action inactivatrice des protéases.

Dans le tableau 4: La trypsine inhibitrice n'influence pas l'activité de la G-6-PDH purifié de levure, de même la phospholipase D. Toutes ces données nous indiquent que l'enzyme érythrocytaire est différente de l'enzyme purifié de levure en ce qui concerne la sensibilité vis à vis de la phospholipase D et de la trypsine inhibitrice. Il reste à purifier l'enzyme érythrocytaire, ainsi que les protéases cytoplasmiques postulées.

Conclusion

Ce travail préliminaire montre que la trypsine inhibitrice est un protecteur puissant de l'enzyme en particulier pendant la conservation prolongée même à 4°C; la phospholipase D détermine une inhibition de l'activité enzymatique soit: en hydrolysant un résidu phospholipidique essentiel pour l'enzyme; ou en libérant dans l'hémolysat depuissants inhibiteurs de l'enzyme. A partir de ces données, nous pourons tester soit l'acide phosphatidique, soit les différentes bases dans leur action sur la G-6-PDH purifié; soit éliminer par dialyse ou par chromatograhie les éventuels inhibiteurs.

Bibliographie

[1] Carson, P. E., C. L. Flanagan, C. E. Ickes et A. S. Alving: Science, N.Y. **124**, 484 (1956)

[2] Rapoport, S., L. Ababei, C. Wagenknecht et S. R. Sarkar: Nature, Lond. **208**, 185 (1965)

[3] Beutler, E., R. J. Dern, C. L. Flanagan et A. S. Alving: J. Lab. clin. Med. **45**, 286 (1955)

[4] Beutler, E. et M. C. Baluda: Lancet **1964**/I, 189

[5] Browne, E. A.: Bull. Johns Hopkins Hosp. **101**, 115 (1957)

[6] Morrison, W. et H. Neurath: J. biol. Chem. **200**, 39 (1953)

[7] Heller, M., P. Edelstein et N. Mayer: Biochim. biophys. Acta **413**, 472 (1975)

[8] Tökes, T. A. et S. Chambers: Biochim. biophys. Acta **389**, 325 (1975)

[9] Lachachai, H., M. Rusu et L. Ababei: Commun. Press. Conf. Annuelle de la Société Med. de l'Quest Algerien (1976)

[10] Kornberg, A. et B. L. Horecker: Meth. Enzym. **1**, 323 (1955)

[11] Löhr, W. et D. Waller en: Methoden der enzymatischen Analyse. H. U. Bergmayer (Ed.). Verlag Chemie, Weinheim. 1962, p. 744

Acta biol. med. germ., Band 36, Seite 753—758 (1977)

Central Institute of Hematology and Blood Transfusion, Moscow, USSR

Characteristics of a new abnormal variant of G-6-PD in human red cells

N. B. CHERNYAK, A. I. BATISCHEV, and YU. N. TOKAREV

Summary

Kinetic and electrophoretic properties were studied in 230—300 fold purified preparations of glucose-6-phosphate dehydrogenase (G-6-PD) from red cells of donors and patients with hemolytic anemia induced by G-6-PD deficiency.
In abnormal variant of G-6-PD isolated from red cells of a patient with hemolytic anemia which had not before been described in the literature was found. The abnormal variant differs from the normal enzyme by a decreased Michaelis constant for G-6-P and NADP, by increased utilization of substrate-analogues (2-deoxy-G-6-P and deamino NADP in particular), by low heat stability, the character of *p*H dependence, and by the appearance of one band of G-6-PD activity during electrophoresis in polyacrylamide gel. The isolated abnormal variant of G-6-PD has been called "Kremenchug" according to the origin of the patient.

The investigation of the properties of glucose-6-phosphate dehydrogenase (G-6-PD, 1.1.1.49) is of great importance in connection with the revealing abnormal enzymes in hemolytic anemias characterized by a decreased activity of this enzyme [2, 3]. This decrease in enzyme activity is transmitted by heredity linked to the *x*-chromosome, and the occurrence of the abnormality is associated with the mutation of the *x*-chromosome locus coding for G-6-PD synthesis.
As a result of mutations an enzyme is formed which differs from the normal one by its activity, and kinetic and electrophoretic characteristics.
The aim of this study was to obtain G-6-PD preparations from a small quantity of human red cells in order to investigate the kinetic and electrophoretic properties of the enzyme in healthy persons and in patients with hemolytic anemia by a decreased G-6-PD activity. The experiments were performed on partially purified (230—300 times) G-6-PD preparations which did not contain hemoglobin and 6-phosphogluconate dehydrogenase (6-PGD 1.1.1.44) the presence of which interferes with calculations of kinetic characteristics. The work was carried out according to the methods suggested by the scientific group of the WHO on standardization of the techniques for the study of G-6-PD with the aim of identifying abnormal variants of the enzyme among the population of the Soviet Union.

Material and methods

G-6-PD was isolated and purified from 15—20 ml of human red cells by the application of our modification of Kirkman's method [6]. The activities of G-6-PD, 6-PGD [7, 8] and the protein content [9] in the obtained prepartation were determined. Enzymatic activity was expressed in international units. The K_m values for G-6-P and NADP were determined by measuring the initial rates at 8 different concentrations of G-6-P (from 20 to 250 μM) and NADP (from 1.5 to 25 μM) respectively. The original G-6-P and NADP concentrations were determined using the enzymatic method. The numerical K_m values were obtained graphically by plotting the ratio Sv versus the final concentration S of the investigated substrate. The utilization of the substrate-analogues (2-deoxy-O-glucose-6-phosphate, deamino-NADP and -NAD) was determined by applying the standard method for the determination of the activity of G-6-PD using the above mentioned analogues instead of G-6-P or NADP in the same concentrations. The activity of G-6-PD was expressed as percentage of the activity rate obtained using G-6-P or NADP [10].

Thermostability was determined by the assay of G-6-PD activity after 10 min incubation of the enzyme in an ultrathermostat "U-10" at 46°, 50°, 52°, 54°, 58 °C and after 5 min incubation at 45°, 48°, 51°, 54°, 57° and 60 °C. The activity of G-6-PD was expressed as percentage of the initial activity.

The *p*H-optimum for G-6-PD was determined in 0.1 M Tris-0.1 M glycine — 0.1 M $NaH_2PO_4 \cdot 2H_2O$ at *p*H 5.5—11.0 with intervals of 0.5 *p*H-units [10]. The activity of the enzyme at various *p*H-values was calculated assuming that the activity at *p*H 8.0 is equal to 100%. The electrophoretic analysis of the obtained preparation of G-6-PD was performed in fine-pore 7.5% polyacrylamide gel (PAG) avoiding the use of 7.5% gel with large pores [11, 12] at the above mentioned conditions of G-6-PD separation and at concentrations of components constituting the reaction mixture for staining electropherograms.

Samples of the G-6-PD preparation with the activity of 0.015—0.020 units/min were mixed with 40% sucrose solution at a ratio of 1:1 and were put in layers under the electrode buffer on the gel using a micropipette in 0.1—0.2 ml aliquots. Electrophoresis was performed applying a 3 mA current to a column for 2—2.5 hrs at 0 °C using 50 mM Tris-glycine buffer, *p*H 8.3, as an electrode buffer. In order to reveal G-6-PD activity, the elctropherograms were placed into the reaction mixture containing 0.1 M Tris-HCl, *p*H 8.0, 3 mM G-6-P, 0.37 mM NADP, 0.98 mM phenacinemetosulphate and 0.12 mM nitrotetrazolium blue. Staining was done in a dark room at 37 °C. Relative electrophoretic mobility (REM) of the zones with G-6-PD activity was calculated according to the generally accepted method [13]. Quantitative evaluation of G-6-PD activity was performed using a densitometer (DMU-2 of "Toyo"-company, Japan) at 620 nm [14].

The electrophoretic mobility of G-6-PD was determined using horizontal electrophoresis in starch gel at 0 °C in Tris-EDTA boric acid (TEB), *p*H 8.6 (30 mA, 150 V, 16 hrs). The mobility of the mutant enzymes was expressed in percent of the migration distance (to the center of the spot) of the control (normal) enzyme which had been parallely subjected to electrophoresis.

Table

Kinetic and electrophoretic properties of G-6-PD of red

Patient	Activity of G-6-PD in hemolysate (% of normal)	Electrophoretic mobility in TEB, *p*H 8.6, (% of normal)	Electrophoretic mobility in PAG (*p*H of separation 8.9)		Ratio of G-6-PD-activity in fractions	
			Rf_I	Rf_{II}	I	II
Norm ($n = 16$)	100	100	0.29	0.36	36.68 ± 3.74 ($n = 6$)	
K—*v* (son)	5.6	100	0,29			
C (mother)	65	100	0.29	0.36		
R—*n* (son)	0	90	0.29	0.36		
R—*n* (mother)	70.8	100	0.29	0.36	37.67	62.33
R—*n* (son)	0		0.29	0.36	37.03	62.97
A (mother)	71.1	100	0.29	0.36		
S—*r* (son)	0	100	0.29	0.36		
S—*r* (mother)	59.2	100	0.29	0.36	31.15	68.5
K—*n*	0	75	0.29	0.36		
B—*y*	11.5	100	0.29	0.36		
B—*v*	84.6	100	0.29	0.36		

Results

This work presents results of the investigation of G-6-PD properties in 7 male patients with hemolytic anemia characterized by this enzyme deficiency, of 4 mothers of the patients and also of 16 donors (Table 1). The investigation of the properties of G-6-PD from red cells of the mothers of the patients with drug induced hemolytic anemia was carried out considering the fact that G-6-PD deficiency is transmitted by heredity via the maternal line. Further results obtained during the investigation of G-6-PD in red cells of the patient Sr and his mother are discussed. In the patient Sr (son) G-6-PD activity in the hemolysate could not be demonstrated. The activity of 6-PGD corresponded to normal values (0.0015 μM NADP/min · mg protein). After purification of the enzyme G-6-PD activity of the obtained preparation was equal to 0.017 μM NADP/min · mg protein (2% of the normal value). In the mother of *S-r* the activity of G-6-PD in the hemolysate was 0.00154 μM NADP/min · mg protein (59.2% of the normal value) while 6-PGD activity corresponded to normal-values (0.0015 μM NADP/min · mg protein). After a 286fold enzyme purification G-6-PD activity in the obtained preparation was 0.44 μM NADP/min · mg protein (51.8% of the normal value). K_m-values for G-6-P and NADP are considerably decreased as compared to normal values both with regard to G-6-PD isolated from the son's red cells as in the case G-6-PD isolated from his mother red cells. Utilization of substrate-analogues 2-deoxy-G-6-P and deamino-NADP was abruptly increased by the son's abnormal enzyme. A less marked rise of deamino-NADP utilization was noted for the enzyme from his mother's erythrocytes. Both mutant enzymes did not utilize NAD. Thermostability of G-6-PD from the son's erythrocytes was greatly decreased — the critical temperature for 10 min incubation in the presence of 10^{-5}M NADP was 42 °C, for the enzyme from red cells of his mother it was higher than 50 °C (Fig. 1). An activity optimum of the enzyme from the son's red cells was observed

1

cells of healthy persons and patients with hemolytic anemia

K_m for G-6-P (μM)	K_m for NADP (μM)	Utilization of 2-deoxy-G-6-P (% of G-6-P level)	Utilization of deamino-NADP (% of NADP level)	Critical t° during 10 min incubation in the presence of 10^{-5} M NADP	pH optimum
35.0 ± 3	4.27 ± 0.3	4	55—60	52°	9.0
25.70	2.27			50°	7.0—9.5
16.07	2.40			50°	9.0
31.80	3.20			45°	7.5 and 9.75
21.75	2.83	5.3	68.8	50°	8.0 and 9.5
26.0	1.82	23	350	42°	7.5 and 10.0
21.20	3.28			50°	9.0
16.44	2.39	70	350	42°	
22.64	2.16	5	63.3	50°	9.5
27.20	2.03			45°	7.0—8.0
33.60	2.32			50°	7.0 and 8.5
27.90	3.34			50°	9.5

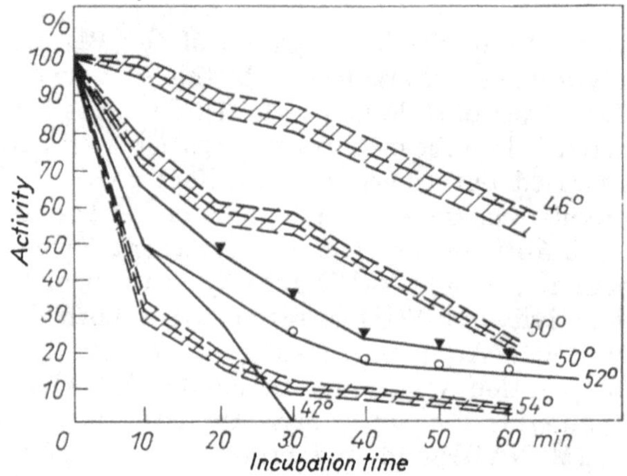

Fig. 1. Thermostability of G-6-PD isolated from red cells of donors, patient *S-r* and his mother *S-r* for 10 min incubation in presence of 10^{-5}M NADP at various temperatures

— — — Change in activity of enzyme isolated from red cells of donors at various temperatures, ▼ change in activity of enzyme isolated from red cells of *S-r*, mother of patient *S-r* at critical temperature; ▿ change in activity of enzyme isolated from red cells of patient *S-r* at critical temperature; ○ change in activity of enzyme isolated from red cells of a donor at critical temperature

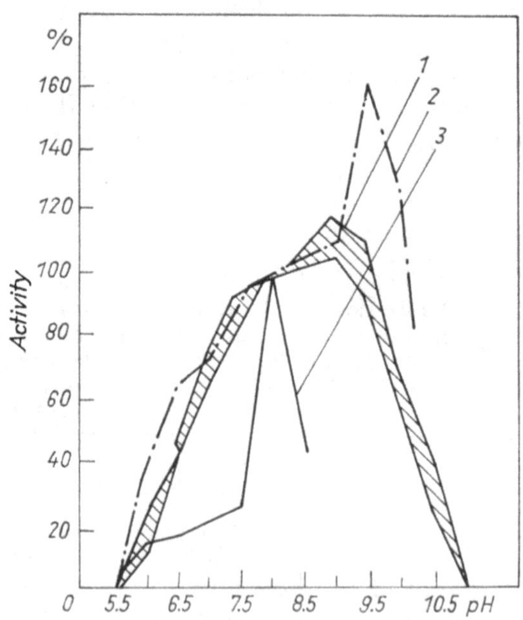

Fig. 2. Influence of *p*H of medium on activity of G-6-PD isolated from red cells of donors, patient *S-r* and his mother *S-r*.

1) Shaded region corresponds to the observed deviations from the average level for 16 experiments (norm); 2) *p*H-dependence of G-6-PD isolated from red cells of mother patient *S-r*; 3) *p*H-different on G-6-PD isolated from red cells of patient *S-r*

at pH 8.0 while at pH 5.5 only 4% activity was registered. With increasing pH up to 8.0 enzyme activity was gradually rising; reaching its maximum and finally inactivation was occurring at pH 9.5 (Fig. 2).

During electrophoresis in polyacrylamide gel abnormal G-6-PD isolated from red cells of patient Sr manifested itself as one band of G-6-PD activity (REM of the band is 0.29) corresponding apparently to a less active tetrameric form of the enzyme (Fig. 3). G-6-PD from red cells of the mother *S-r* manifested itself as two bands of activity (REM of band I was 0.29; of band II 0.36) corresponding to dimeric and tetrameric forms of the enzyme, with relative G-6-PD activities (as compared to total enzyme activity) of 31.5% and 68.5% respectively.

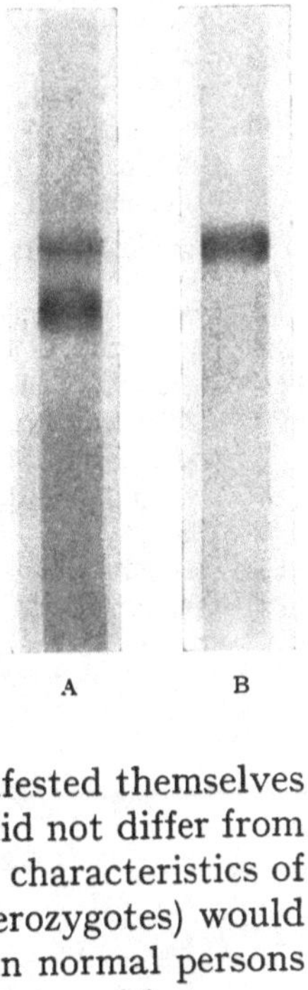

Fig. 3. Electropherogram of G-6-PD isolated from red cells of a donor and patient *S-r* after disk-electrophoresis in polyacrylamide gel.
A) Electropherogram of G-6-PD isolated from red cells of the donor;
B) Eletrophoreogram of G-6-PD isolated red cells of patient *S-r*

During electrophoresis in starch gel both mutant enzymes manifested themselves as one band of G-6-PD activity, and electrophoretic mobility did not differ from normal. It would be suggested that kinetic and electrophoretic characteristics of G-6-PD isolated from red cells of mothers of the patients (heterozygotes) would have intermediate values between properties of the enzyme in normal persons and mutant forms isolated from red cells of the sons (hemizygotes of heterozygote mothers because with regard to G-6-PD deficiency there is a normal and defective population in erythrocytes of the mothers. But the data obtained by us have not proved or in any case have not completely proved this suggestion (Table 1). Apparently it will be possible to explain the observed phenomenon after finding a suitable technique for the separation of erythrocyte populations and the investigation of G-6-PD properties in them.

Discussion

Concerning the results of the investigation of G-6-PD isolated from red cells of the patient *S-r* we may conclude that the comparison of the properties of the enzyme isolated from erythrocytes of this patient has demonstrated that with regard to deficiency in activity this G-6-PD corresponds to the Mediterranean form

of abnormality and according to investigated characteristics differs from enzymes described in the literature.
The disclosed abnormal variant of G-6-PD from red cells of the patient *S-r* has been called "Kremenchug" according to the origin of the patient.

References

[1] YOSHIDA, A., E. BEUTLER, and A. G. MOTULSKY: Bull. Wld Hlth Org. **45**, 243 (1971)
[2] YOSHIDA, A.: Mei Lin. **41**, 877 (1973)
[3] YOSHIDA, A.: Science, N.Y. **179**, 532 (1973)
[4] Standardization of procedures for the study of glucose-6-phosphate dehydrogenase. Report of a WHO Scientific Group. Wld Hlth. Org. Techn. Rep. Ser. No. 366, 1967
[5] KIRKMAN, H. N.: J. biol. Chem. **237**, 2364 (1962)
[6] BATISCHEV, A. I., N. V. LAMZINA, and N. B. CHERNYAK: Vop. med. Khim. **22**, 351 (1976)
[7] BEUTLER, E. (Ed.): Red cell metabolism. A manuel of biochemical methods. Grune and Stratton, New York, London 1971
[8] MOTULSKY, A. G., and A. YOSHIDA in: Red cell genetics. Academic Press, New York 1969, p. 52
[9] LOWRY, O. H., N. J. ROSEBROUGH, A. L. FARR, and R. J. RANDALL: J. biol. Chem. **193**, 265 (1951)
[10] BUETLER, E., C. K. MATHAI, and J. E. SMITH: Blood **31**, 131 (1968)
[11] DAVIS, B. I.: Ann. N.Y. Acad. Sci. **121**, 404 (1964)
[12] ORNSTEIN, L.: Ann. N.Y. Acad. Sci. **121**, 321 (1964)
[13] SAFONOV, V. I., and M. P. SAFONOVA in: Electrophoresis — Polyacrylamidnom Gele i jego primeneniye — biologiu. Selskom Khozyajstve, Meditsine i Pischevoy Promeshlennosti, Trudi Vsesoyuznogo Seminara, October 1971, p. 16
[14] NEVALDINE, B. H., C. M. HYDE, and H. R. LEVY: Archs Biochem. Biophys. **165**, 398 (1974)
[15] PORTER, I. H., S. H. BOYER, J. WATSON et al. **1**, 895 (1964)
[16] Treatment of haemoglobinopathies and allied disorders. Report of WHO scientific group. Wld Hlth. Org. Techn. Rep. Ser. No. 509, 1972

Acta biol. med. germ., Band 36, Seite 759—763 (1977)

Department of Biological Chemistry, University of Torino, Torino (Italy)
Department of Internal Medicine(I), University of Torino, Torino (Italy)
Department of Pediatrics(II), University of Torino, Torino (Italy)

Regulation of NAD and NADP synthesis in human red cell

G. P. PESCARMONA, A. BRACONE, O. DAVID, M. L. SARTORI, and A. BOSIA

Summary

NAD is synthesized in red cell from nicotinic acid and PRPP through the formation of nicotinate mononucleotide and desamido-NAD. Synthesis of one mole of NAD requires two moles of ATP. NADP comes from NAD phosphorylation by NAD-kinase (EC. 2.7.1.23). NAD and NADP analysis on a population with ATP level ranging from 800 to 2500 nmoles/ml red cells showed a close correlation between ATP and pyridine cofactors. Moreover, NADP level appeared to be dependent of the redox-state of NADP/NADPH couple. Subjects with low NADPH (G-6-PD) deficient red cells, Hb Köln) showed lower $NAD_{tot}/NADP_{tot}$ ratio, suggesting a NAD-kinase equilibrium shift toward NADP related to lower levels of the negative effector NADPH, as already described in rat liver.

Introduction

NAD is synthesized in red cells (RC) from nicotinic acid and phosphoribosyl pyrophosphate (PRPP), through the formation of nicotinate mononucleotide and desamido-NAD. Synthesis of one mole of NAD requires two moles of ATP. NADP comes from NAD phosphorylation by NAD-kinase (EC. 2.7.1.23) (Fig. 1). Accurate determination of reduced and oxidized forms of pyridine cofactors in red cells requires complex and rather time-consuming techniques, owing to the instability of the reduced forms in acid, and the oxidized forms in alkali [1—3]. Total NAD ($NAD^+ + NADH$) and NADP ($NADP^+ + NADPH$) levels are, on the other hand, easily determined by standard spectrophotometric methods. NAD_{tot}

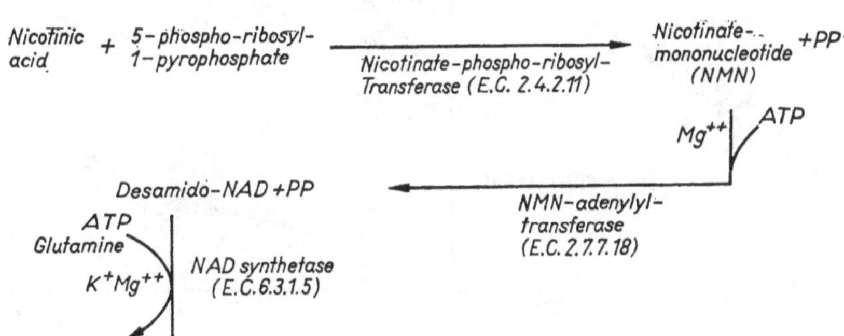

Fig. 1. Synthesis of nicotinamide adenine dinucleotide (NAD) and nicotinamide adenine dinucleotide phosphate (NADP) in human red cell

and $NADP_{tot}$ changes have been described in some cases of hereditary hemolytic anemia (G-6-PD and PK deficiency) and Hb Köln disease [3—5]. Correct interpretation of these data requires further information on the factors affecting the level of total pyridine coenzymes, such as the substrates and inhibitors of the enzymes involved in the synthetic pathway.

Materials and methods

NAD_{tot}, $NADP_{tot}$ and ATP were determined by standard spectrophotometric methods [6] on whole blood perchloric acid extracts. One volume of heparinized blood freshly drawn, or stored according to Beutler [7] (no more than 5 days at 25 °C, or 20 days at 4 °C) was added to 3 volumes of cold 6% perchloric acid, mixed thoroughly and centrifuged for 10 min 5000 × *g*. Supernatant was buffered with 1 M TRAP and neutralized with 10 N KOH, taking care to avoid alkalinization. Hematocrit was measured for each blood sample and cofactors level related to ml RC. $NADP_{tot}$ and NADPH assay with Lowry's fluorometric method [2] was used to obtain the $NADP^+$ values. Alcohol dehydrogenase, G-6-PD, hexokinase were purchased from Boehringer, glucose-6-phosphate disodium salt and NADP from Sigma Chem. Other reagents were from Merck.

Results

In normal donors, $NADP_{tot}$, NAD_{tot} and ATP levels show little variability: $ATP = 1347 \pm 35$, $NAD_{tot} = 79 \pm 3.1$, $NADP_{tot} = 37 \pm 3.7$ nmoles/ml RC. No correlation can be found between ATP and both NAD_{tot} and $NADP_{tot}$ level on such a population. In a group of 48 persons composed of patients with unidentified hemolytic anemia and their relatives, we observed a larger scattering of data ($ATP = 1591 \pm 386$, $NAD_{tot} = 78 \pm 16$, $NADP_{tot} = 41 + 8.8$ nmoles/ml RC). This probably depends on a number of uncertain factors, such as red cell age or unknown primary defects. This group, plus 12 normal donors, with ATP levels ranging from 0.8 to 2.5 µmoles/ml RC, was used to see whether a positive correlation exists between ATP and both NAD_{tot} and $NADP_{tot}$ level. Hb electrophoresis

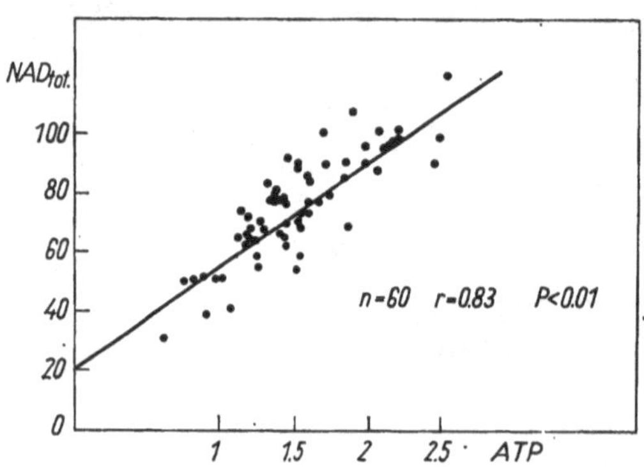

Fig. 2

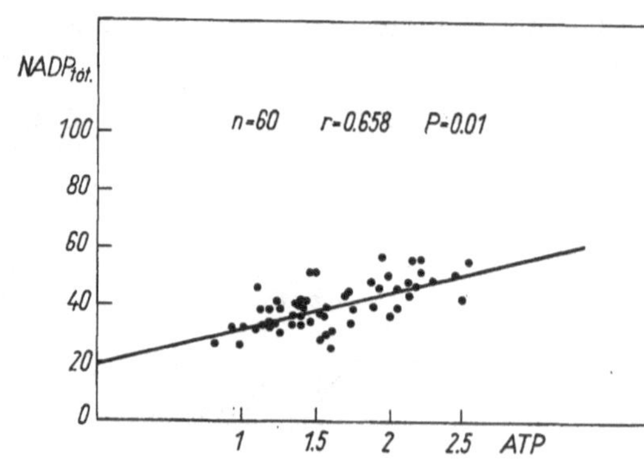

Fig. 3

Fig. 2. Correlation between ATP and NAD_{tot} in human red cells. ATP is expressed as µmoles per ml RC, NAD_{tot} as nmoles/ml RC. r = correlation coefficient

Fig. 3. Correlation between ATP and $NADP_{tot}$ in human red cells. ATP is expressed as µmoles/ml RC, $NADP_{tot}$ as nmoles /ml RC. r = correlation coefficient

was routinely performed to exclude β-thalassemia, since this is associated with low ATP levels, even in the heterozygous form [8].

Both NAD_{tot} and $NADP_{tot}$ showed a close correlation with ATP (Fig. 2 and 3). NADPH is one of the factors that can affect $NADP_{tot}$ level. The association of low NADPH with high $NADP_{tot}$ level in G-6-PD deficient red cells (Mediterranean variant) has been previously described by Kirkman et al. [3].

Table 1 shows ATP, NAD_{tot}, $NADP_{tot}$ and NADPH levels in three different populations: controls, homozygous G-6-PD deficiency and Hb Köln disease. In both G-6-PD deficiency and Hb Köln disease, we observed a low NADPH and a high $NADP_{tot}$ level. The increase of the latter cannot be justified by the ATP level, which was normal in G-6-PD deficiency and even lower in Hb Köln disease.

If we assume that, as in rat liver [9], NADPH is a negative effector of NAD-kinase, the substrate/product ratio of this enzymatic step will vary according to NADPH level, as can be expected from the cross-over theorem [10]. Assuming 50% NADH, we can calculate $NAD^+/NADP^+$ ratio in low NADPH (G-6-PD deficient) and high NADPH (normal) red cells. Table 2 shows that in low NADPH cells (G-6-PD deficient) the $NAD_{tot}/NADP_{tot}$ ratio is very low and quite similar to the $NAD^+/NADP^+$ ratio. In normal red cells, where NADPH is virtually completely reduced, the $NAD_{tot}/NADP_{tot}$ ratio is much lower than the $NAD^+/NADP^+$ ratio, but always significantly higher than in G-6-PD deficient red cells. The ability of $NAD_{tot}/NADP_{tot}$ ratio to identify the cross-over point at the NAD-kinase step is confirmed by Fig. 4, where G-6-PD deficient cells are compared with controls and the unidentified hemolytic anemia group.

Table 1

ATP, NAD_{tot}, $NADP_{tot}$ and NADPH levels in normal, G-6-PD deficient (Mediterranean variant) red cells and in hemoglobin Köln disease

	ATP	NAD_{tot}	$NADP_{tot}$	NADPH
Normals ($n = 12$)	1347 ± 35	79 ± 3.1	37.2 ± 3.7	35.3 ± 4.2
G-6-PD def. ($n = 12$)	1322 ± 216	64.3 ± 8.8	63.3 ± 6.7	14.4 ± 4.5
Hb Köln ($n = 2$)	992 ± 93	81 ± 6	55 ± 3	23 ± 2

Mean values are expressed as nmoles/ml RC (±S.D.)

Table 2

NADPH level, $NAD^+/NADP^+$ and $NAD_{tot}/NADP_{tot}$ ratios in normal and G-6-PD deficient (Mediterranean variant) red cells

	G-6-PD deficients	Normals
NADPH	14.4 ± 4.5	35.3 ± 4.2
$NAD^+/NADP^+$	0.65 $P < 0.001$	20
$NAD_{tot}/NADP_{tot}$	1.01 $P < 0.001$	2.2

Mean values are expressed as nmoles/ml RC (±S.D.)

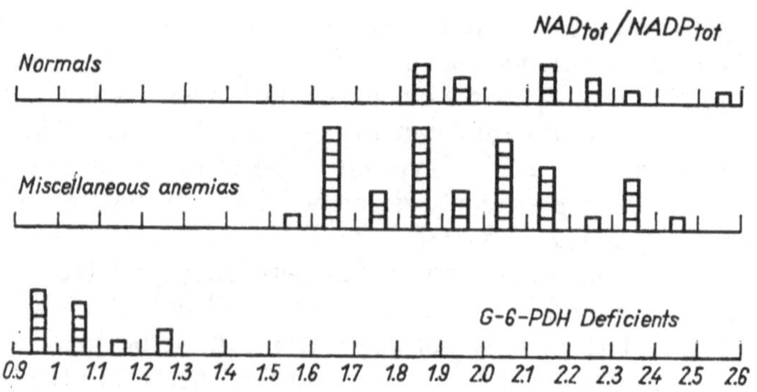

Fig. 4. $NAD_{tot}/NADP_{tot}$ ratio in normal controls, miscellaneous hemolytic anemias and G-6-PD deficient (Mediterranean variant red cells)

Discussion

Analysis of steady-state metabolite level of an enzymatic chain is mostly performed in terms of the cross-over theorem [10]. HEINRICH, RAPOPORT et al. [11, 12] have criticized this model and have proposed a different procedure for the identification of the interaction of an outer effector with an enzymatic chain. This requires the determination of the flux through the chain, the concentration of the substrates and products of the enzymatic step under consideration, and the rate law by which an inner effector, if present, influences the reaction rate at this step. This treatment may be applied, with some assumptions, to red cell glycolysis, whose flux and metabolite level are easily assayed, but is not easy in the case the metabolic chain starting from PRPP and nicotinic acid and leading to NADP. Net synthesis of NADP from nicotinamide has been achieved only once in the literature [13]. Moreover, this occurred in long term incubation (20 hrs) and in the absence of a steady-state, as ATP was reduced to nil in the synthesis of NAD and NADP.

Owing to lack of data on flux in steady-state conditions in this metabolic pathway, the Heinrich-Rapoport treatment could not be applied and we reverted to the cross-over theorem. A cross-over point was identified at the NAD-kinase step. NADPH has been, in fact, identified as one of the negative effectors for NAD-kinase in rat liver: the enzyme has been purified 70fold and its activity shows marked inhibition by NADH and NADPH (K_i of NADH was approximately $1 \cdot 10^{-4}$ M and that of NADPH $5 \cdot 10^{-5}$ [9]).

Kinetic constants of purified human red cell enzyme are lacking, but reliable information on the regulatory role of NADPH in vivo may be drawn from the NAD-kinase steady-state equilibrium. ATP and ADP are not included in the equilibrium, as they are involved in may ATP consuming (Na^+-K^+-dependent ATPase, shape conservation) or ATP producing (PGK and PK steps in glycolysis) processes. Any of these can modify the ATP/ADP ratio much more drastically than NAD-kinase reaction. The striking difference in $NAD^+/NADP^+$ ratio (Table 2) in high and low NADPH cells points to NADPH as a strong negative effector of red cell NAD-kinase in vivo: high NADPH level shifts the equilibrium to the left with increase of $NAD^+/NADP^+$ ratio, low NADPH level shifts the equilibrium to the right with decrease of $NAD^+/NADP^+$ ratio.

For practical purposes, the cross-over point at the NAD-kinase step may be identified by using $NAD_{tot}/NADP_{tot}$ ratio instead of $NAD^+/NADP^+$ ratio, which requires rather cumbersome measuring techniques and freshly drawn blood, factors that restrict its use on large populations. $NAD_{tot}/NADP_{tot}$ ratio decrease specifically indicates a metabolic situation leading to low NADPH level as the pentose phosphate pathway impairment in G-6-PD deficiency, or oxidative stress in the presence of unstable hemoglobins (Hb Köln [14]). The ratio is unchanged (Fig. 4) when $NADP_{tot}$ increase depends on high ATP level, which raises both NAD_{tot} and $NADP_{tot}$ without any modification of NAD-kinase equilibrium.

The work was supported by the Consiglio Nazionale delle Ricerche (Rome, Italy).

References

[1] Lowry, O. H., J. V. Passonneau, and M. K. Rock: J. biol. Chem. **236**, 2756 (1961)
[2] Burch, H. B., M. E. Bradley, and O. H. Lowry: J. biol. Chem. **242**, 4546 (1967)
[3] Kirkman, H. N., G. D. Gaetani, E. H. Clemons, and C. Mareni: J. clin. Invest. **55**, 875 (1975)
[4] Loder, P. B., and G. C. de Gruchy: Br. J. Haemat. **11**, 21 (1965)
[5] Pescarmona, P. P., P. Arese, A. Bosia, and U. Mazza: Boll. Soc. ital. Biol. sper. **48**, 3 (1972)
[6] Klingenberg, M. in: Methoden der enzymatischen Analyse. H.-U. Bergmeyer (Ed.). Verlag Chemie, Weinheim 1970, p. 1979
[7] Beutler, E. (Ed.): Red Cell Metabolism. A Manual of Biochemical Methods. Grune & Stratton, New York, London 1971, p. 99
[8] Brewer, G. J.: Biochem. Genet. **1**, 25 (1967)
[9] Oka, H., and J. B. Field: J. biol. Chem. **243**, 815 (1968)
[10] Chance, B., and G. R. Williams: Adv. Enzym. **17**, 65 (1956)
[11] Heinrich, R., and T. A. Rapoport: Eur. J. Biochem. **42**, 97 (1974)
[12] Rapoport, T. A., R. Heinrich, G. Jacobasch, and S. Rapoport: Eur. J. Biochem. **42**, 107 (1974)
[13] Tsuboi, K. K., J. F. Allan, and K. Fukunaga: J. biol. Chem. **241**, 1616 (1966)
[14] Winterbourn, C., and W. Carrell: J. clin. Invest. **54**, 678 (1974)

Acta biol. med. germ., Band 36, Seite 765—771 (1977)

Institut für Physiologische und Biologische Chemie der Humboldt-Universität Berlin, 104 Berlin, DDR

Glukose-6-phosphat-Dehydrogenase-Mangel roter Blutzellen in der DDR

A. Guckler, M. Grieger, G. Jacobasch und K. Bier

Zusammenfassung

34 G-6-PD-Defektträger wurden diagnostiziert und die pathologischen Enzymvarianten der roten Blutzellen entsprechend den Empfehlungen der WHO charakterisiert. Die Untersuchungen erfolgten an gereinigten G-6-PD-Präparaten.

Auf Grund der unterschiedlich hohen Restaktivitäten in den roten Blutzellen der Probanden und ihrem unterschiedlichen Verhalten bei der kinetischen und physikochemischen Charakterisierung, wird auf eine Vielfalt seltener pathologischer G-6-PD-Varianten in der DDR geschlossen.

Anhand der bestimmten Enzymparameter war es nicht in allen Fällen möglich, die nachgewiesenen G-6-PD-Varianten mit denen bereits in der Literatur beschriebenen Fällen direkt zu vergleichen, da die gewählten Parameterkombinationen nicht immer übereinstimmen.

Einleitung

G-6-PD-Enzymopathin roter Blutzellen zählen in der DDR, ähnlich wie in einigen anderen Ländern, zu den häufigsten Ursachen angeborener nichtsphärozytärer hämolytischer Anämien.

Aus den Ergebnissen bisher durchgeführter Charakterisierungen des Enzyms von Defektträgern geht hervor, daß im Gegensatz zu den Ländern des Mittelmeerraumes in Mittel- und Nordeuropa nicht die $Gd^{mediterran}$-Variante des Enzyms überwiegt, sondern verschiedene Mutanten mit einer annähernd gleichen Häufigkeit nachweisbar sind. Das Ziel der vorliegenden Arbeit war es deshalb, anhand vergleichender Untersuchungen des Enzyms an G-6-PD-Defektträgern aus der DDR sowie von Patienten aus Afrika, Vietnam und Griechenland

1) Aussagen über den Polymorphismus dieser Enzymopathie in unserem Lande machen zu können und
2) Eigenschaften dieser pathologischen Enzymvarianten, die vom Mittelmeertyp abweichen, herauszustellen.

Diese Problematik berührt eine aktuelle wissenschaftliche Frage der Humangenetik und ist zum anderen eine wichtige Grundlage für eine gezielte medizinische Beratung der Patienten.

Material und Methoden

G-6-PD-Aktivitätsbestimmungen wurden im stromafreien Hämolysat und gereinigten Enzympräparat durchgeführt. Heparinisiertes Blut, abgenommen aus der Kubitalvene, wurde nach Abtrennung des Plasmas 3mal mit 0,9%iger NaCl-Lösung gewaschen. Anschließend erfolgte eine osmotische Hämolyse und Abtrennung des Stromas. Dieses stromafreie Hämolysat wurde sowohl zur G-6-PD-Aktivitätsbestimmung als auch als Ausgangsmaterial zur Reinigung des Enzyms verwandt.

Die Reinigung der G-6-PD erfolgte nach den Empfehlungen der WHO [1]. Aus ca. 30 ml Hämolysat wurde das Hämoglobin mittels DEAE-Zellulose abgetrennt. Als Elutionspuffer wurde 5 mM Kalium-Phosphat-Puffer, *p*H 6,4, mit Zusatz von ÄDTA (1 mM) und 2-Merkaptoäthanol (1 mM) benutzt. Die Abtrennung des Enzyms erfolgte mit 0,5 M Kalium-Phosphat-Puffer, *p*H 6,4, mit Zusatz von ÄDTA (1 mM), 2-Merkaptoäthanol (1 mM), NaCl (0,5 M) und NADP (0,02 mM). Das Enzymeluat wurde bei Raumtemperatur gewonnen, die anschließenden Fällungen erfolgten bei 4 °C.

1. $(NH_4)_2SO_4$-Fällung bei 57% Ammoniumsulfat-Sättigung (kaltgesättigte $(NH_4)_2SO_4$-Lösung, *p*H 6,4); der Niederschlag wurde in 0,1 M Kalium-Phosphat-Puffer, *p*H 6,4, mit Zusatz von ÄDTA (1 mM), 2-Merkaptoäthanol (1 mM) und NADP (0,02 mM) aufgenommen.
2. $(NH_4)_2SO_4$-Fällung bei 20% Ammoniumsulfat-Sättigung zur Entfernung von Fremdeiweißen; aus dem Überstand G-6-PD-Fällung bei 50% $(NH_4)_2SO_4$-Sättigung; der Niederschlag wurde in 0,1 M Kalium-Phosphat-Puffer, *p*H 6,4, mit Zusatz von ÄDTA (1 mM) und 2-Merkaptoäthanol (1 mM) aufgenommen.

Das so gewonnene Enzympräparat entsprach einer ca. 200fachen Reinigung, die spezifische Aktivität von Kontrollbluten wurde von 1 μMol $NADPH_2$/mg Eiweiß bei 37 °C auf 200 μMol $NADPH_2$/mg Eiweiß bei 37 °C bei einer Ausbeute von 50% der Anfangsenzymaktivität erhöht.

Die Bestimmung der G-6-PD-Aktivität erfolgte in Anlehnung an BERGMEYER [2] im optischen Test bei 334 nm und 37 °C. Der Meßansatz enthielt (Endkonzentrationen): 41 mM Tra-Puffer, *p*H 7,6, 3 mM $MgSO_4$, 360 μM NADP, 625 μM G-6-P und Hämolysat bzw. Enzympräparat. Für die Ermittlung der relativen G-6-PD-Aktivität mit Analogen wurde dem Meßansatz Gal-6-P bzw. 2-D-G-6-P in einer Konzentration von jeweils 625 μM zugesetzt.

Der G-6-PD-Zytotest nach TÖNZ [3] wurde mit frischem Blut und einer Inkubationsdauer von 3 Std. bei 37 °C durchgeführt.

Die Bestimmung der Michaeliskonstanten von G-6-P erfolgte sowohl durch Einzelbestimmungen bei verschiedenen G-6-P-Konzentrationen im Bereich zwischen $5 \cdot 10^{-6}$ bis $6{,}25 \cdot 10^{-4}$ M als auch durch Progresskurven [4] mit einer Startkonzentration für G-6-P von $2{,}2 \cdot 10^{-4}$ M.

Die Bestimmung der Michaeliskonstante für NADP erfolgte nur mittels Progresskurven bei einer Startkonzentration für NADP von $6{,}0 \cdot 10^{-5}$ M.

Die Bestimmung der Michaeliskonstanten erfolgte rechnerisch nach WILKINSON [5].

Die elektrophoretische Wanderungsgeschwindigkeit wurde auf Cellogel (*p*H 8,8, 50 mM Tris-HCl-Puffer mit ÄDTA-Zusatz) bei 250 V/cm, 6 mA nach 30 min bestimmt. Die Färbung erfolgte mit Nitroblautetrazoliumsalz. Bei Abweichung der Laufgeschwindigkeit von der normalen Aktivitätsdarstellung der G-6-PD wurde zusätzlich eine Stärkegel-Elektrophorese durchgeführt (gleicher Puffer, 30 V/cm, 14 Std.) [1], um eine Einschätzung der WHO-Vorgaben vornehmen zu können. Die Thermostabilität der G-6-PD wurde bei 46 °C im Hämolysat und Enzympräparat bis zu 60 min Inkubation geprüft.

Das *p*H-Optimum wurde im Bereich von *p*H 4,0–10,2 mit 50 mM Zitrat-, Tris- und Phosphat-Puffer bestimmt. Als Bezugswert (100%) diente der Meßwert, der bei *p*H 7,6 in 50 mM Tra-Puffer gewonnen wurde.

Folgende Reagenzien wurden verwendet:

NADP (VEB AWD); Glukose-6-phosphat (G-6-P) als Dinatriumsalz, 6-Phosphoglukonat (6-P-G) als Natriumsalz und Galaktose-6-phosphat (Gal-6-P) als Monobariumsalz (Boehringer); 2-deoxy-Glukose-6-phosphat (2-d-G-6-P) als Natriumsalz (Ferak Chemikalien); DEAE SS-Zellulose (Kapazität 0,61 mÄq/l) von Serva, Heidelberg.

Ergebnisse

Bei 28 Patienten (10 männliche und 18 weibliche) aus 18 Familien und bei 6 Ausländern (Afrika, Vietnam, Griechenland) wurde ein G-6-PD-Mangel roter Blutzellen nachgewiesen. Von den bisher in der DDR bekannten G-6-PD-Mangelträgern haben 20 eine chronische Anämie, die bei 11 Patienten seit der Geburt besteht (7 männliche, 4 weibliche). Durch Medikamente kam es in 4 Fällen zur Auslösung hämolytischer Schübe.

Im Tönz-Test zeigten 10 Hemizygote zwischen 10% und mehr als 90% „leerer Zellen" bei hell gefärbter Gesamtpopulation roter Blutzellen. Die 18 heterozygoten Merkmalsträger ließen bei 3% bis 90% der Zellen einen Defekt erkennen. Von den 3 Patienten ohne G-6-PD-Aktivität der roten Blutzelle war auch ein totaler Defekt der G-6-PD in den Leukozyten nachweisbar. Auch bei G-6-PD-Defekten verschiedener Zelltypen handelt es sich wahrscheinlich nicht um einen Ausfall der Enzymsynthese, wie in einem der analysierten Fälle bestätigt werden konnte (siehe Patient N. Wo.).

Die Enzymaktivitäten im stromafreien Hämolysat von Kontrollpersonen ($n = 6$) ergab einen Mittelwert von 174,8 ± 13,1 μMol $NADPH_2$/ml Zellen · 1 Std. bei 37 °C. Die Enzymaktivität der roten Blutzellen der Patienten war sehr unterschiedlich und erstreckte sich von nicht meßbaren Werten (N. Wo.) bis zu Aktivitäten, die weit über denen von Kontrollpersonen lagen (s. Tab. 1 und 3).

Tabelle 1

Verhalten der G-6-PD-Aktivität roter Blutzellen bei verschiedenen Defektträgern

% der Normal-aktivität	Zahl der Patienten	
	weiblich	männlich
0–20	–	4
30–50	4	2
100	4	–
>100	5	2

Tabelle 2

Spezifische Aktivität der G-6-PD in Hämolysaten und Enzympräparaten einiger Patienten

Patient	Hämolysat[2]	Enzympräparat[2]	Ausbeute
H. Br.	0,8	144,0	68,0
M. Be.	0,47	30,0	30,0
N. Wo.	0	0	0
B. Gu.	1,0	160,0	41,0
H. Ph.[1]	0,76	91,0	20,0

[1] Patient aus Vietnam, alle anderen aus der DDR;
[2] in μMol $NADPH_2$/mg Eiweiß

Die Charakterisierung der G-6-PD roter Blutzellen erfolgte an gereinigten Enzympräparaten. Für die untersuchten Blute von G-6-PD-Mangelträgern lag die spezifische Aktivität im Hämolysat zwischen 0 und 1,0 μMol $NADPH_2$/mg Eiweiß und für das Enzympräparat (s. Tab. 2) zwischen 0 und 160 μMol $NADPH_2$/mg Eiweiß. Die Aktivitätsausbeute lag dabei zwischen 25–60% bezogen auf das Hämolysat.

Die Bestimmung der Michaeliskonstanten für G-6-P ergab bei einem Vergleich der Progresskurvenmethode und der Einzelbestimmung unter Variation der G-6-P-Konzentration gut übereinstimmende Ergebnisse. Die Progresskurven bieten dabei gegenüber der Direktbestimmung eine wesentliche Rationalisierung und

Tabelle 3

Kasuistik einiger G-6-PD-Mangelpatienten

	Enzymaktivität [μMol $NADPH_2$/ml Zellen · Std. bei 37 °C]	Tönz [% leere] Zellen]	K_m G-6-P [μM]	K_m NADP [μM]	elektrophoretische Wanderung [cm]	Thermostabilität	pH-Optimum	Phänotyp	Genotyp
Normalpersonen	174,8 ± 13,1	0	42,8 ± 6,4	11,7 ± 1,6	7,5	normal	7,6		
H. Br.	187,0	30	64,2	12,0	7,5	↓	zweigipflig 7,6; 9, 8	c. n. s. h. A[1]	vermutlich heterozygot
M. Be.	47,2	30	54,5	12,8	6,0	↓	4,8	angeborene c. n. s. h. A	heterozygot
N. Wo.	0	90	—	—	7,5	—	—	angeborene c. n. s. h. A	hemizygot
B. Gu.	196,5 variabel	24	52,9	12,7	—	⇊	7,6	c. n. s. h. A	heterozygot
H. Ph.	134,5	12	42,2	13,1	9,0	normal	—	schwere c. n. s. h. A	hemizygot

[1] chronische nichtsphärozytäre hämolytische Anämie

eine hohe Genauigkeit bei der Arbeit, so daß dieser Methode der Vorzug in der weiteren Arbeit gegeben wurde. Bei den charakterisierten Enzympräparaten wurde bei einem Patienten (H. Ph.) sowohl für den K_m-Wert für G-6-P als auch für den für NADP eine normale Substrataffinität festgestellt. Bei 2 Patienten (M. Be., B. Gu.) entsprachen die NADP-Affinitäten der Norm, die K_m-Werte für G-6-P lagen im oberen Grenzbereich. Bei einem weiteren Patient (H. Br.) der ebenfalls eine der Norm entsprechende NADP-Abhängigkeit zeigte, war die G-6-P-Affinität herabgesetzt. Der ermittelte K_m-Wert von 64 μM liegt außerhalb des 3s-Bereiches (s. Tab. 3).

Die elektrophoretische Wanderungsgeschwindigkeit der G-6-PD betrug unter den angegebenen Testbedinungen bei Kontrollpersonen 7,5 cm. Zwei der vier überprüften Patientenpräparate weisen in der elektrophoretischen Wanderungsgeschwindigkeit Abweichungen im Vergleich zur Norm auf. Der Patient M. Be. zeigt eine langsamer, der vietnamesische Patient H. Ph. eine schneller wandernde Aktivitätsbande (s. Tab. 3).

Die Temperaturstabilität ist ein sehr empfindlicher Parameter bei der Differenzierung pathologischer Enzymvarianten. Bei Kontrollpersonen ist die G-6-PD-Aktivität roter Blutzellen über einen längeren Zeitraum stabil. Nach 20 min Inkubation bei 46 °C waren noch 88% und nach 60 min noch 80% der Ausgangsaktivität zu verzeichnen. Die G-6-PD des Patienten B. Gu. erwies sich als sehr instabil; nach 20 min Inkubation waren nur noch 39% und nach 60 min keinerlei Aktivität der G-6-PD mehr zu beobachten. Auch die G-6-PD der Patienten H. Br. und M. Be. zeigte eine Instabilität bei höheren Temperaturen. Die Aktivität der G-6-PD betrug nach 20 min noch 55%, sank jedoch nach 60 min auf 0 bis 6% der Ausgangsaktivität herab. Die Mutante des Patienten H. Ph. war dagegen auch bei 46 °C stabil (s. Tab. 3).

Einige G-6-PD-Mutanten zeigten von der Norm abweichende *p*H-Abhängigkeiten. Die Messung der Enzymaktivität bei unterschiedlicher H^+-Ionenkonzentration ergab bei Kontrollpersonen eine eingipflige Aktivitätskurve mit einem *p*H-Optimum bei *p*H 7,6 (Tris-, Phosphat-, Tra-Puffer). Identische Resultate ergab auch die Überprüfung der *p*H-Abhängigkeit des Enzyms bei dem Patienten B. Gu. Bei der G-6-PD-Mutante des Patienten M. Be. war dagegen das *p*H-Optimum zum sauren Bereich auf *p*H 4,8 (Zitrat-Puffer) hin verschoben. Ein zweigipfliges Aktivitätsverhalten der G-6-PD wurde bei dem Patienten H. Br. festgestellt. Ein Gipfel entsprach den Kontrollwerten, ein 2. Optimum lag bei *p*H 9,8 (Tris-, Phosphat-Puffer) (s. Tab. 3).

Substratanaloga wurden auf ihre Umsatzrate bei G-6-PD-Präparaten von 2 Patienten getestet. Die G-6-PD-Aktivität des Patienten H. Br. zeigte eine relative Aktivitität mit Gal-6-P von 10% und mit 2-D-G-6-P von 3,9%, die den Kontrollen entsprachen. Erhöhte Umsatzraten von Gal-6-P mit 37% und 2-D-G-6-P mit 64% wurden dagegen für die G-6-PD des Patienten B. Gu. nachgewiesen.

Diskussion

Die in dieser Arbeit aufgeführten Beispiele von ausgewählten G-6-PD-Defektträgern veranschaulichen, daß sehr unterschiedlich hohe Restaktivitäten in den roten Blutzellen der Probanden vorliegen (Tab. 1). Aus dieser Tatsache leitet sich die Schlußfolgerung ab, daß eine Vielfalt seltener pathologischer G-6-PD-Varianten in der DDR anzunehmen ist. Eine alleinige G-6-PD-Aktivitätsbestimmung ist auf Grund der vorliegenden Befunde zur Identifizierung von G-6-PD-Defektträgern nicht ausreichend. Da eine Reihe von Patienten mit hohen G-6-PD-Restaktivitäten thermolabile Enzymvarianten aufweisen, ist neben der routinemäßigen G-6-PD-Aktivitätsbestimmung im frisch hergestellten stromafreien Hämolysat die Aktivitätsüberprüfung nach 20 min Inkubation des Hämolysates bei 46 °C zu empfehlen.

Die kinetische und physikochemische Charakterisierung einer pathologischen Enzymvariante erfordert eine ausreichend hohe G-6-PD-Aktivität des Untersuchungsmaterials, die mindestens der des Hämolysates von Kontrollpersonen entsprechen sollte.

Die Reinigung des Enzyms auch aus kleinen Blutproben von ca. 3 ml Zellen von G-6-PD-Defektträgern ist deshalb die Methode der Wahl. In Fällen, wo keine G-6-PD-Aktivität in den roten Blutzellen feststellbar ist, muß geprüft werden, ob es sich um einen kompletten Ausfall dieser Enzym-Synthese handelt oder, was häufiger ist, eine inaktive G-6-PD-in der roten Blutzelle vorkommt (siehe Patient N. Wo.), die als Hinweis auf eine extrem instabile Enzymvariante gewertet werden kann. Im letzteren Fall sollte bestimmt werden, inwieweit die Stabilität der Leukozyten-G-6-PD eine Charakterisierung der pathologischen Enzymvariante erlaubt.

Die Zuordnung einer pathologischen G-6-PD-Variante zu einer der vielen in der Literatur beschriebenen Mutanten ist ohne den Aufwand einer chemischen Analytik des Enzymeiweißes, für die nur selten die notwendige Enzymmenge erhalten werden kann, nicht möglich. Überwiegend stützt man sich deshalb bei der Charakterisierung auf den Vergleich der Ergebnisse aus einer Kombination mehrerer Enzymparameter. Unterschiede in der Affinität zu den unmittelbaren Reaktionspartnern der G-6-PD werden dabei nur selten festgestellt. Häufiger sind abweichende Umsatzgeschwindigkeiten mit den Substratanalogen 2D-G-6-P und Gal-6-P (siehe Patient B. Gu.) und die Beeinflussung durch Inhibitoren. Weitere Parameter, die zum Standardprogramm bei der Charakterisierung pathologischer G-6-PD-Varianten zählen, sind die Thermostabilität, *p*H-Abhängigkeit und elektrophoretische Wanderungsgeschwindigkeit. Die Anwendung der zuletzt genannten Parameter erlaubt nicht nur Hinweise auf einen möglichen Aminosäureaustausch im Enzymeiweiß, sondern diese Parameter sind auch geeignet, um den Verdacht eines vorliegenden Zellmosaiks bzw. doppelt heterozygoten Merkmalsträgers abzusichern. Von den fünf in dieser Arbeit aufgeführten Patienten entspricht die G-6-PD des Patienten H. Br. am ehesten der als „El-Karga" beschriebenen G-6-PD-Variante [6], die sowohl durch eine normale Enzymaktivität als auch unveränderte elektrophoretische Wanderungsgeschwindigkeit und nahezu normale K_m-Werte für G-6-P und NADP gekennzeichnet ist. Sie ist jedoch thermoinstabil und hat 2 *p*H-Optima im alkalischen *p*H-Bereich.

Die G-6-PD des Patienten H. Be. ist der seltenen „Berlin-Variante" zuzuordnen [7]. Es ist eine thermoinstabile G-6-PD-Mutante mit einer eindeutig von der Norm abweichenden *p*H-Abhängigkeit und verringerten elektrophoretischen

Wanderungsgeschwindigkeit. Zwei charakteristische Merkmale der G-6-PD-Varianten des Patienten B. Gu., die hohe Umsatzrate der Substratanalogen 2D-G-6-P und Gal-6-P sowie die schnelle Denaturierbarkeit des Enzymproteins bei normaler G-6-PD und NADP-Affinität, lassen die Schlußfolgerung zu, daß diese Variante Ähnlichkeiten zu der von BEUTLER et al. [8] als „G-6-PD-Duarte" beschriebenen Form aufweist. Um eine GdA-Variante mit hoher Restaktivität handelt es sich auf Grund der schnelleren elektrophoretischen Wanderungsgeschwindigkeit im Falle des Patienten H. Ph.

Literatur

[1] World Health Organization: Standardization of procedure for the study of glucose-6-phosphate dehydrogenase. Wld Hlth Org. Techn. Rep. Ser. **366** (1967)
[2] BERGMEYER, H. U. (Hrsg.): Methoden der enzymatischen Analyse. **1**, 417, 605. Akademie-Verlag, Berlin 1970
[3] TÖNZ, O.: Ann. Paediat. **204**, 24 (1965)
[4] HÖHNE, W. E., T. RAPOPORT u. P. HEITMANN: Acta biol. med. germ. **29**, 841 (1972)
[5] WILKINSON, G. N.: Biochem. J. **80**, 324 (1961)
[6] MCCURDY, P. R., K. KAMEL u. O. SELIM: J. Lab. clin. Med. **84**, 673 (1974)
[7] STREIFF, F., u. C. VIGNERON: Nouv. Rev. fr. Hématol. **11**, 279 (1971)
[8] BEUTLER, E., C. K. MATHAI u. J. E. SMITH: Blood **31**, 131 (1968)

Summary

A. GUCKLER, M. GRIEGER, G. JACOBASCH, and K. BIER: G-6-PD deficiency of red cells in the G.D.R.

34 persons with G-6-PD deficiency were diagnosed, and the pathological enzyme-variants of red blood cells were characterized according to the recommendations of WHO.
We conclude from the differing residual G-6-PD-activities in red blood cells of the propositi and the differing reactivity of the enzyme in kinetic and physicochemical characterizations that a multiple variety of rare pathological G-6-PD variants exists in the GDR.
Using the estimated enzymeparameters it was not possible in all cases to compare directly the newly demonstrated G-6-PD variants with cases already described in the literature. In addition, the differing combinations of parameters render a classification more difficult.

Acta biol. med. germ., Band 36, Seite 773—777 (1977)

Institut für Physiologische und Biologische Chemie der Humboldt-Universität, Berlin
104 Berlin, DDR

Die Problematik der Erfassung heterozygoter Glukose-6-phosphat-Dehydrogenase-Mangelträger

M. GRIEGER und G. JACOBASCH

Zusammenfassung

Die Diagnostik heterozygoter Merkmalsträger ist sowohl für Populationsstudien als auch für die Erfassung und Beratung hämolysegefährdeter G-6-PD-Mangelträger unerläßlich. Das Auffinden heterozygoter Merkmalsträgerinnen ist problematisch, wenn in der Familie kein hemizygot Kranker als Anhaltspunkt für die Zuordnung existiert.

1. Die Aktivitätsbestimmung kann nur bei Ergebnissen unterhalb des Referenzbereiches zur Differenzierung in verdächtige Heterozygote und Homozygote beitragen; heterozygote G-6-PD-Mangelträger mit normaler Aktivität bleiben unerkannt.
2. Die Darstellung des Mosaiks (Tönz-Test) gesunder und defekter Zellen beweist auch bei normaler Enzymaktivität den G-6-PD-Defekt.
3. Die Prüfung der Temperaturstabilität bei 46 °C mit und ohne NADP als Stabilisator führt bereits nach 20 min zum Nachweis hitzelabiler Enzymvarianten.
4. Die Cellogel-Elektrophorese dient zum Nachweis einer Ladungsänderung des mutierten Enzyms.
5. Familienuntersuchung mit Aufstellung eines Stammbaumes sichert den Erbgang und damit den Verdacht auf Heterozygotie.

Zur Verbesserung der Heterozygotenerfassung wird eine Parameterkombination empfohlen.

Der Genort der G-6-PD befindet sich im X-Chromosom. Daraus leiten sich verschiedene Möglichkeiten des Geno- und Phänotyps ab (Abb. 1). Die Gruppe heterozygoter Merkmalsträgerinnen für einen G-6-PD-Mangel ist schwer zu identifi-

Geschlecht	*Gonosomenmuster*	*Genotyp*	*Phänotyp*
♂	X Y	*normal*	*gesund*
♂	X̸ Y	*hemizygot*	*Anämie (abh. v. Variante)*
♀	X X	*normal*	*gesund*
♀	X̸ X	*heterozygot*	*gesund oder Anämie*
♀	X̸ X̸	*homozygot*	*Anämie abh. v. Variante*

X--- normales X-Chromosom
X̸---- G-6-PD-Mangel-Chromosom

Abb. 1. Vererbungsmöglichkeiten einer G-6-PD-Mangel-Mutante

zieren; denn es besteht nicht immer eine Enzymaktivitätsverminderung auf die Hälfte der Norm und die Enzymaktivität korreliert nicht mit dem Schweregrad der Anämie [1].

Eine Ursache für die Vielfalt ist in der Geninaktivierung zu sehen, die bereits ab 14. Tag post conceptionem beginnt und entweder das gesunde oder das pathologische X-Chromosom in den Zellen des Föten unterdrückt [2, 3]. Der Zeitpunkt der Inaktivierung ist für einzelne Gewebe unterschiedlich [4, 5]. Es ist unbekannt, wonach die Auswahl des Gens erfolgt, das im jeweiligen Zellclon dominiert.

Darüber hinaus erschwert die Vielzahl bisher beschriebener pathologischer G-6-PD-Varianten die Erkennung und Zuordnung einer Mutation [6].

Bisher gibt es keine verbindliche Festlegung zum Nachweis heterozygoter G-6-PD-Mangelträger. In vielen Fällen ist eine sichere Unterscheidung zwischen homo- und heterozygoten Merkmalsträgerinnen unmöglich. Die Notwendigkeit Heterozygote zu erkennen ist für die genetische Erforschung von Familien mit hemizygot Kranken und für die Erfassung und Beratung hämolysegefährdeter Heterozygoter von großer Wichtigkeit. Populationsstudien erhalten darüber hinaus in einem Gesundheitswesen, das auf die Prophylaxe orientiert ist, besondere Bedeutung.

Als Parameter zur Heterozygotenerfassung dient in erster Annäherung die Enzymaktivität. Die Aktivitätsbestimmung der G-6-PD aus roten Blutzellen ist immer das Ergebnis der Mittelung von Einzelaktivitäten aller Zellen. Reifungsunterschiede und ein Mosaik defekter und intakter Zellen schränken die Aussage zusätzlich ein. Die G-6-PD-Aktivität Heterozygoter kann zwischen 30 und 100% der Norm betragen. Daraus ergibt sich eine Einengung der Aussage empfohlener Siebtests, die auf Entfärbung von DCPIP, Brilliantkresylblau oder Methylenblau durch reduziertes Koenzym aus der G-6-PD-Reaktion beruhen [7—9]. Auch der einfache Fluoreszenztest von Beutler [10] erbringt nur bei Heterozygoten mit deutlich vermindeter G-6-PD-Aktivität den Nachweis eines G-6-PD-Mangels. Die Ergebnisse der Aktivitätsmessung an heterozygoten G-6-PD-Mangelträgern in der DDR zeigt Abb. 2. Auf der Grundlage dieses Parameters wäre nur etwa ein Drittel der Gefährdeten erkannt worden. Durch die Aktivitätsbestimmung ist damit keine Abgrenzung heterozygoter Merkmalsträgerinnen von Homozygoten möglich.

Größere Sicherheit beim Auffinden Heterozygoter bietet ein Test zur Erfassung des Mosaiks von G-6-PD-Mangelzellen neben Zellen normaler G-6-PD-Aktivität, wie er von Tönz [11] angegeben wurde (Abb. 3). Er basiert auf der NADPH-Bildung der Einzelzelle, das oxydiertes Hämoglobin in die reduzierte Form überführt. Bei G-6-PD-Mangelzellen wird aufgrund des resultierenden NADPH-Defizits und herabgesetzter Pseudoperoxidaseaktivitat das Hb^{3+} eluiert. Unter der angewendeten Inkubationsdauer von 3 Std. besteht ab 3% „leerer" Zellen der Verdacht auf einen G-6-PD-Mangel [8]. Mit Hilfe dieses Tests konnten in Populationsstudien 2100 Studenten untersucht und bei 20 von ihnen ein Zellmosaik nachgewiesen werden. Außerdem wurden in einer Risikogruppe von 1000 Anämiepatienten 18 miteinander nicht verwandte G-6-PD-Mangelträger diagnostiziert.

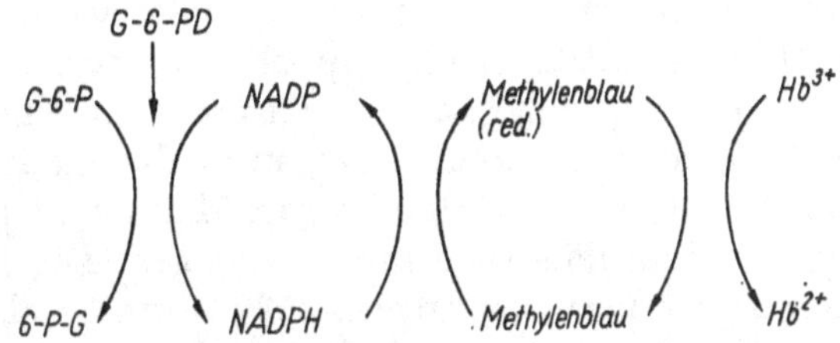

Abb. 3. Reaktionsschema des Tönz-Testes

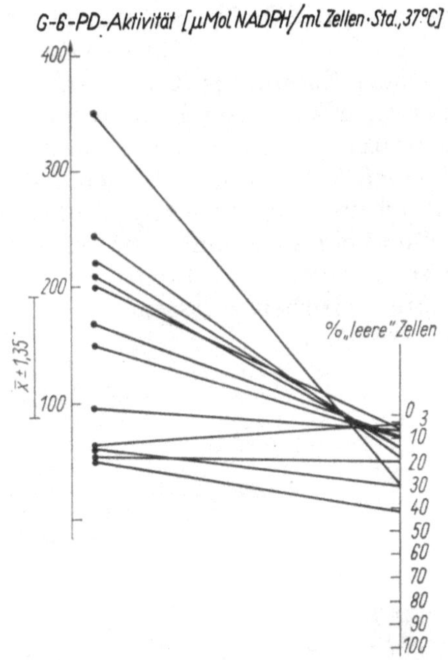

Abb. 2. Beziehung zwischen Enzymaktivität unter optimalen Meßbedingungen und Ergebnis des Tönz-Testes bei heterozygoten G-6-PD-Mangelträgern

Eine Differenzierung homozygoter und heterozygoter Defektträger ist in vielen Fällen allein mit dieser Methode nicht möglich. In Kombination mit der Aktivitätsbestimmung läßt sich jedoch die Aussagemöglichkeit erweitern.

Zur Sicherung der Diagnose eines G-6-PD-Defektes ist der Temperaturstabilitätstest ein besonders geeigneter Parameter (Tab. 1). Das SFH von Normalpersonen zeigt nach 20 und 60 min Inkubation bei 46 °C noch 88 bzw. 80% der Ausgangsaktivität, NADP steigert diesen Anteil nicht. Eine verminderte Stabilität ist ein zuverlässiges Kriterium für das Vorliegen einer G-6-PD-Variante (Abb. 4), normale Stabilität kann jedoch ein pathologisches Enzym nicht ausschließen.

Als weiterer Parameter läßt die Elektrophorese durch den Nachweis einer veränderten Ladung des Eiweißmoleküls Rückschlüsse auf eine Mutation zu. Mit der Cellogel-Elektrophorese ist eine einfache und schnelle Methode gegeben, die

Tabelle 1

Temperaturstabilität der G-6-PD heterozygoter Merkmalsträger nach Inkubation des stromafreien Hämolysats bei 46 °C

	% der Ausgangsaktivität	
	nach 20 min	nach 60 min
Kontrollen ($n = 5$)	88	80
Fr. S.	85	20
Pö. K.	40	0
Koa. A.	30	10
Wo. I.	25	0
Kör. S.	50	12

eine gleichzeitige Überprüfung mehrerer Enzymproben erlaubt. Sie ermöglicht auch die Identifizierung doppelt heterozygoter Fälle [4]. In allen Fällen, in denen mit den empfohlenen vier Parametern eine sichere Zuordnung zum Genotyp nicht eindeutig möglich ist, muß diese auf der Grundlage von Familienuntersuchungen erfolgen. Hemizygote Männer oder heterozygote Frauen können heterozygote Merkmalsträgerinnen erzeugen. Diese können 25% normale Jungen, 25% hemizygot kranke Jungen, 25% normale Mädchen und 25% heterozygote Mädchen als Nachkommen haben [12]. Die Aufstellung eines Stammbaumes sichert den Erbgang (Abb. 5). Aufgrund der Mendel'schen Gesetze wird dabei die Untersuchung der Eltern eines Patienten die größte Sicherheit zulassen.

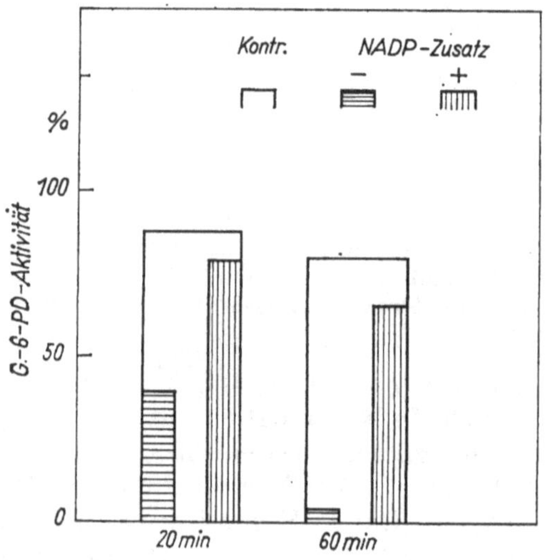

Abb. 4

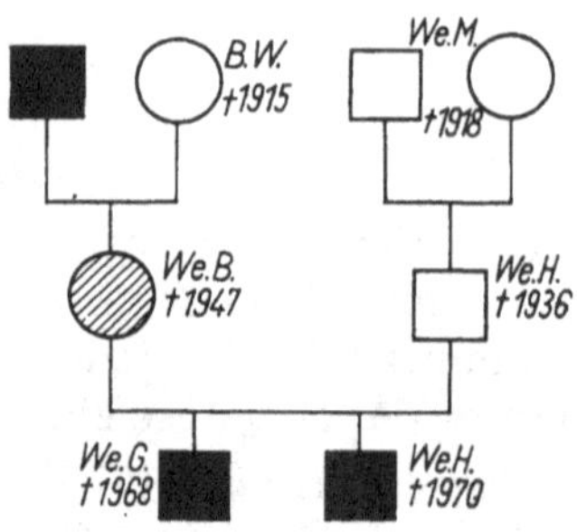

Abb. 5

Abb. 4. Temperaturstabilität der G-6-PD nach Inkubation des stromafreien Hämolysates mit und ohne NADP (1 µM) bei 46 °C

Abb. 5. Familienstammbaum eines Patienten mit G-6-PD-Mangel

Die vier von uns vorgeschlagenen Parameter: Tönz-Test, G-6-PD-Aktivität, Temperaturstabilität und Elektrophorese garantieren in ihrer Kombination die sichere Abrenzung Gesunder von G-6-PD-Defektträgern. Sie erlauben darüber hinaus die Diagnostik homozygoter Merkmalsträger, wenn der Tönz-Test mehr als 50% leere Zellen zeigt und die G-6-PD-Aktivität an der Erfassungsgrenze liegt.

Literatur

[1] Harris, H.: Biochemische Grundlagen der Humangenetik. Akademie-Verlag, Berlin 1974, S. 116
[2] Lyon, M.: Nature, Lond. **190**, 372 (1962)
[3] Beutler, E., M. Yett u. V. Fairbanks: Proc. natn. Acad. Sci. U.S.A. **48**, 9 (1962)
[4] Chan, T., u. M. Lai: J. med. Genet. **8**, 149 (1971)
[5] Steele, M., u. A. Migeon: Biochem. Genet. **9**, 164 (1973)

[6] YOSHIDA, A., E. BEUTLER, u. A. MOTULSKY: Bull. Orgmond. Sante Bull. Wld Hlth Org. **45**, 243 (1971)
[7] SCHEUCH, D., u. H. KUTSCHER: Z. med. Labortechn. **3**, 22 (1962)
[8] FRISCHER, H., u. P. CARSON: Am. J. Med. **41**, 743 (1966)
[9] MOTULSKY, A.: Acta genet. Statist. med. **17**, 465 (1967)
[10] BEUTLER, E.: Blood **28**, 533 (1966)
[11] TÖNZ, O.: Ann. paediat. **204**, 24 (1965)
[12] MENDEL, G.: Versuche über Pflanzenhybriden. Verh. Naturf. Ver. Brunn **43**, (1865)

Summary

M. GRIEGER and G. JACOBASCH: Detection of female heterozygous G-6-PD defiency

Diagnostics of heterozygotes are required for population studies, for the detection and consultation of persons with G-6-PD deficiency prone to hemolysis.
The diagnostics of heterozygous females with the corresponding trait are problematic in families without hemizygous patients.

1. The determination of the activity is only applicable to the differentiation between heterozygotes and homozygotes if the activities are below the reference range. Heterozygous G-6-PD deficiency with normal activity cannot be identified by this method.
2. Existence of G-6-PD defects is demonstrated by mosaicism even in case of normactivity (Tönztest).
3. Incubation with and without NADP of stroma-free hemolysates involving heat labile enzyme mutants results in a marked decrease of activity within 20 min at 46 °C.
4. Electrophoresis on Cellogel demonstrates changes of charge in the mutated enzyme.
5. Family examination verifies suspicion of the heterozygous trait.

A combination of parameters is recommended to obtain an improvement in the detection of persons with the heterozygous trait.

Acta biol. med. germ., Band 36, Seite 779—772 (1977)

Zaklad Biochemii Instytutu Fizjologiczno-Biochemicznego WAM,
Instytut Pediatrii AM and Zaklad Biofizyki Instytutu Biochemii i Biofizyki Uniwersytetu Lodzkiego, Lodz, Poland

The level of superoxide dismutase in erythrocytes of children with Down syndrome (trisomy G and unbalanced translocation G 21/22)

J. Kędziora, J. Jeske, H. Witas, G. Bartosz, and W. Leyko

Introduction

In the last few years special attention has been paid to inborn metabolic disturbances in various chromosomal aberrations [1—5].

The connection between Down syndrome and chromosomal anomaly (trisomy G 21) found by Lejeune et al. [4] has given the basis to investigations aiming at determining those chromosomal markers, the genes loci of which would be placed in acrocentric chromosomes of the group G.

One of the first reports confirming the finding of Tan et al. [6] was the discovery that high level of activity of indophenol oxidase (IPO)-A occurs in children with Down syndrome (trisomy G-21) [7]. Further investigations in that direction done by Sinet et al. showed a characteristic increase of the level of superoxide dismutase (SOD-1) both in erythrocytes [8] and blood platelets [9].

The aim of our investigations was to indicate how SOD-1 activity would behave in the extremely rare aberration unbalanced translocation G 21/22, centrifusic type. What gave rise to this investigation was the discovery that IPO-A activity level in children with trisomy G 21 was approximately 30% higher than in normals [10].

Material and methods

Venous blood for the tests was taken from patients (10 female and 15 male) between 16 and 26 years of age. Erythrocyte SOD-1 activity was studied in three groups. The first group consisted of individuals with trisomics for chromosome G 21 (6 females and 7 males aged from 16 to 26 years). The second group contained a brother and a sister with unbalanced translocation (t/21q22q/mat.) and a full appearance of Down phenotype. The third group consisted of 19 healthy individuals aged from 19 to 22 years. All the persons examined were Caucasians. The subjects had no infectious disease at the time of the tests.

Superoxide dismutase activity in hemolysates of erythrocytes was determined according to the method of Winterbourn et al. [11]. Erythrocytes were hemolysed by addition of 1.5 volume of cold distilled water. Then 0.5 ml hemolysate was added to 3.5 ml cold distilled water, 1 ml ethanol and 0.6 ml chloroform. The obtained mixture was shaken for 60 s and centrifuged. The enzyme activity was determined in clear hemoglobin-free supernatants. Six different quantities of the supernatant (5—50 μl) were put into tubes and added with: 0.2 ml 0.1 M EDTA (containing 1.5 mg KCN per 100 ml), 0.1 ml 1.5 mM nitroblue tetrazolium (NBT), 0.05 ml of 0.12 mM riboflavine and 1/15 M sodium phosphate buffer, *p*H 7.8, up to 3 ml. Riboflavin was added as the last compound and the tubes were then placed on a revolving table and irradiated for 30 min using a halogen source. The absorbance was determined at 560 nm. In the time interval of the exposure the increase of absorbance in control test tubes containing no SOD was a linear function of the time of exposure.

From diagrams of dependence of the percent inhibition of NBT reduction on the supernatant volume, the volume of supernatant causing 50% inhibition of the reduction seen in control tubes was determined. SOD activity standardized with the use of a commercial enzyme preparation (Sigma) was expressed as units SOD per g hemoglobin. The concentration of hemoglobin in hemolysates was determined spectrophotometrically in the cyanmet form, on the assumption that the millimolar absorption coefficient equals to 11 at 540 nm for a molecular weight of 16100 daltons.

Results

The investigations have indicated that an about 80% increase of SOD-1 activity could be observed in the group with trisomy G 21 and an about 60% increase is

Table 1

SOD activity in erythrocytes of control subjects (units per g hemoglobin)

No.	Sex	SOD activity × 10^{-3}
1	F	2.62
2	F	3.20
3	F	4.01
4	M	1.99
5	M	2.47
6	M	2.39
7	M	1.83
8	M	2.27
9	M	3.54
10	M	2.88
	Mean ± S.D.	2.72 ± 0.70

Table 2

SOD activity in erythrocytes of patients with trisomy G 21 and unbalanced translocation G 21/22 (units per g hemoglobin)

No.	Sex	SPD activity · 10^{-3}
	Trisomy	
1	F	5.24
2	F	5.62
3	F	3.85
4	F	4.84
5	F	5.21
6	F	4.22
7	M	4.95
8	M	4.17
9	M	4.58
10	M	6.78
11	M	7.50
12	M	4.06
13	M	3.13
Mean ± S.D.		4.93 ± 1.21
	Unbalanced translocation	
1	F	4.82
2	M	4.00
Mean		4.41

typical of the group with unbalanced translocation G 21/22 (Tab. 1 and 2). Nevertheless, the fact that in patients with unbalanced translocation G 21/22 a tendency to lower the level of SOD-1 activity exists must not be disregarded.

Discussion

The obtained data prove that the average activity of SOD-1 in erythrocytes of our control group ($2.72 \cdot 10^3$ units per g hemoglobin) corresponds with the value obtained by WINTERBOURN et al. [11] ($2.9 \cdot 10^3$ units per g hemoglobin). It should be also stressed that the level of SOD-1 activity seems to be independent of sex what is confirmed in the results of both WINTERBOURN et al. [11] and ours.

The results obtained in our investigation performed on patients with trisomy G 21 correspond with previous observations [8,12] showing that the level of SOD-1 activity in patients with Down syndrome is significantly increased. The average ratio of SOD activity in erythrocytes of trisomic and of controls equals to 1.81 in our study. Although this value is higher than the theoretically foreseen value of 1.50, the observed deviation has no statistical significance ($p > 0{,}2$) as estimated by means of one-tailed Student's test. In two patients with translocation G 21/22 an increased SOD-1 level has been observed, which approximated to the level typical of patients with trisomy G21.

This finding is astonishing indeed, considering the fact that the previous data obtained from tests performed on the same group of patients with unbalanced translocation indicate that the IPO-A activity level is on the borderline of normal values, whereas in patients with trisomy G 21 a distinct increase of IPO-A activity levels was observed [10].

The data contained in recent references indicate that IPO-A and SOD-1 have identical gene loci [6, 13]. In the interpretation of the present results, possible inadequacy of our methods of estimation IPO-A and SOD-1 must be taken into consideration. It seems clear enough, however, that the above mentioned results may contradict the thesis putting an equality mark between IPO-A and SOD-1.

It is worthwhile considering that in our previous investigation we have found that copper in plasma of patients with trisomy G 21 is not different from the level typical of healthy individuals [14].

In spite of this, the level of erythrocuprein (SOD-1) in erythrocytes of those patients is significantly higher, which confirms the thesis of ALEXANDER and BENSON [15] based on the estimation of SOD level and ceruloplasmin in Wilson disease, that SOD-1 level in erythrocytes is independent of copper concentration in plasma.

References

[1] HARRIS, H.: The principles of human biochemical genetics. North-Holland Publishing Co., Amsterdam, London 1971
[2] HSIA, D. Y. Y., T. INOWYE, P. WONG, and A. SOUTH: New Engl. J. Med. **270**, 1985 (1964)
[3] HSIA, D. Y. Y.: Inborn errors of metabolism. Year Book Medical Publishers Inc., Chicago 1966
[4] LEJEUNE, J., N. GAUTIER, and R. TURPIN: Compl. Acad. Sci. **248**, 1721 (1959)
[5] LEJEUNE, J.: Pediatrics **32**, 326 (1963)
[6] TAN, Y. H., J. TISCHFIELD, and F. H. RUDDLE: J. exp. Med. **137**, 317 (1973)
[7] SICHITIU, S., P. M. SINET, J. LEJEUNE, and J. FREZAL: Humangenetik **23**, 65 (1974)

[8] SINET, P. M., D. ALLARD, J. LEJEUNE, and H. JERÔME: C. r. hebd. Séanc. Paris **278**, 3267 (1974)
[9] SINET, P. M., F. LAVELLE, A. M. MICHELSON, and H. JERÔME: Biochem. biophys. Res. Commun. **67**, 904 (1975)
[10] KĘDZIORA, J., D. ROŻYNKOVA, M. KOPFF, and J. JESKE: Hum. Genet. (in press)
[11] WINTERBOURN, C. C., R. E. HAWKINS, M. BRIAN, and R. W. CARRELL: J. Lab. clin. Med. **85**, 337 (1975)
[12] FRANTS, R. R., A. W. ERIKSSON, P. H. JONGBLOET, and A. J. HAMERS: Lancet **1975**/II, 42
[13] FRIDOVICH, I.: Adv. Enzym. **41**, 35 (1974)
[14] WACHOWICZ, B., and J. KĘDZIORA: Endokryn. Pol. **25**, 9 (1974)
[15] ALEXANDER, N. M., and G. D. BENSON: Life Sci. **16**, 10025 (1975)

Acta biol. med. germ., Band 36, Seite 783—791 (1977)

Marzinovskij-Institut für medizinische Parasitologie und tropische Medizin, Moskau, UdSSR

Funktionelle Besonderheiten normaler und pathologischer Erythrozyten

F. F. Soprunov, Je. I. Benkowitsch und T. I. Kazarinskaja

Zusammenfassung

Die Abgabe von Sauerstoff mittels polarographischer Titration von Blut und konzentrierten Hb-Lösungen wurde in einer geschlossenen Zelle untersucht. Der Einfluß der Temperatur, von Polyäthylenglykol und verschiedener Metaboliten auf das desoxygenierte Tetramer wurden spektrometrisch und polarographisch registriert. Besonderheiten des O_2-Transports und der O_2-Abgabe bei β-Thalassemie, funktioneller Hypoxie und bei Cooley-Krankheit werden beschrieben.

Einleitung

Die Regulationsmechanismen, die die genaue Adaptation der Erythrozyten an die zu erfüllende Funktion gewährleisten, sind wenig bekannt, im wesentlichen deshalb , weil keine einfache und empfindliche Methode zur Untersuchung des Sauerstofftransports besteht. Für die Erfoschung der Mechanismen, die die Abgabe von Sauerstoff in der normalen Zelle, bei β-Thalassämie, Glukose-6-phosphat-Dehydrogenase-Defekten und anderen Erkrankungen regulieren, wurde die früher von uns vorgeschlagene Methode der polarographischen Titration benutzt [1—3].

Material und Methodik

Verwendet wurden Erythrozyten oder konzentrierte Hämoglobinlösungen (8—20 mM Häm) von Spendern oder Kranken. Die Titration wurde bei konstantem *p*H-Wert in einer geschlossenen Zelle mit Dithionit, Hydrazin oder unter Verwendung von Hefen bzw. eines Systems Mitochondrien + Atmungssubstrat durchgeführt. Die polarographische Titrationskurve wurde mit 1 ml Blut im Verlaufe von etwa 10 min aufgezeichnet. Auf der Grundlage der Kurve, die einer Oxygenierungskurve mit 50—100 Gleichgewichtszuständen entspricht, wurden die Affinitätskonstanten K_1 und K_4 sowie P_{50} errechnet und die Steigung in der mittleren Sättigungszone bestimmt. Anstelle des Hill-Koeffizienten wurde zur Auswertung das Verhältnis K_4/K_1 berechnet.

Der gelöste Sauerstoff stellt nur einen kleinen Bruchteil des gesamten Sauerstoffs des Blutes dar. Wird die Affinität des Hämoglobins zum Sauerstoff verändert, so ändert sich die Menge des gelösten Sauerstoffs in einer geschlossenen Zelle sehr wesentlich: bei einer Änderung der Hämoglobinkonzentration um 2% wird die Menge des gelösten Sauerstoffs um etwa 100% erhöht. Die polarographische Titration in einer geschlossenen Zelle ist im Vergleich zu den gebräuchlichen spektrophotometrischen oder volumetrischen Methoden ein schnelles und sehr empfindliches Verfahren.

Die Absorptionsspektren wurden mit dem Specord UV-VIS, die Streuungs- und Fluoreszensspektren mit dem Hitachi MPF-3 aufgezeichnet. Die Absorptionsspektren wurden in Bereich von 330—775 nm, die Anregungsspektren im Bereich von 580—800 nm, d. h. in dem Bereich, in dem der Tetrapyrrolring fluoresziert, registriert. Zur chromatographischen Trennung wurden Proben der Hämoglobinlösung aus der Titrationszelle ohne Sauerstoffzutritt entnommen und auf Säulen (2 × 50 cm) mit DEAE-Zellulose aufgetragen, die zuvor mit einer Lösung äquilibriert wurden, deren pO_2 bekannt war. Der Partialdruck des Sauerstoffs, der sich in den einzelnen Säulen einstellte, wurde polarographisch in der ein- und auslaufenden Flüssigkeit kontrolliert.

Ergebnisse

In frisch entnommenem Blut sind die Werte für die erste und vierte Affinitätskonstante fast gleich (Abb. 1), was einer Vergrößerung der Affinität des Hämoglobins um das 16fache bei seiner Oxygenierung entspricht. In frisch entnommenem Blut von 30 gesunden Spendern wurde gefunden:

$K_1 = 1{,}7 \pm 0{,}5 \cdot 10^4 \cdot M^{-1}$; $K_4 = 1{,}6 \pm 0{,}3 \cdot 10^4 \cdot M^{-1}$.

In 2 Tage konserviertem Blut wurden folgende Werte bestimmt:

$K_1 = 1{,}07 + 0{,}10 \cdot 10^4 \cdot M^{-1}$ und $K_4 = 2{,}08 \pm 0{,}13 \cdot 10^4 \cdot M^{-1}$.

Bei Aufbewahren des Blutes bei einem pH-Wert < 7,6 unter praktisch anaeroben Bedingungen verringert sich der Wert der ersten Konstante stark, was sich durch eine Anhäufung von 2,3-DPG erklären läßt.

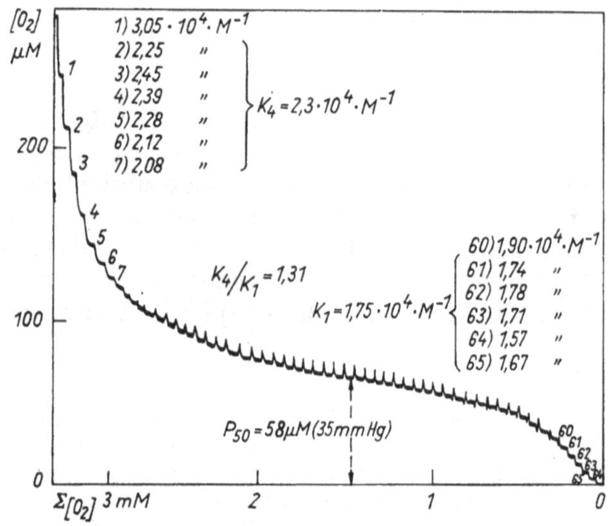

Abbl. 1. Polarographische Titrationskurve des Blutes eines Donors.
Die Zelle enthielt 3,4 ml heparinisiertes Blut, 2 : 1 mit physiologischer NaCl-Lösung verdünnt; pH = 7,3; pH der gepufferten Dithionit-Lösung = 7,35; Temperatur 21 °C

Bei Temperaturerhöhung verändern sich die scheinbaren Affinitätskonstanten des Blutes der gesunden Probanden; die Ergebnisse sind in Tab. 1 dargestellt.

Tabelle 1
Einfluß der Temperatur auf die Sauerstoffaffinität des Hämoglobins

	23 °C	28 °C	34 °C	41 °C
K_1 (1—2 Std. nach Blutentnahme)	2,28	1,55	1,19	1,10 · 10⁴ · M⁻¹
K_1 (nach 24 Std.)	0,85	0,74	0,54	0,48 · 10⁴ · M⁻¹
K_4	1,21	1,23	1,59	1,84 · 10⁴ · M⁻¹
K_4/K_1	0,53	0,8	1,3	1,7
$P_{50(\text{mmHg})}$	20	22	28	32

Wir fanden keinen Einfluß der Temperatur auf die scheinbaren Affinitätskonstanten, wenn der *p*H-Wert des Blutes von 7,2 auf 7,6 erhöht wurde.
Die Affinitätskonstanten und die Form der polarographischen Titrationskurve wurden bei 50 Blutspendern und bei einigen Gruppen von Kranken analysiert (Tab. 2). Die vierte Affinitätskonstante wird durch den Einfluß äußerer Faktoren im Gegensatz zur ersten Affinitätskonstante kaum verändert. Letztere dagegen verändert sich unter Hypoxiebedingungen und bei Aufbewahrung des Blutes.

Tabelle 2
Scheinbare Affinitätskonstanten des Blutes unter normalen und pathologischen Bedingungen

Anzahl der untersuchten Blutproben		K_1 (10^{-1} M)	K_4	K_4/K_1
Gesunde Blutspender	50	1,07 ± 0,10 (24 Std. nach Blutentnahme)	2,08 ± 0,13	1,9
mit Brandwunden	15	1,20 ± 0,13	2,19 ± 0,19	1,8
mit Neoplasmen	15	1,51 ± 0,15	1,99 ± 0,16	1,3
mit funktioneller Hypoxie	8	0,61 ± 0,08	2,16 ± 0,12	3,5
Hämoglobin F	4	3,58	2,94	0,8

$t°$ war 22—26 °C; *p*H = 7,3

Es ist bekannt, das DPG, ATP und andere Effektoren mit der räumlichen Struktur des Tetramers, die dem desoxygenierten Zustand des Hb entspricht, in Wechselwirkung stehen. Kann aber diese räumliche Struktur unter physiologischen Bedingungen, wenn das Hämoglobin im Durchschnitt immer über 55% oxygeniert ist, entstehen? Um diese Möglichkeit zu prüfen, beschickten wir DEAE-Zellulosesäulen mit halboxygenierten Hb-Lösungen, die der geschlossenen Titrationszelle entnommen wurden. Oxyhämoglobin durchläuft die oxygenierte Säule als einheitliche Bande. Ebenso durchläuft Desoxyhämoglobin in Abwesenheit von Sauerstoff die Säule. Teilweise oxygeniertes Hämoglobin durchläuft die Säule, die bei niedrigem Partialdruck des Sauerstoffs ins Gleichgewicht gebracht wird, in zwei Hauptbanden. Vereinzelt war das Auftreten von mehreren kleineren Fraktionen zu beobachten, die zwischen den beiden stark gefärbten Hauptfraktionen lagen, welche ihrerseits sogar bei einem Abstand von 5 cm voneinander durch eine schwach rosa gefärbte Übergangszone verbunden blieben.
Proben, die einen verschiedenen Oxygenierungszustand aufweisen, wurden weiterhin zur Überprüfung der Streuungs-, Fluoreszens- und Absorptionspektren benutzt.
Im Bereich von 600—800 nm waren Fluoreszensspektren für (O_2, CO, CN-, NO-) ligandierte Proben nachweisbar, deren Intensität vom Sättigungsgrad abhängig war. Eine quantitative Beurteilung war jedoch durch die intensive Absorption und Streuung erschwert.
Die Veränderungen des Absorptionsspektrums wurden aus den in 5—10 Punkten im Prozess der Oxy- oder Desoxygenierung des Blutes oder Hämoglobins auf-

gezeichneten Spektren beurteilt. Für die quantitative Bestimmung wurden die siosbestischen Punkte und die Serien vorangegangener Standaufzeichnungen bei wachsender Konzentration von Oxy- und Desoxyhämoglobin genutzt. In Tab. 3 sind die Resultate von Berechnungen dargestellt, die das Ergebnis der polarographischen Titration in der geschlossenen Zelle darstellen. Sie sind das Resultat einer geometrischen Analyse der Veränderungen der Absorptionskurven im Bereich der α- und β-Peaks, die auf der Grundlage eines geschätzten Wertes der Intensität der Banden bei der Säulchenchromatographie erhalten wurden. Die im großen und ganzen gute Übereinstimmung deutet auf eine geringe Wahrscheinlichkeit des Vorhandenseins von Tetrameren im halboxygenierten Zustand hin.

Tabelle 3
Veränderungen der räumlichen Struktur des Tetramers im Laufe der Desoxygenierung

pO_2	Analyse der polarographischen Titrationskurven:		Analyse der Veränderung der Absorptionsspektren der α- und β-peaks:		Ergebnisse der Säulenchromatographie auf DEAE-Zellulose: Hb^{II}/Hb^{I}
	y% erhalten	Hb^{II}/Hb^{I} kalkuliert	y% kalkuliert	Hb^{II}/Hb^{I} erhalten	
49	94	24	93	20	
33,6	78	3,8	80	2,5	>2
27	62,4	1,6	66	1,5	
24,5	59,1	1,4	57,5	1,2	≈1
17	38,3	0,65	37,5	0,6	
11,2	21,8	0.28	23	0,25	<0,3
5,1	5,6	0.01	7,5	0,05	

Auf der Grundlage der Gleichgewichtszustände (Adair's Polynom) wurde die Veränderung der Oxygenierungszustände des Tetramers bei wachsendem pO_2 in einer geschlossenen Zelle mit der Hilfe eines Rechners modelliert. Die experimentell erhaltene Werte der Affinitätskonstanten sind in Übereinstimmung mit der Annahme von $Hb_4\,(O_2)_2$ in halboxygeniertem Zustand des Tetramers (Abb. 2, III). Wie bekannt ist, sind die Tetramere im Erythrozyten dicht gepackt (der Abstand zwischen den Tetrameren im Kristall ist etwa 15 Å, im Erythrozyten 27 Å) und der Übergang von einer räumlichen Struktur in die andere ist mit intermolekularen Wechselwirkungen, in denen die Hydrathülle der Tetramere eine wesentliche Rolle spielt, verbunden [4].

Wir untersuchten deshalb welchen Einfluß die Beseitigung der Hydrathülle und die Lockerung der Wasserstoffbrückenbindungen des Tetramers auf die optischen Eigenschaften und Ligandenaffinität hat. Die Erythrozyten des Menschen wurden in physiologischer Kochsalzlösung gewaschen, in einem minimalen Volumen destillierten Wassers hämolysiert und das erhaltene Hämoglobin sofort auf Sephadex G-50 oder Akrylex P-10 absorbiert, um das Wasser und die niedermolekularen Verbindungen zu entfernen. Hämoglobin wurde mit einem geringen Volumen Wasser eluiert und so eine konzentrierte Lösung erhalten. Zur Auf-

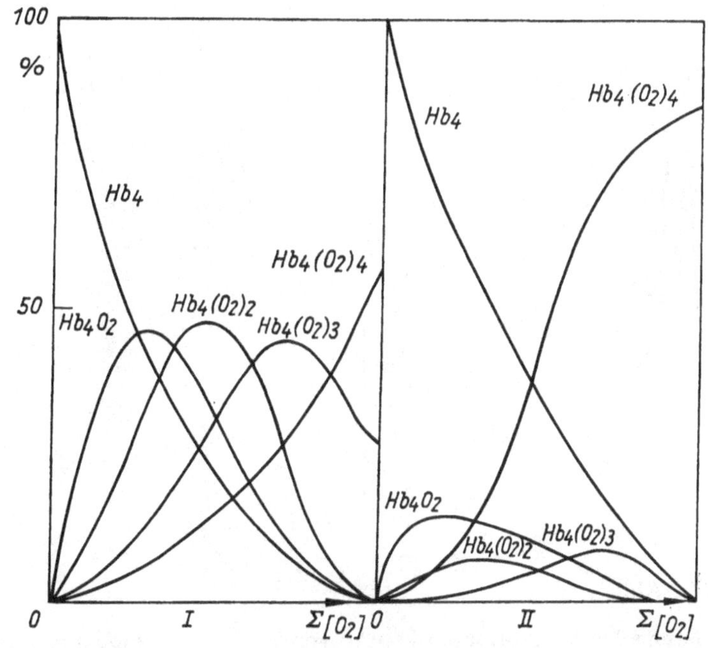

Abb. 2. Rechnerische Modellierung der Veränderung des Oxygenierungszustandes des Tetrames bei der Oxygenierung des Blutes in einer geschlossenen Zelle.
I = Keine Häm-Häm Wechselwirkung; II = Berücksichtigung experimentell erhaltener Affinitätskonstanten

spaltung von Wasserstoffbrückenbindungen, wurden Polyäthylenglykol und Harnstoff verwendet. Es wurden Polyäthylenglykole von Diäthylenglykol bis PEG mit einem Molekulargewicht von 6000 erprobt. Harnstoff wurde in 4molarer Konzentration verwendet.

Mit steigender PEG-Konzentration geht eine Veränderung in der Struktur des Tetramers vor sich, die man an der Veränderung des Hämspektrums verfolgen kann. In der ersten Phase sinkt die Affinität des Tetramers zu Sauerstoff um 30 bis 40%. Dieser Prozeß ist vollständig reversibel, das Tetramer behält die Fähigkeit der Bindung und Abgabe des Sauerstoffs in vollem Umfange bei. Bei höheren Konzentrationen von PEG (40—50%) verändert sich das Absorptionsspektrum stark: es verschwindet die breite Bande bei 559 nm und es erscheinen zwei deutliche Banden bei 533 nm (E_{mM} = 10,2) und 562 nm (E_{mM} = 38,7). Die Lage der Soret-Bande ändert sich um 3 nm. Das Eisen liegt dabei in zweivalenter Form vor, der *p*H-Wert lag oberhalb 7,0. Eine Erhöhung des *p*H-Wertes bewirkt das Auftreten von 2 Absorptionsbanden ebenso wie eine Erhöhung der Hämoglobinkonzentration, der Ionenstärke und der Temperatur. Bei einer langsamen Veränderung eines dieser Faktoren, die auf das Gleichgewicht zwischen 2 Hämoglobinstrukturen einwirken, kann man Übergangszustände mit entsprechenden Spektren erhalten (Abb. 3).

Nach vorrausgegangener Oxydation des Hämeisens mit Hilfe von $K_3(CN)_6Fe$ oder $NaNO_2$, nach Überführung des Hämoglobins in CN-Hb oder CO-Hb, kann PEG das Erscheinen des charakteristischen Spektrums mit den Maxima bei

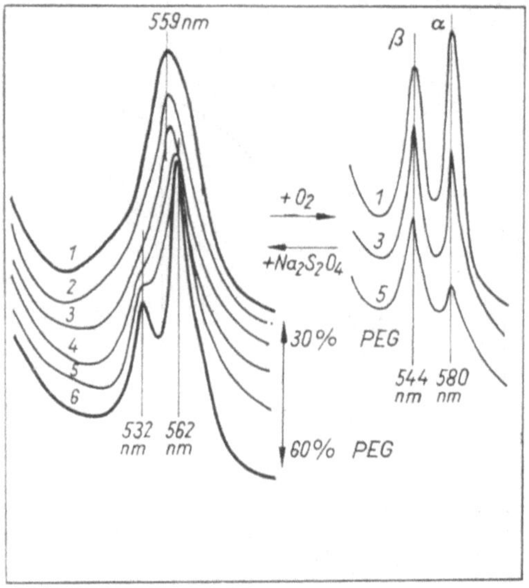

Abb. 3

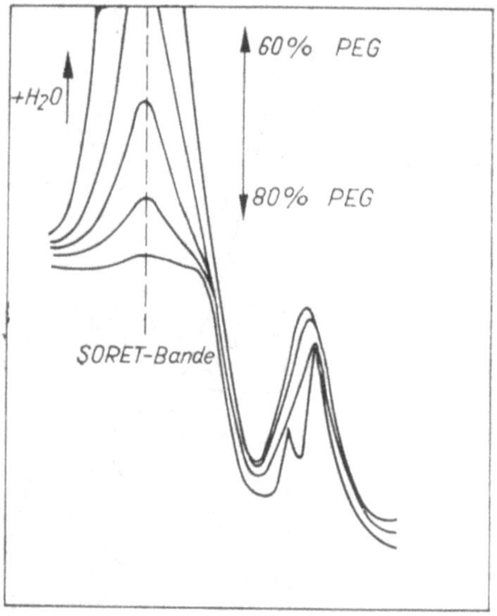

Abb. 4

Abb. 3. Veränderung des Absorptionsspektrums von Desoxyhämoglobin in Anwesenheit von Polyäthylenglykol (PEG) bei *p*H 8

Abb. 4. Veränderung des Absorptionsspektrum von Desoxyhämoglobin bei der Bildung supramolekularer Aggregate bei *p*H 7,6

533 nm und 562 nm nur unter reduzierenden Bedingungen d. h. bei Zugabe von $Na_2S_2O_4$ und NH_2-NH_2H_2O hervorrufen [5, 6].

Es ist wahrscheinlich, daß unter dem Einfluß von PEG eine teilweise Despiralisierung stattfindet, die zu einer Verschiebung des E-Abschnittes der Globinkette führt. Infolgedessen nähert sich das distale Histidin E_7(58), welches mit der Seitenkette des Häms in der α-Kette unter normalen Bedingungen zusammenwirkt, dem Fe^{2+} und konkurriert mit den Liganden um die sechste Valenz des Eisens. Das Zusammenwirken von Histidin E_7 mit Fe^{2+} ist im physiologischen *p*H- und Temperaturbereich umkehrbar, was sich in der umkehrbaren Veränderung der Affinität des Hämoglobins zu Liganden ausdrückt.

Wenn ein Überschuß von PEG zugegeben wird, werden die Veränderungen des Tetramers irreversibel. Es sinkt die Löslichkeit des Hämoglobins in Wasser und die Affinität zu Sauerstoff und anderen Liganden geht vollständig verloren. Wir nehmen an, daß eine starke Bindung zwischen Histidin E_7 und Fe^{2+} entstanden ist und daß die Tetramere supramolekulare Aggregate bilden, was durch die Entfernung der Hydrathülle erleichtert, wird. Die Bildung der supramolekularen Aggregate kann man an der Intensitätsminderung der Soret-Bande verfolgen (Abb. 4).

Unter physiologischen Bedingungen wirken Metabolite, die im Erythrozyten enthalten sind, als Effektoren, die die Dissoziationskurve des Hämoglobins verändern. Das kann leicht in einer geschlossenen Zelle mit konzentrierter Hämoglobinlösung in halboxygeniertem Zustand gezeigt werden. Werden 2,3-DPG,

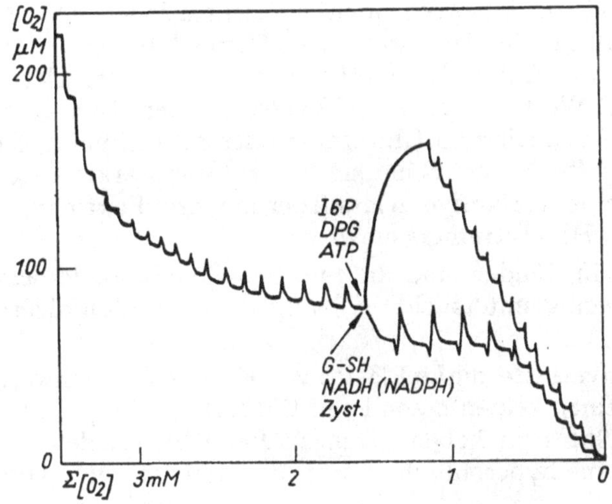

Abb. 5. Veränderung der Titrationskurve einer konzentrierten Hämoglobinlösung nach Zugabe von Metaboliten bei *p*H 7,3

Tabelle 4

Einwirkung von Faktoren auf die Struktur des Desoxyhämoglobins und auf seine Affinität zu Liganden

Veränderung des Redox-Potential des Milieu	← $Na_2S_2O_4$ ← $NH_2–NH_2 \cdot H_2O$			→ O_2	
Eh (V)	−0,6	−0,4	0	+0,4	+0,8
Zusätzliche Faktoren	*p*H, t^0, Ionenstärke, Hb-Konzentr., PEG, Harnstoff				
Biologische Faktoren	2,3-DPG, ATP, NADH/NAD, G-SH/GSSG, t^0, *p*H, Mg^{++}				
Absorptionsspektrenveränderung im Bereich 500—600 nm	Spektrum des Hämochrom	Übergangsspektrum	Spektrum des Desoxyhämoglobin	Übergangsspektrum	Spektrum des Oxyhämoglobin
Oxygenierungszustand	Aggregation der Tetrameren bei PEG-Überschuß	Hb_4	Hb_4	$Hb_4(O_2)_2$	$Hb_4(O_2)_2$
Veränderung der Affinität zu Sauerstoff	0%	3%	6,2%	40%	100%
Wechselwirkung mit Liganden	CN	CO	O_2		CN–Hb CO–Hb O_2–Hb

ADP, ATP, G-6-P in entsprechenden Konzentrationen den Proben zugesetzt, findet eine Abspaltung des Sauerstoffs vom Hämoglobin statt mit einem entsprechenden Anstieg des freien Sauerstoffs in der Zelle. Der gleiche Effekt ist bei Senkung des *p*H-Wertes oder einer Erhöhung der Temperatur nachweisbar. Nach Zugabe von reduziertem Glutathion oder NADH nahm die Konzentration des freien Sauerstoffs ab, der „Schaukel-Effekt" war stärker ausgeprägt (Abb. 5). Tab. 4 enthält eine vorläufige Klassifizierung der Faktoren, die auf Struktur und Affinität des Hb_4-Tetramers einwirken.

Bei krankhaften Zuständen sind die funktionellen Besonderheiten der Erythrozyten noch zu wenig untersucht, um allgemeine Schlußfolgerungen ziehen zu können.

Wasser und Ionenverluste, die im Plasma von Kranken mit schweren Brandwunden vorkommen, scheinen keinen besonderen Einfluß auf die polarographische Titrationskurve des Blutes zu haben (Tab. 2). Bei HbF (Cooley-Krankheit) hat die Titrationskurve eine hyperbolische Form; die erste Affinitätskonstante hat einen hohen Wert.

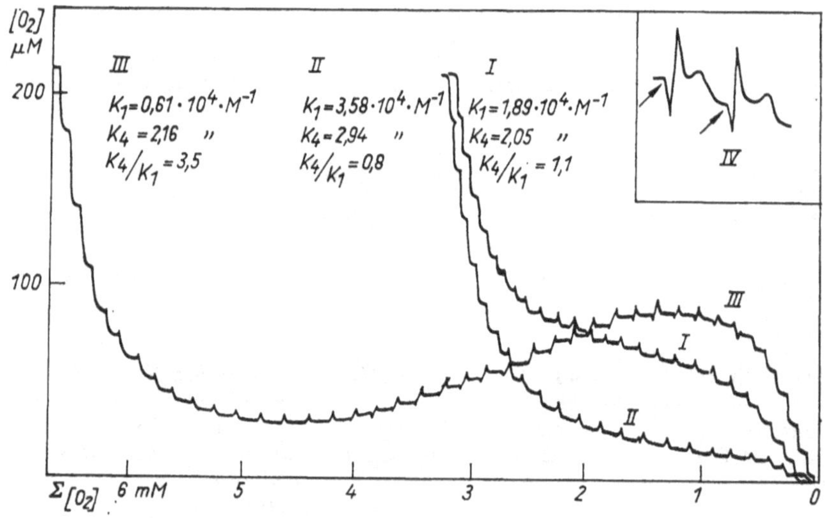

Abb. 6. Polarographische Titrationskurven von Patientenbluten.
I = Kontrollblut; II = Hb F (Cooley-Krankheit); III = schwere funktionelle Hypoxie; IV = Effekt bei Lungenfibrose und Bronchektasie.
$t = 29$ °C, *p*H 7,3

Zwei Fälle von Lungenfibrose und Lungensilikose bei jungen Probanden wurden genauer untersucht. In beiden Fällen war ein schwerer Sauerstoffmangel mit Hyperhämoglobinämie — 23 und 24 g% Hb — und mit respiratorischer Alkalose vorhanden. In beiden Fällen hatte die polarographische Titrationskurve eine paradoxe Form: die Sättigung des Blutes mit Sauerstoff erfolgte bei einem niedrigerem pO_2 als die Sauerstoffabgabe (Abb. 6).

Literatur

[1] SOPRUNOV, F. F.: Biofizika **1**, 153 (1973)
[2] SOPRUNOV, F. F.: Biochemie **56**, 355 (1974)
[3] SOPRUNOV, F. F. in: VII. Internationales Symposium über Struktur und Funktion der Erythrozyten, Abhandlungen der Akademie der Wisschenschaften der DDR, Jahrgang 1973, Akademie-Verlag, Berlin, 1975, S. 205
[4] GARY-BOBO, C. M., u. A. K. SALOMAN: J. gen. Physiol. **57**, 283 (1971)
[5] VON ZEYNEK, R.: Hoppe-Seyler's Z. physiol. Chem. **70**, 224 (1910)
[6] MYER, Y. P., u. H. A. HARBURG: Ann. N.Y. Acad. Sci. **206**, 685 (1973)

Summary

F. F. SOPRUNOV, Je. I. BENKOWITSCH, and T. I. KAZARINSKAJA: Oxygen transport regulation in erythrocytes under physiological conditions and in disease

Polarographic titration in a closed cell of blood samples and concentrated Hb solutions was used to study the regulation of the release of oxygen by the red cell. The effect of temperature, polyethylenglycol and metabolites normally found in the erythrocyte was determined by spectrophotometric and polarographic methods. Some pecularities of O_2-transport and release in β-thalassemia, functional hypoxia and Cooley-disease are described.

Acta biol. med. germ., Band 36, Seite 793—795 (1977)

Forschungsabteilung der Kinderklinik des Bereichs Medizin der Humboldt-Universität zu Berlin[1], DDR,
Medizinisch-Chemisches Institut der Universität Bern[2], Schweiz
Poliklinik für konservierende Stomatologie[3] und
Kinderklinik der Karl-Marx-Universität Leipzig[4], DDR

Charakterisierung der Katalase roter Blutzellen eines Patienten mit den Symptomen einer Takahara-Krankheit

J. Gross[1], B. Scherz[2], Sonja Wyss[2], W. Künzel[3], H. J. Maiwald[3], A. Hartwig[1] und H. Polster[4]

Zusammenfassung

Die Katalaseaktivität von roten Blutzellen des Patienten R. G., der 1969 als Akatalasie beschrieben worden war, zeigte bei einer 4—6 Jahre später durchgeführten Nachuntersuchung normale Werte.
Die bei der Nachuntersuchung vorgenommene Charakterisierung des Enzyms hinsichtlich Hitzestabilität, Veränderungen der Enzymaktivität während Reifung und Altern der roten Blutzellen und immunchemischer Reaktion gegenüber Antikatalase ergab unauffällige Befunde. Als einzige Abweichung fanden wir eine gegenüber der Norm verminderte elektrophoretische Wanderungsgeschwindigkeit, ähnlich der Allokatalasie. Die erhobenen Befunde sprechen dafür, daß es sich bei dem Patienten z. Zt. nicht um eine Akatalasie im klassischen Sinne handeln kann. Es wird diskutiert, daß z. Zt. der Erstbeschreibung ein Katalasedefekt vorlag und daß im Laufe des Lebens katalasehaltige Zell-Populationen auftraten.

Einleitung

1969 berichteten Polster et al. [1] über eine Akatalasie eines 4jährigen Jungen (R. G.), der durch eine charakteristische klinische Symptomatik im Sinne der Takahara-Krankheit aufgefallen war.
Bei einer Nachuntersuchung im Jahre 1971 wurde überraschenderweise bei unverändertem klinischen Bild eine normale Katalaseaktivität des Blutes nachgewiesen.
Um mögliche Besonderheiten der biochemischen Eigenschaften der Erythrozytenkatalase dieses Patienten auszuschließen, wurde das Enzym auf seine Hitzestabilität, die elektrophoretische Wanderungsgeschwindigkeit und die Veränderungen der Enzymaktivität während Reifung und Altern der Zellen untersucht.

Methoden und Material

Die Bestimmung der Katalaseaktivität (E. C. 1.11.1.6) im Hämolysat erfolgte mit der titrimetrischen Methode mit $KMnO_4$ oder mit der spektrophotometrischen Methode [2].
Die Prüfung der Hitzestabilität erfolgte im Hämolysat (Hämolyse in Aqua dest., Hämoglobinkonzentration ca. 5 g/100 ml) und an einem mittels DEAE-Zellulose gereinigten Katalasepräparat [2]. Die Trennung der roten Blutzellen nach Reifegrad und Alter erfolgte im kontinuierlichen 50 ml Dextrandichtegradienten [3]. Zur Charakterisierung der getrennten Zellpopulationen wurde die Retikulozytenzahl und die Aktivität der Glukose-6-phosphat-Dehydrogenase (E.C. 1.1.1.49) bestimmt.

Die Stärkegel-Elektrophorese wurde nach der Vorschrift von Smithies [4] in Tris-Borat-ÄDTA-Puffer, *p*H 8,5 ausgeführt.
Die Doppel-Immuno-Diffusion nach Ouchterlony [5] wurde in Oxoid Ionagar Nr. 2, Natrium-Veronal-Puffer (0,05 M), *p*H 8,6, durchgeführt. Anti-Katalase-Serum wurde durch Immunisierung von Kaninchen mit gereinigtem Katalasepräparat gewonnen [6].

Ergebnisse und Diskussion

In Tab. 1 sind die Ergebnisse der Nachuntersuchungen der Katalaseaktivität zusammengestellt. Es wird deutlich, daß sowohl der Patient G. als auch die Familienmitglieder normale Katalaseaktivitäten aufweisen. Zum Ausschluß anderer erythrozytärer Enzymdefekte wurden noch die Glukose-6-phosphat-Dehydrogenase (G-6-PD), die 6-Phosphoglukonat-Dehydrogenase, die $NADPH_2$-abhängige Glutathionreduktaseaktivität und die Pyruvatkinase bestimmt, die bis auf eine gering verminderte G-6-PD-Aktivität unauffällige Werte zeigten.
Um das Vorliegen einer labilen Enzymvariante auszuschließen, untersuchten wir die Hitzestabilität des Enzyms. Bekanntlich kann die Prüfung der Hitzestabilität als empfindlicher Test für die Aufdeckung von Katalasevarianten herangezogen werden.

Tabelle 1
Katalaseaktivität in roten Blutzellen von R. G. und Familienangehörigen

Proband	Katalaseaktivität (mg Katalase/gHb)
R. G. (Patient)	2,32
W. C. (Vater)	1,58
C. G. (Mutter)	1,72
U. G. (Schwester)	1,61
Normalwerte	2,38 ± 0,36

Tabelle 2
Inaktivierung von Erythrozytenkatalase nach 10min Inkubation bei 63 °C

Proband	Katalase-Aktivität (mg Katalase/g/Hb		Inaktivierung (%)
	vor Inkubation	nach Inkubation	
R. G.	2,47	1,40	56,7
Kontrollgruppe ($n = 9$)	1,92 ± 0,21	1,17 ± 0,15	61,5 ± 10,32

Tab. 2 zeigt, daß die Inaktivierung des Enzyms von R. G. nach Inkubation des Hämolysats bei 63 °C für 10 min etwa 57% beträgt und damit kein Unterschied zur Inaktivierung von Erythrozytenkatalase gesunder Normalpersonen besteht.
Die kompletten Hitzestabilitätskurven des partiell gereinigten Enzyms zeigten bei R. G. und bei Kontrollpersonen einen identischen Verlauf.
Auch die Untersuchung des Verhaltens der Katalaseaktivität während Reifung und Altern der roten Blutzellen gab keinen Hinweis auf Besonderheiten des Enzyms, da die Katalaseaktivität in jungen und alten Zellpopulationen nahezu identische Enzymaktivitäten aufwies [7].

Hingegen konnte bei dem Patienten eine deutlich verminderte elektrophoretische Wanderungsgeschwindigkeit beobachtet und in wiederholten Versuchen bestätigt werden [7]. Eine kausale Beziehung zur Erkrankung nehmen wir nicht an. Vielleicht handelt es sich um eine Enzymvariante ähnlich der von BAUER [8] beschriebenen Allokatalasie.

Gegen Anti-Katalase-Serum wurde eine komplette immunchemische Identität der Antigene gefunden. Alle vorliegenden biochemischen Befunde sprechen dafür, daß es sich bei dem Patienten R. G. z. Zt. nicht um eine Akatalasie im klassischen Sinne handeln kann.

Es ist möglich, daß z. Zt. der Erstbeschreibung ein Katalasedefekt vorgelegen haben kann. Mit zunehmendem Alter von R. G. traten dann im peripheren Blut katalasehaltige Zellpopulationen auf. MINTZ [9] nimmt an, daß rote Blutzellen klonaler Herkunft sind und im Laufe des Lebens verschiedene Klone aktiv werden.

Mitteilungen über einen Anstieg der Enzymaktivität im Laufe des Lebens bei einem erythrozytären Enzymdefekt sind uns bisher aus der Literatur nicht bekannt. JACOBASCH [10] beobachtete einen Patienten mit einem atypischen Pyruvatmangel, der im Säuglings- und Kleinkindesalter häufig niedrigere Maximalaktivitäten zeigte als jenseits des 10. Lebensjahres.

Sowohl die in der ersten als auch in der zweiten Dentitionsperiode nachgewiesenen klinischen Veränderungen unterschieden sich in keiner Weise von der Symptomatologie infantiler bzw. juveniler Parodontolysen anderer Genese, beispielsweise von den oralen Symptomen eines Ceratoma palmo-plantare. Das erste klinische Bild sowie der spätere Verlauf des parodontalen Zerstörungsprozesses entsprechen jedoch am ehesten der Takahara-Erkrankung.

Literatur

[1] POLSTER, H., K. BEYREISS u. R. J. NOSTITZ: Kinderärztliche Praxis **36**, 367 (1968)
[2] AEBI, H., SONJA R. WYSS, B. SCHERZ u. F. SKVARIL: Eur. J. Biochem. **48**, 137 (1974)
[3] GROSS, J., S. ROSENTHAL u. J. SYLLM-RAPOPORT: Acta biol. med. germ. **26**, 643 (1971)
[4] SMITHIES, O.: Biochem. J. **71**, 585 (1959)
[5] OUCHTERLONY, O.: Progr. Allergy **5**, 1 (1958)
[6] BEN-YOSEPH, Y., u. M. SHAPIRA: J. Lab. clin. Med. **81**, 133 (1973)
[7] GROSS, J., B. SCHERZ, S. WYSS, W. KÜNZEL, J. J. MAIWALD, A. HARTWIG u. H. POLSTER: Kinderärztliche Praxis (im Druck)
[8] BAUER, E. W.: Science, N.Y. **140**, 816 (1963)
[9] MINTZ, B.: Fedn Proc. Fedn Am. Socs exp. Biol. **30**, 935 (1971)
[10] JACOBASCH, G.: persönliche Mitteilung

Acta biol. med. germ., Band 36, Seite 797—800 (1977)

Institut für Physiologische und Biologische Chemie der Humboldt-Universität Berlin, 104 Berlin, DDR und
Institut für Laboratoriumsdiagnostik, 1115 Berlin-Buch, DDR

Screening auf Galaktose-1-phosphat-uridyl-transferase-Mangel (klassische Galaktosämie) bei Neugeborenen

I. Ahlbehrendt und C. Wagenknecht

Zusammenfassung

In der vorliegenden Arbeit wird ein Screening-Test beschrieben, der durch direkten Nachweis des Enzyms Galaktose-1-phosphat-uridyltransferase (E.C. 2.7.7.10) und Verwendung von Nabelvenenblut als Probenmaterial die Diagnose der Erkrankung bereits kurz nach der Geburt ermöglicht. Unter ca. 52000 Neugeborenen fanden wir 2 Fälle von klassischer Galaktosämie.
Seit Juni 1975 ist auch eine Differenzierung zwischen normaler und verminderter Enzymaktivität möglich. Bei 21 von ca. 15000 Probanden wurde bisher eine verminderte Enzymaktivität nachgewiesen.

Einleitung

Die klassische Galaktosämie ist eine angeborene Stoffwechselstörung mit autosomal-rezessivem Erbgang. Der biochemische Defekt besteht in dem Fehlen der Galaktose-1-phosphat-uridyltransferase (E.C. 2.7.7.10) in den roten Blutzellen und in der Leber und somit in der Unfähigkeit, Galaktose in Glukose umzuwandeln. Die Folge ist ein Anstau von Galaktose und besonders von Galaktose-1-phosphat, da die Kinasereaktion bekanntlich der einzige praktisch irreversible Reaktionsschritt in diesem Stoffwechselweg ist. Galaktose-1-phosphat ist ein Zellgift und wahrscheinlich vorwiegend für die klinische Symptomatik — Erbrechen, Gedeihstörungen, Reflexstörungen, Hepato(spleno)megalie, hämorrhagische Diathesen, Katarakte, Leberzirrhose — verantwortlich. Das Krankheitsbild ist nicht einheitlich, und man kann zwei Verlaufsformen unterscheiden. Bei der schweren Verlaufsform tritt die klinische Symptomatik ab 6. Lebenstag spontan auf, und innerhalb der ersten zwei Lebenswochen tritt der Tod ein [1, 2]. Kinder mit der sog. leichteren Form versterben im 2. oder 3. Lebensjahr an Leberzirrhose.
Durch laktose- bzw. galaktosefreie Ernährung kann die Manifestation des Krankheitsbildes verhindert werden. Voraussetzung ist die präsymptomatische Erfassung, so daß die Therapie möglichst vor Ausbildung der Symptome beginnt.
Auf Grund dieser Problematik haben wir einen Screening-Test entwickelt, mit dem der Enzymdefekt direkt nachgewiesen und durch Verwendung von Nabelvenenblut als Probenmaterial die Diagnose bereits kurz nach der Geburt gestellt werden kann. Der Enzymnachweis wird mit dem für das KAPA-System modifizierten Beutler-Test geführt. [3]

Abkürzungen: KAPA = Kammer-Papier-Analyse; NADP, $NADPH_2$ = Nikotinsäureamid-Adenin-Dinukleotid-Phosphat (oxydierte und reduzierte Form); UDPG = Uridindiphosphat-Glukose

Material und Methode

Das Prinzip des Tests beruht auf dem fluorometrischen Nachweis von $NADPH_2$, das in Gegenwart von Galaktose-1-phosphat, UDPG und Galaktose-1-phosphat-uridyltransferase aus NADP gebildet wird.

Probenmaterial ist ein Tropfen Nabelvenenblut, der bei der Geburt auf FN 7-Filterpapier (VEB Spezialpapierfabrik Niederschlag, DDR) aufgesaugt wird. Die Proben aller Neugeborenen Berlins gelangten täglich (außer samstags und sonntags) über den „Zentralen Hol- und Bringedienst" in das Untersuchungslabor. Aus den eingetrockneten Blutflecken werden 2 Blutplättchen von 1,8 mm Durchmesser in die KAPA-Kammer eingestanzt. Nach Zugabe von 10 µl Reagensgemisch (Galaktose-1-phosphat, UDPG, NADP; Leerwert ohne Substrat) wird 3 Std. bei 37 °C und unter Schütteln inkubiert. Die Reaktion wird durch Abklatschen des Reaktionsgemisches auf FN 3-Filterpapier und anschließende Trocknung unter Warmluft beendet. Unter UV-Licht (366 nm) müssen die Hauptwerte der normalen Proben eine strahlende Fluoreszenz zeigen (Abb. 1).

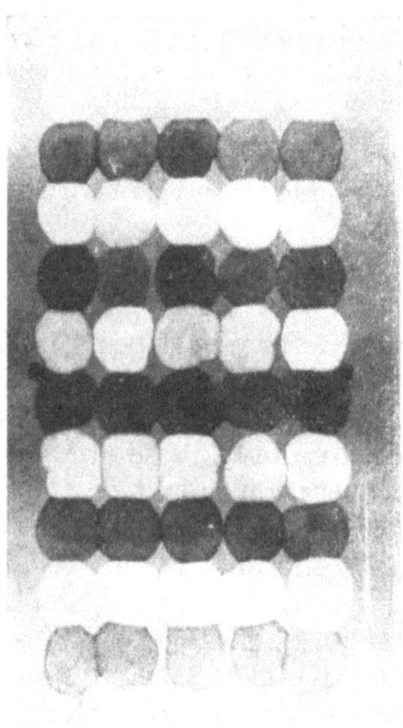

Abb. 1

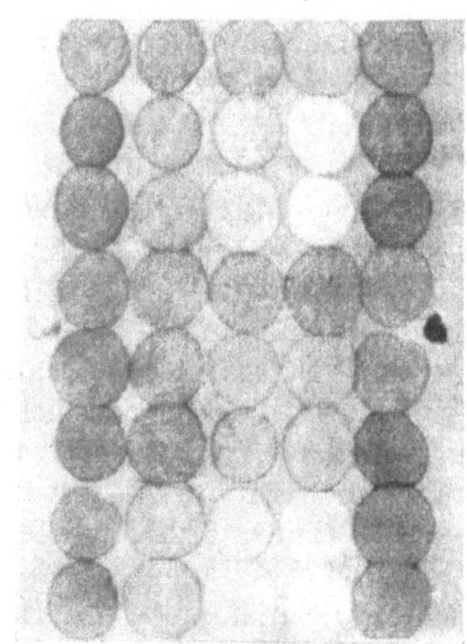

Abb. 2

Abb. 1. Filterpapier-Abklatsch des Screening-Tests nach 3 Std. Inkubation. HW = Hauptwert; LW = Leerwert

Abb. 2. Filterpapier-Abklatsch des gestaffelten Screening-Tests nach 1, 2 und 3 Std. Inkubation.
1 = verminderte Enzymaktivität; 2, 3, 7 und 8 = normale Enzymaktivität; 4 = keine nachweisbare Enzymaktivität; 5 und 6 = Eltern von 4, verminderte Enzymaktivität. LW = Leerwert

Weist eine Probe nach 3 Std. Inkubation eine verminderte Fluoreszenz auf (das entspricht einer verminderten Enzymaktivität), so wird in einem Wiederholungsansatz die Fluoreszenzbildung in 3 verschiedenen Blutplättchen nach 1, 2 und 3 Std. Inkubation bewertet. Während die normale Probe bereits nach 1 Std. eine deutliche Fluoreszenz zeigt, ist in Proben mit verminderter Enzymaktivität nur eine geringe bzw. keine Reaktion nachweisbar (Abb. 2).

Ergebnisse

Seit April 1972 wurden im systematischen Screening 52578 Neugeborene des Raumes Berlin auf klassische Galaktosämie untersucht. In diesem Zeitraum konnten 2 Fälle von klassischer Galaktosämie nachgewiesen werden (Tab. 1). Das erste Kind verstarb am 10. Lebenstag, da zu diesem Zeitpunkt die Testung noch aus Kapillarblut vom 5. Lebenstag erfolgte und die Probe zu spät ins Unter-

Tabelle 1

Ergebnisse des Neugeborenen-Screenings auf Galaktose-1-phosphat-uridyltransferase-Mangel

Untersuchungszeitraum:	April 1972—August 1976
Probandenzahl:	52578
Nicht nachweisbare Enzymaktivität (klassische Galaktosämie):	2
Untersuchungszeitraum:	Mai 1975—August 1976
Probandenzahl:	15179
Verminderte Enzymaktivität	21

Tabelle 2

Galaktose-1-phosphat-uridyltransferase-Aktivitäten von im Screening auffälligen Neugeborenen, ihren Eltern und Geschwistern

Proband	Mutter	Vater	Geschwister
1. ♀ 0, †	⇓	↓	♂ =
2. ♀ 0	⇓	⇓	—
3. ♂ ↓	=	↓	—
4. ♂ ↓	↓	=	—
5. ♀ ⇓	↓	⇓	—
6. ♀ ⇓	⇓	=	♂ ⇓ S. Nr. 19
7. ♂ ⇓	⇓	=	—
8. ♀ ⇓	=	⇓	♀ =
9. ♀ ⇓	⇓	=	—
10. ♀ ⇓	=	⇓	—
11. ♀ ⇓	⇓	=	—
12. ♂ ⇓	⇓	=	—
13. ♂ ⇓	⇓	=	♂ ⇓
14. ♂ ⇓	—	—	—
15. ♀ ⇓	—	—	—
16. ♀ ⇓	↓	⇓	♀ ↓
17. ♀ ⇓	⇓	↓	—
18. ♂ ⇓	⇓	↓	—
19. ♂ ⇓	⇓	=	♀ ⇓ s. Nr. 6
20. ♂ ⇓	↓	—	—
21. ♂ ⇓	—	—	—
22. ♀ ⇓	⇓	=	—
23. ♂ ↓	=	↓	—

0: keine nachweisbare Enzymaktivität; ⇒: stark verminderte Enzymaktivität; →: verminderte Enzymaktivität; =: normale Enzymaktivität; —: nicht untersucht.

suchungslabor kam. Ab Mai 1974 wurde die Testung generell aus Nabelvenenblut durchgeführt, so daß bei dem zweiten Kind das Ergebnis bereits am 3. Lebenstag (Wochenende) vorlag und präsymptomatisch mit der diätetischen Behandlung begonnen werden konnte.
Seit Juni 1975 ist eine Differenzierung zwischen normaler und verminderter Enzymaktivität möglich. Unter 15179 Neugeborenenen fanden wir 21 Probanden mit verminderter Enzymaktivität (Tab. 1). Diese Befunde ließen sich mindestens auch bei einem Elter nachweisen (Tab. 2). Die Geschwister der Neugeborenen wurden ebenfalls in die Untersuchung mit einbezogen.

Diskussion

Der Nachweis der klassischen Galaktosämie aus Nabelvenenblut mittels einer einfachen enzymatischen Methode bietet den Vorteil der präsymptomatischen Erfassung dieser Erkrankung und damit einer chancenreichen Therapie. Bei allen Methoden, die auf dem Metabolitnachweis (Galaktose) beruhen, liegt das Testergebnis erst nach Ausprägung der klinischen Symptomatik vor, so daß die therapeutischen Maßnahmen weniger erfolgversprechend sind oder bereits zu spät kommen [4].
Zur Klärung genetischer Zusammenhänge werden bei allen Probanden mit fehlender bzw. verminderter Enzymaktivität auch Blutproben der Eltern und Geschwister untersucht. Die Eltern beider Kinder mit klassischer Galaktosämie zeigten deutlich verminderte Galaktose-1-phosphat-uridyltransferase-Aktivitäten. Auch bei den Kindern mit verminderter Enzymaktivität konnte mindestens bei einem Elter ebenfalls eine verminderte Enzymaktivität nachgewiesen werden.
Der Test hat sich bei unseren Screening-Untersuchungen gut bewährt. Wie jedoch die Erfahrung zeigt, liegt die Schwierigkeit einer optimalen Testung nicht in der Analytik sondern vielmehr in der Organisation des Screenings. Wir arbeiten mit allen Geburtshilflichen Kliniken sehr eng zusammen. Zweimal jährlich erfolgt eine allgemeine Auswertung und eine Belehrung über die Ausführung und Durchführung des Tests.

Literatur

[1] STANBURY, J. B., J. B. WYNGAARDEN u. D. S. FREDRICKSON: The metabolic basis of inherited disease, 3rd ed., MacGraw-Hill Book Company, A Blakiston Publication, 1972
[2] SCHEIBENREITER, S. u. O. THALHAMMER: Wien. klin. Wschr. **84**, 93 (1972)
[3] URNER, U., I. AHLBEHRENDT, H. FRIEDEMANN u. C. WAGENKNECHT: Dte GesundhWes. **28**, 49 (1973)
[4] SCHÖN, R.: Wien. klin. Wschr. **88**, 274 (1976)

Acta biol. med. germ., Band 36, Seite 801—804 (1977)

Institut für Hygiene und Ernährung,
Medizinische Akademie, 1431 Sofia, Bulgarien

Uroporphyrinogen-I-Synthetase in den Erythrozyten bei akuter intermittierender Porphyrie

E. D. Iwanov, D. Adjarov und R. Tsanev

Zusammenfassung

Bei der akuten intermittierenden Porphyrie (AIP) liegt ein partieller Mangel an Uroporphyrinogen-I-Synthetase vor. Im Rahmen einer genealogischen Untersuchung bekannter Familien mit vorliegenden Erkrankungen dieser Porphyrie wurde die Uroporphyrinogen-I-Synthetase (EC 4.3.1.8) in Erythrozyten nach der Methode von Mandel et al. [8] bestimmt.

Außer bei drei Erkrankten wurde der Enzymdefekt auch bei drei Familienangehörigen festgestellt, bei denen die Anomalie latent ausgeprägt ist. Die Uroporphyrinogen-I-Synthetase-Bestimmung erwies sich als ein wertvoller Parameter zum Nachweis der latenten AIP.

Einleitung

Die akute intermittierende Porphyrie ist die häufigste angeborene Leberporphyrie. Sie wird autosomal dominant vererbt und verläuft gewöhnlich als latente Anomalie im Porphyrinstoffwechsel symptomlos. Bei Applikation von Barbituraten, Sulfonamiden, Steroiden u. a. entwickelt sich das klinische Bild einer akuten Porphyrie mit Bauch-, Neuro-, Psycho-, kardiovaskulären und anderer Symptomatik bei hoher Mortalität. Der genetische Grunddefekt stellt einen partiellen Mangel an Uroporphyrinogen-I-Synthetase dar [1—4]. Diese Störung geht einher mit einer erhöhten Aktivität der δ-Aminolävulinsäure-Synthetase (ALS-S) [5]. Die erwähnten Arzneimittel lösen einen akuten Anfall aus, indem sie die ALS-S induzieren, die Biosynthese der Porphyrinvorstufen ist erhöht und der Enzymblock tritt deutlicher in Erscheinung. Bis vor kurzem waren die stark erhöhten PBG- und ALS-Werte im Harn die einzigen labordiagnostischen Parameter zum Nachweis der AIP. Da das Uro-S-Defizit auch in den Erythrozyten existiert, wurde durch die Entwicklung neuer Untersuchungsverfahren [6—8] eine Möglichkeit für den direkten Nachweis von Defektträgern geschaffen [9—12]. Wir stellten uns als Aufgabe, bekannte Familien mit Erkrankungen an AIP in Bulgarien genealogisch mittels Uro-S-Bestimmung in Erythrozyten zu untersuchen und die entdeckten latenten Fälle unter Beobachtung zu nehmen. Die vorliegende Mitteilung stellt erste Resultate vor.

Material und Methodik

Untersucht wurden 9 Personen: 3 Kranke von verschiedenen Familien mit bestätigter AIP, die durch typische Anfälle und erhöhte PBG-, ALS- und Uroporphyrinausscheidung im Harn

Abkürzungen: AIP = akute intermittierende Porphyrie; Uro-S = Uroporphyrinogen-I-Synthetase; ALS = δ-Aminolävulinsäure; PBG = Porphobilinogen; ALS = δ-Aminolävulinsäure-Synthetase

charakterisiert sind, sowie 6 Angehörige dieser Familien. Während der Untersuchung waren die Kranken anfallsfrei.
Uro-S wurde in den Erythrozyten nach MANDEL et al. [8] bestimmt[1]. Die erhaltenen Werte wurden mit den Daten 20 gesunder Frauen und Männer verglichen. Im Harn wurden ALS und PBG nach MAUZERALL und GRANICK [13], Uroporphyrin nach GUTNIAK und DANCEWICZ [14] untersucht.

Ergebnisse und Diskussion

Die Aktivität der Uro-S in den Erythrozyten ist bei den 3 Erkrankten und bei 3 ihrer Familienmitglieder vermindert (Tab. 1 und Abb. 1). Die Vererbung des Enzymdefizits ist autosomal dominant (Abb. 2), wobei die entdeckten Latentträger des pathologischen Gens — der Vater der einen Kranken und zwei achtjährige Kinder der anderen Kranken — symptomfrei waren

Tabelle 1
Ergebnisse der Familienstudien zum Uroporphyrinogen-I-Synthetase-Defekt

Untersuchte Familien	Parameter			
	Erythrozyten	Harn		
	Uroporphyrinogen-I-Synthetase [pMol Porphobilinogen/min/10/ Ery.] Normalbereich 1050—1800	ALS [µg/24 Std.] Normalbereich 700—4000	Porphobilinogen [µg/24 Std.] Normalbereich 0—2200	Uroporphyrin [µg/24 Std.] Normalbereich 0—20
I. W. P., ♀, 18 J. erkrankt	843	50810	76576	70
M. P., ♀, 8 J. Tochter, latente Trägerin	551	2785	186	0
II. G. B., ♂, 31 J. erkrankt	957	32712	30696	18
M. B., ♀, 8 J. Tochter, latente Trägerin	839	1330	490	nicht untersucht
Z. B., ♂, 7 J. Sohn gesund	1614	1200	450	nicht untersucht
III. W. M., ♀, 25 J. erkrankt	714	30245	35120	154
N. M., ♂, 61 J. Vater, latenter Träger	923	1560	7920	0
P. M., ♀, 50 J. Mutter, gesund	1314	528	627	0
D. M., ♂, 29 J. Bruder, gesund	1064	1952	2428	2

[1] PBG wurde uns dankenswerterweise von Dr. T. WITH, Svendborg (Dänemark) überlassen.

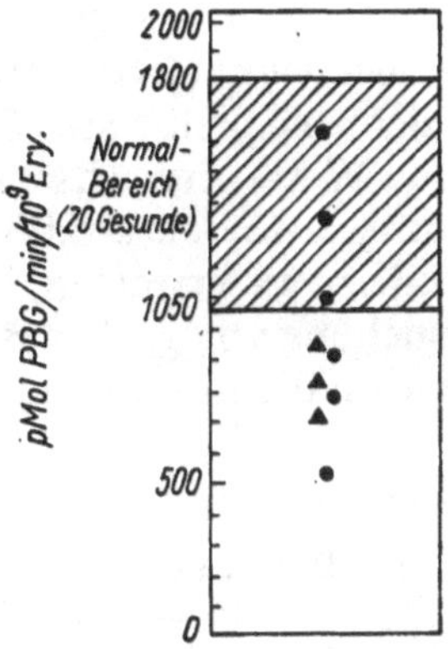

Abb. 1. Uroporphyrinogen-I-Synthetase in Erythrozyten.
▲ = Kranke mit ausgeprägter akuter intermittierender Porphyrie;
● = untersuchte Familienangehörige.

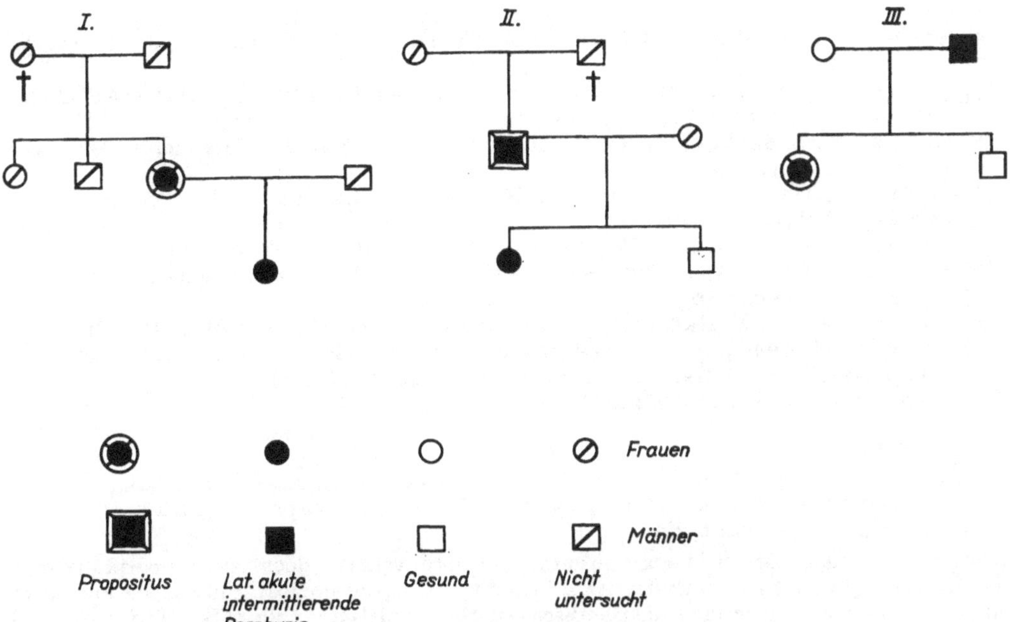

Abb. 2. Verbreitung des Enzymmangels bei den untersuchten Familien

Bei allen 3 Erkrankten war die Ausscheidung von PBG und ALS im Harn stark und von Uroporphyrin leicht erhöht, obwohl sie sich in einer anfallfreien Periode befanden. Dieselben Parameter waren bei den 2 Kindern mit niedriger Uro-S in den Erythrozyten im Normalbereich, bei W. M. s Vater wurde aber eine mäßig erhöhte PBG-Ausscheidung im Harn gefunden.

Die gefundene Aktivitätsminderung der Uro-S in den Erythrozyten bei Kranken mit ausgeprägter AIP beträgt in Übereinstimmung mit den Angaben anderer Autoren [4, 11] ca. 40—50% der Werte der Gesunden. Bei Kranken mit ausgeprägter AIP liefert der Uro-S-Aktivitätsnachweis in Erythrozyten keine wesentlichen zusätzlichen dignostischen Informationen gegenüber der PBG- und ALS-Bestimmung im Harn. Die Bedeutung des Uro-S-Nachweises in den Erythrozyten liegt vor allem darin, daß Latentträger des pathologischen Gens mit normaler Ausscheidung von PBG und ALS erkannt werden können, was auf andere Weise nicht möglich wäre. Da das Enzym bei Gesunden schon im frühen Kindheitsalter die Werte Erwachsener erreicht [11], ist dessen Nachweis besonders wertvoll für das Auffinden kindlicher Latentträger.

Es ist zu erwarten, daß die Ausführung der geplanten genealogischen Untersuchung von großer Bedeutung für die Prophylaxe der oft fatalen akuten Porphyrieanfälle in unserem Land sein wird.

Literatur

[1] HEILMEYER, L., u. R. CLOTTEN: Klin. Wschr. **47**, 71 (1969)
[2] STRAND, L. J., B. FELSHER, A. REDEKER u. H. MARVER: Proc. natn. Acad. Sci. U.S.A. **67**, 1315 (1970)
[3] MEYER, U., L. J. STRAND, M. DOSS, A. REES u. H. MARVER: New Engl. J. Med. **286**, 1277 (1972)
[4] STRAND, L. J., U. MEYER, B. FELSHER, A. REDEKER u. H. MARVER: J. clin. Invest. **51**, 2530 (1972)
[5] TSCHUDY, D., M. PERLROTH, H. MARVER, A. COLLINS, G. HUNTER u. M. RECHCIGL, Jr.: Proc. natn. Acad. Sci. U.S.A. **53**, 841 (1965)
[6] GRANICK, S., S. SASSA, J. GRANICK, R. LEVERE u. A. KAPPAS: Proc. natn. Acad. Sci. U.S.A. **69**, 2381 (1972)
[7] SASSA, S., J. GRANICK, S. GRANICK, A. KAPPAS u. R. LEVERE: Fedn Proc. Fedn Am. Socs exp. Biol. **32**, 565 (1973)
[8] MANDEL, P., C. KOEHL u. J. WERTENSCHLAG: Annls Biol. clin. **32**, 353 (1974)
[9] SASSA, S., S. GRANICK, D. BICKERS, R. LEVERE u. A. KAPPAS: Enzyme **16**, 326 (1973)
[10] MEYER, U.: Enzyme **16**, 334 (1973)
[11] GRANDCHAMP, B., M. GRELIER u. Y. NORDMANN: Nouv. Presse méd. **4**, 1801 (1975)
[12] KOEHL, C., J. ABECASSIS u. J. WERTENSCHLAG: Nouv. Presse méd. **4**, 2735 (1975)
[13] MAUZERALL, D., u. S. GRANICK: J. biol. Chem. **219**, 435 (1956)
[14] GUTNIAK, O., u. A. DANCEWICZ: Medycyna Pr. **9**, 1 (1958)

Summary

E. D. IVANOV, D. ADJAROV and R. TSANEV: Red cell uroporphyrinogen I-synthetase in acute intermittent porphyria

The primary genetic defect in acute intermittent porphyria is a decreased uroporphyrinogen I-synthetase [EC. 4.3.1.8] activity. As a beginning of a genealogical study of the known families with members suffering from this disease in the People's Republic of Bulgaria, the red cell uroporphyrinogen I-synthetase was determined in 3 families by the method of MANDEL et al [8]. Except for the three propositi, an enzyme deficiency was established in 3 latent carriers of the pathological gene, two of whom had normal values of the urinary δ-aminolevulinic acid and porphobilinogen. The determination of red cell uroporphyrinogen I-synthetase proved to be a valuable parameter for revealing the latent AIP.

Acta biol. med. germ., Band 36, Seite 805—808 (1977)

Medizinische Universitäts-Poliklinik Münster i. W., BRD

Klinische Bedeutung des Folsäuregehaltes im Plasma und in Erythrozyten

E. Heilmann und E. Bönninghoff

Einleitung

In der Differentialdiagnose megaloblastärer Anämien sind Vitamin B_{12}- und Folsäuremangel voneinander abzugrenzen. Direkte Bestimmungen dieser Vitamine im Plasma bzw. Serum ermöglichen die Diagnose. Mikrobiologische Verfahren mittels verschiedener Testorganismen haben wegen der aufwendigen Methodik keinen Eingang in die klinische Routinediagnostik gefunden [1, 2, 4]. Die ersten Mitteilungen über einen Radioassay zur Vitamin-Bestimmung stammen aus dem angloamerikanischen Schrifttum [7, 8, 18]. In früheren Studien berichteten wir über ein routinemäßig durchführbares Verfahren zur Folsäurebestimmung im Plasma und in Erythrozyten [5, 6]. In der nachfolgenden Untersuchung gingen wir der Frage nach, welche klinische Bedeutung dieser Untersuchung beizumessen ist.

Patientengut und Methodik

Zur Untersuchung gelangten 20 männliche (m. N.), 20 weibliche gesunde Probanden (w. N.), 20 Frauen unter der Einnahme von Ovulationshemmern (Ov.), 20 chronische Alkoholiker (Al.), 6 Patienten mit terminaler Nierensuffizienz vor (a. Dia.) und nach 8stündiger Hämodialyse (p. Dia.) und 4 Patienten mit idiopathischer Sprue (Sp.).

Die Folsäurebestimmungen wurden mittels eines Radioassays durchgeführt, der auf dem Prinzip der kompetitiven Proteinbindungsmethoden beruht. Zur Anwendung gelangte das Folsäure-Radioassay-Testsystem der Firma Schwarz/Mann (Division of Becton, Dickinson and Company, Heidelberg). Den nüchternen Probanden wurde je 5 ml Venenblut entnommen und in Vakutainer-Röhrchen gefüllt, die mit ÄDTA beschickt waren. Zur Ermittlung des Folsäuregehaltes in den Erythrozyten wurde 100 ml Blut zu 1,9 µl 1%ige Askorbinsäure gegeben. Diese Proben wurden 30 min lang bei Zimmertemperatur belassen und anschließend bis zur Bestimmung bei —20 °C gelagert.

Für die Bestimmung des Folsäuregehaltes im Plasma wurde das Venenblut nach der Entnahme bei 1000 U/min zentrifugiert und die gewonnene Plasmaprobe ebenfalls bei —20 °C gelagert. Für die Untersuchung wurde 50 µl Plasma benötigt.

Ergebnisse

Abb. 1 zeigt die Ergebnisse der Folsäurebestimmungen im Plasma bei den untersuchten Probanden. Wir fanden bei den männlichen und weiblichen Normalpersonen einen Mittelwert des Folsäuregehaltes im Plasma von 6,9 ± 3,3 bzw. 6,7 ± 3,4 ng/ml. Bei Frauen unter der Einnahme von Ovulationshemmern lag der Wert mit 7,13 ± 2,8 ng/ml nur geringfügig höher. Bei den chronischen Alkoholikern lag der Mittelwert mit 4,85 ± 2,02 ng/ml signifikant ($p < 0{,}01$) niedriger, ebenso bei den Hämodialyse-Patienten. Nach Durchführung der Hämodialysen erfolgte ein weiterer Abfall des Folsäure-Gehaltes im Plasma ($p < 0{,}05$). Die niedrigsten Folsäurewerte wiesen die Patienten mit „idiopathischer Sprue" auf (3,67 ± 2,1 ng/ml).

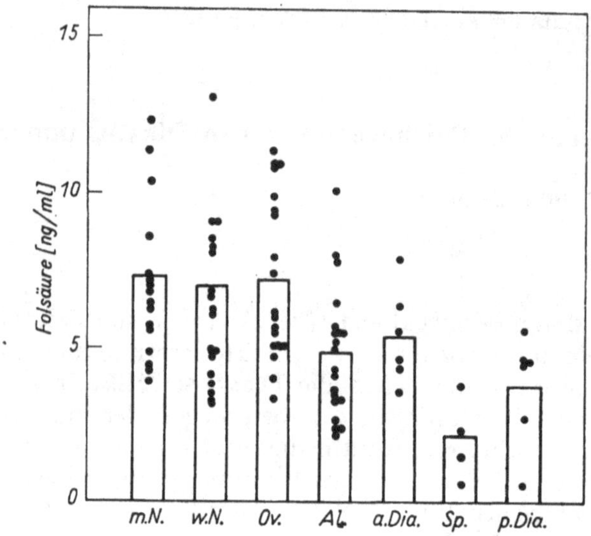

Abb. 1. Folsäuregehalt im Plasma bei verschiedenen Kollektiven (Einzelheiten siehe Text)

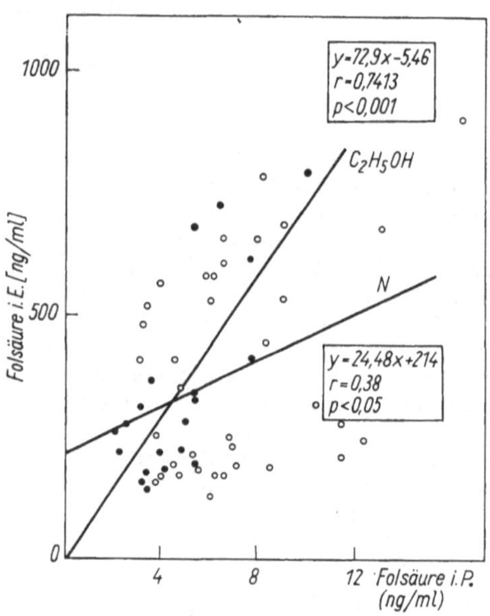

Abb. 2

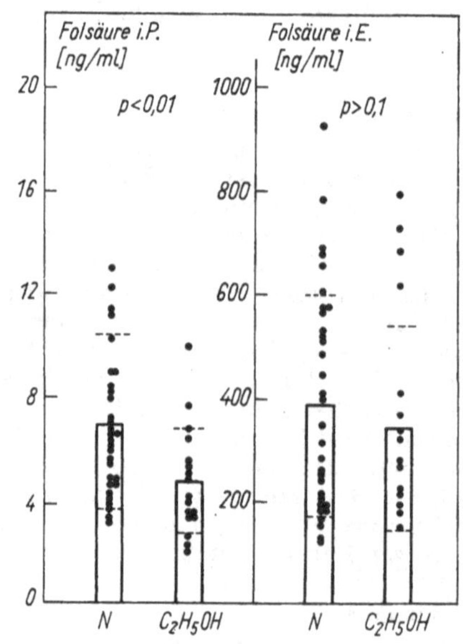

Abb. 3

Abb. 2. Beziehungen zwischen Folsäuregehalt im Plasma (i. P.) und in Erythrozyten (i. E.) bei chronischen Alkoholikern (●) und Normalpersonen (N = o)

Abb. 3. Beziehungen zwischen Folsäuregehalt im Plasma (i. P.) und in Erythrozyten (i. E.) bei chronischen Alkolikern (C_2H_5OH) und Normalpersonen (N)

Die Beziehungen zwischen dem Folsäuregehalt im Plasma und in den Erythrozyten bei chronischen Alkoholikern und Normalpersonen gehen aus Abb. 2 und 3 hervor. Bei den Alkoholikern fand sich ein höherer Korrelationskoeffizient mit einem $p < 0{,}001$ als bei den Normalpersonen ($p < 0{,}05$). Bei einer Gegenüberstellung der Ergebnisse des Folsäuregehaltes im Plasma und in den Erythrozyten der Normalpersonen und chronischen Alkoliker ergab sich lediglich für den Gehalt im Plasma ein signifikanter Unterschied ($p < 0{,}01$). Der Mittelwert des intraerythrozytären Folsäuregehaltes betrug bei den Normalpersonen 396,5 ± 215,1 ng/ml, bei den chronischen Alkoholikern 374,4 ± 198,9 ng/ml. Bei den Frauen unter der Einnahme von Ovulationshemmern ergab sich ein Mittelwert von 407,0 ± ±192,4 ng/ml. Die größten Schwankungen der intraerythrozytären Folsäurespiegel wiesen die Dialysepatienten auf (100—800 ng/ml), die niedrigsten Werte ergaben sich bei den Patienten mit „idiopathischer Sprue" (221,6 ± 186,2 ng/ml).

Diskussion

Für die Differentialdiagnose megaloblastärer Anämien hat sich die Bestimmung des Folsäuregehaltes im Plasma bzw. Serum mittels Radioassay bewährt [5, 6, 8, 11, 13, 17, 18]. Die klinische Häufigkeit eines Folsäuremangels ergibt sich daraus, daß die mit der Nahrung zugeführte, resorbierbare Menge mit etwa 270 ng pro Tag relativ gering ist gegenüber dem täglichen Verbrauch von 100—200 ng [10]. Da der Organismus nur über wenige Folsäurereserven verfügt, ergibt sich, daß unter verschiedenen Krankheitsbedingungen rasch ein Folsäuredefizit entstehen kann. Dabei ist mit Bestimmung des Folsäuregehaltes im Plasma bzw. Serum die Erfassung eines latenten Folsäuredefizits möglich, während die intraerythrozytären Bestimmungen einen manifesten Folsäuremangel aufdecken können [9].
Die von uns ermittelten Werte bei Normalpersonen stimmen mit den Mitteilungen anderer Autoren [7, 8, 13, 17] gut überein. Ein Folsäuremangel wird im allgemeinen bei Werten unter 4 ng/ml angenommen. Bei den intraerythrozytären Werten liegt der untere Grenzwert bei 150 ng/ml. Ein Folsäuremangel wird auch in der Literatur bei chronischen Alkoholikern häufig beschrieben [5, 6, 10, 16]. Über den Folsäuregehalt im Plasma bei Frauen unter der Einnahme von Ovulationshemmern finden sich in der Literatur unterschiedliche Angaben [12, 14]. Über erniedrigten Folsäuregehalt bei Hämodialyse-Patienten wird in der Literatur berichtet, wobei auf die Dialysierbarkeit dieses wasserlöslichen Vitamins besonders hingewiesen wird [15]. Der manifeste Folsäuremangel bei Patienten mit „idiopathischer Sprue" wird auf die gestörte Resorption dieses Vitamins im Dünndarm zurückgeführt [3].
Die Möglichkeit, direkte Folsäurebestimmungen im Plasma und in den Erythrozyten durchführen zu können, ermöglicht somit einen Einblick in Störungen des Stoffwechsels dieses Vitamins bei verschiedenen Erkrankungen zu gewinnen. Darüber hinaus ergibt sich die Möglichkeit, eine entsprechende Substitutionstherapie zu kontrollieren.

Literatur

[1] BAKER, H., O. FRANK, I. PASHER, H. SOBOTKA u. S. HUTNER: Nature, Lond. **191**, 78 (1961)
[2] BAKER, H., V. HERBERT, O. FRANK, I. PASHER, S. H. HUTNER, L. R. WASSERMAN u. H. SOBOTKA: Clin. Chem. **5**, 275 (1959)
[3] DOIG, A., u. R. H. GIRDWOOD: Q. Jl. Med. **29**, 333 (1960)

[4] GIRDWOOD, R. H. in: Ciba Foundation Symposium on Chemistry and Biology of Pteridines. G. E. W. WOLSTENHOME u. M. P. CAMERON (Hrsg.). Churchill, London 1954
[5] HEILMANN, E., u. E. BÖNNINGHOFF: Diagnostik **9**, 347 (1976)
[6] HEILMANN, E., u. E. BÖNNINGHOFF: Schweiz. med. Wschr. (im Druck)
[7] HERBERT, V. in: Modern Nutrition in Health and Disease. 5. Aufl. R. S. GOODHART u. M. E. SHILS (Hrsg.). Lea und Febiger, Philadelphia 1973, S. 221
[8] HERBERT, V., A. R. LARRABEE u. J. N. BUCHANAN: J. clin. Invest. **41**, 1134 (1962)
[9] HINES, J. D., A. V. HOFFBRAND u. D. H. MOLLIN: Am. J. Med. **43**, 555 (1967)
[10] HUSER, H.-J.: Schweiz. med. Wschr. **105**, 1823 (1975)
[11] KAMEN, B. A., u. J. D. CASTON: J. Lab. clin. Med. **83**, 164 (1974)
[12] PAINE, C. J., W. D. GRAFTON, V. L. DICKSON u. E. R. EICHNER: J. Am. med. Ass. **231**, 731 (1975)
[13] ROTHENBERG, S. P., M. DaCOSTA u. Z. ROSENBERG: New Engl. J. Med. **286**, 1335 (1972)
[14] SHOGANIA, A. M., G. HORNADY u. P. H. BARNES: Lancet **1968**/II, 1376
[15] SKONTAKIS, V. A., S. R. ACCHIARDO, M. C. MEYER u. F. E. HATCH: Clin. Pharmac. Ther. **18**, 200 (1975)
[16] STRAUB, P. W.: Schweiz. med. Wschr. **105**, 1072 (1975)
[17] TAJUDDIN, M., u. H. A. GARDYNA: Clin. Chem. **19**, 125 (1973)
[18] WAXMAN, S., u. C. SCHREIBER: Blood **42**, 281 (1973)

Acta biol. med. germ., Band 36, Seite 809 (1977)

Discussion

The discussion on genetic defects was mainly concerned with the factors that determine the instability of mutant enzymes and additional parameters to classify variants. There is no complete correspondence between the stability under in vitro conditions and in vivo as exemplified by the fact that G-6-PD A^- is less stable than normal B^+ in vivo but not in vitro. There were suggestions to include a fractionation of cells according to density, which correlates well with age (SYLLM-RAPOPORT) and to add the effects of inhibitors such as NADPH, NADH-X, and of analogues such as thio-NADP and acetyl-NADP, and also to employ immunological methods. The discussion on PK was concerned mainly with the question whether the mutant enzyme forms dimers.

The report of STERN was supplemented by the information that the work was done with 0.2% red cell suspensions at *p*H 7.4 in 50 mM phosphate buffer. Extended spectrophotometric studies indicated that a large proportion of the hemoglobin is converted to non-oxygen-carrying products. While scavengers of OH· radicals are protective with respect to hemolysis, they do not inhibit hemoglobin breakdown in red cells exposed to dihydroxyfumaric acid. In the discussion to the report of AEBI it was emphasized that catalase hybrids apparently are much more stable than those of hemoglobin.

Acta biol. med. germ., Band 36, Seite 811—813 (1977)

Workshop

The following topics were discussed:

1) Diagnostics of pathological glucose-6-phosphate dehydrogenase.
2) Phosphofructokinase enzymopathy of red cells.
3) Causes of the instability of enzyme mutants.
4) Mechanisms of hemolysis in connection with the superoxide dismutase.

Ad 1: Dr. JACOBASCH asked for opinions on the place and value of immunologic methods in the determination and discrimination of G-6-PD mutants. Dr. YOSHIDA doubted that immunological methods cold distinguish effectively point mutations which form the bulk of enzymopathies. In support he cited the fact that enzymes of man and apes cannot be discriminated immunologically. Dr. AEBI agreed. He referred to own work on catalase which showed that it may be impossible to find differences in the sequence of amino acids by immunological studies. Dr. MARTI pointed out that improvement of immunological methods may be hoped for. In the field of hemoglobins recently the determination of HbA_2 and HbF has become feasible, these methods now compete with the chemical ones but do require great care. Such immunological determinants are mostly located in the surface, it may be advantageous to study the immunological determinants in the unfolded state (RAPOPORT).

Such experiments were carried out by AEBI who tried to distinguish dimers and tetramers of catalase immunologically. In the discussion of screening methods it was pointed out that difficulties arose from both the maturation dependency of the activities of G-6-PD as well as from the existence of a cell mosaicism, which may occur either in heterozygotes or can be caused by the enzyme instability. Thus normal or even elevated ones — all enzyme activities may be found. It was proposed to supplement the usual determination of enzyme activities by one at 45° and to perform the cytological metHb-reduction test according to TÖNZ. This proposal was agreed to by all participants. In support Dr. GRIEGER reported that of the 28 cases of hemolytic anemia caused by a G-6-PD defect found in the GDR, 10 had enzyme activities in the normal range. They could be established only by a combination of tests for temperature stability, electrophoresis and the Tönz-test. These variants were then characterized according to the recommendations of the WHO. Dr. MARTI reported similar experiences. He pointed out that the Tönz-test requires careful standardization of the elution time and may not be suitable for field work in all countries. He recommended in such cases the tube test of BREWER and that for thermostability. Even the simplest characterization should encompass electrophoresis and the kinetic parameters including inhibition studies.

A particular problem in mediterranean countries are patients prone to hemolytic crises. Dr. ARESE reported that 20% of the population of Sardinia exhibit a G-6-PD defect, mostly the mediterranean type B^-, which may be subdivided into several variants. But only 10—15% react with violent hemolytic crises to the ingestion of fava beans or to some drugs. So far no correlation has been established between the type of 6-6-PD variant or the occurrence of icterus in the new-born on the one hand and the proneness to hemolytic crises on the other. Additional environmental or genetic factors have been postulated. Dr. GRIEGER proposed studies with drugs on immature cell populations which may be obtained by density gradient centrifugation. She had the impression that severe anemias may be correlated with low values of the G-6-PD in the reticulocytes. BOSIA observed a connection between changes of the NADP/NADPH ratio and the occurrence of hemolytic anemias. He found lower ratios in anemias caused by G-6-PD deficiency as compared with those of other origin. Finally the discussion turned to the question of the formation of inhibitor NADP-X. The question was raised whether declines of the NADPH may be caused by it (ARESE).

YOSHIDA pointed out that NADH-X and NADPH-X may be formed both in vitro and in vivo. At neutral *p*H and 37 °C NADPH is transformed irreversibly to NADPH-X in the absence of phosphate.

Ad 2: ETIEMBLE reported a case of PFK deficiency causing hereditary nonspherocytic hemolytic anemia. Similar to other reported cases the enzyme from red cells did not crossreact with antiserum to human muscle PFK. In contrast however there were no muscular symptoms while the anemia was pronounced. Kinetic studies showed higher sensitivity of the PFK to inhibition by ATP whereas the effect of citrate was reduced. On electrophoresis in starch-gel the muscle PFK of the patient was unstable, whereas the mobility was higher than normal. It is assumed that the subunit corresponding to the muscle enzyme is defective. Anemia is presumably caused by the strong inhibitory effect of ATP, the maximal activity was only reduced to 60%. A high Ca^{2+} concentration was noted in the red cells. The influence of K^+ and NH_4^+ (JACOBASCH) had not been tested. The absence of myopathy may be caused by a high rate of synthesis of the enzyme.

Ad 3: There seems to exist a lack of systematic studies on this question. The scattered reports indicate a variety of causes for instability, one factor may be dissociation of active tetramers to less active or inactive dimers, as found with catalase (AEBI). It is suggested that mutations in the catalytic center affect the interactions between enzyme and substrate, while changes in hydrophobic regions, particularly if related to contact regions between subunits may cause instability. Dr. YOSHIDA found counter-examples. The enzyme variants "Capetown" and "Manchester" are unstable, but occur almost exclusively as tetramers. They are quite resistant to dissociation. He proposes aggregation as a possible cause of instability.

Ad 4: STERN reported on his unsuccessful search for SOD mutants with lowered activity in patients with hemolytic anemias after menadion application. Isoenzymes were found which differ in their metal component. For these reasons studies on the possible significance of SOD have been restricted to model experiments. It was found that dihydroxy-fumaric acid leads to the formation of superoxide anions with subsequent hemolysis. In other experiments the SOD was inhibited by dithiocarbamate. This treatment without additional insult led also to hemolysis. AEBI also has never found a reduction of SOD activity. He mentioned the work of BEEHMAN who described a variant of SOD with aberrant mobility but without alteration of its activity. Possibly SOD variants with low activity are lethal mutants. TANNERT considered the possible importance of a decline of SOD for the aging process of cells.

The discussion turned to the toxic effects of various radicals and peroxides. Lipid peroxides apparently accumulate only if the activity of glutathione peroxidase is diminished or inhibited.

Addition of 1,4-naphthoquinone-2-sulfonate inhibits glutathione-peroxidase. At the same time there are formed radicals in a reaction of the quinone with hemoglobin. This leads to its breakdown with the formation of green pigments, which cause an increased permeability of the cell membrane. Sulfhydryl groups are not oxidized to a significant extent. Another interesting effect is an increased hypotonic hemolysis produced by lipid peroxidation. Dr. STERN holds phenyl radicals to be more important than lipid peroxides in the action of phenylhydrazine. The phenyl radicals can react both with hemoglobins and the cell membrane and produce large amounts of superoxide radicals, which is presumably the reactant species with Hb.

Section IV

Membrane Processes

Acta biol. med. germ., Band 36, Seite 817—821 (1977)

Max-Planck-Institut für Biophysik, Frankfurt/Main, GFR

Anion transport across the red blood cell membrane and the protein in band 3

H. Passow

Summary

The paper reviews existing evidence for the participation of the protein in band 3 (nomenclature of Steck, [1]) in anion transport across the red cell membrane and discusses the possible role of common binding sites on band 3 for 1-fluoro-2,4-dinitrobenzene, 2-(4'-aminophenyl)-6-methylbenzenethiazol-3',7-disulfonic acid and dihydro 4,4'-diisothiocyanato stilbene-2,2'-disulfonic acid in the transport process.

One of the intrinsic glycoproteins of the red cell membrane migrates in band 3 of SDS polyacrylamide gel electropherograms (nomenclature of Steck [1]). This glycoprotein has a molecular weight of about 100,000 and spans the membrane [1—3]. It has been proposed by Cabantchik and Rothstein [4], Passow et al. [5] and Ho and Guidotti [6] that it participates in anion transport across the red blood cell membrane. It is the purpose of this paper to review briefly the evidence which justifies this proposition, to describe some of the problems associated with it, and to discuss the possible function in anion transport of the binding sites for certain inhibitors on that protein.

Essentially, the evidence is based on 6 types of studies:

(1) Selective labeling with irreversibly binding inhibitors

Since the early work from this laboratory it is known that anion transport can be inhibited by amino reactive reagents [7—9]. Further work has shown that the site of action are amino groups on membrane proteins rather than the membrane lipids [10, 11]. Among the many inhibitory amino reactive reagents, the non-penetrating 4'-isothiocyanato-4'-acetamido stilbene-2,2'-disulfonic acid (SITS), first synthesized by Maddy [12a], acquired a prominent place since this compound inhibits without reacting with the many proteins at the inner membrane surface [12]. Systematic variations of the molecule revealed [4] that its diisothiocyanate derivative labeled two [4] or three [13] bands only, the protein in band 3, glycophorin, and an additional unidentified band in the region of bands 4.2—4.5, called satellite band. Most of the binding takes place at the protein in band 3. There exists a linear relationship between inhibition of sulfate equilibrium exchange and binding of DIDS or its dihydroderivative, H_2DIDS, to the protein in band 3 [13]. Although binding to membrane constituents outside band 3 could

Abbreviations: APMB = 2-(4'-aminophenyl)-6-methylbenzenethiazol-3',7-disulfonic acid; DNFB = 1-fluoro-2,4-dinitrobenzene; DIDS = 4,4'-diisothiocyanato stilbene-2,2'-disulfonic acid; H_2DIDS = dihydro DIDS; SITS = 4-acetamido 4'isothiocyanato stilbene-2,2'-disulfonic acid

not be excluded as site of inhibition by H_2DIDS or DIDS, these experiments were clearly compatible with some role of the protein in band 3 in anion transport. Additional evidence for some role of band 3 in anion transport was derived from work with an isothiocyanate derivative of a benzene sulfonic acid [6].

(2) Demonstration of common binding sites for different irreversibly binding inhibitors of anion transport

The penetrating inhibitor DNFB labels a large variety of amino groups on many membrane proteins. If, however, red cells are first treated with SITS and subsequently exposed to DNFB, the dinitrophenylation of band 3 is greatly reduced while that of all other membrane proteins remains uneffected [5]. Similar results were obtained with another pair of inhibitors, DIDS and pyridoxal phosphate [14].

(3) Protection by reversibly acting inhibitors against irreversible modification by DNFB of sulfate transport system and binding sites on the protein in band 3

Certain reversibly acting inhibitors of anion transport, including stilbene derivatives like 4,4′-diacetamido stilbene-2,2′-disulfonic acid and the thiazole derivative APMB were shown to have common binding sites with DNFB [5, 15] and H_2DIDS [16] on the protein in band 3. This was inferred from the observation that dinitrophenylation in the presence of these agents prevented the modification of band 3. After removal of the reversibly acting inhibitors and of the unreacted DNFB by washing, the rate of anion transport far exceeded the rate observed after dinitrophenylation in the absence of reversibly acting inhibitors. With APMB at least 85% of both anion transport and the common binding sites with DNFB and H_2DIDS on band 3 can be protected against irreversible modification by DNFB [16].

(4) Photoaffinity labeling with a competitive inhibitor of anion transport

Rothstein et al. [14, 17, 18] are currently engaged in attempts to demonstrate that the anionic NAP-taurine is a competitive inhibitor of anion transport and that the irreversible fixation of this compound by illumination can be modified by the presence of the anions to be transported. Preliminary results are compatible with the participation of the protein in band 3 in anion transport, but they are not yet conclusive.

(5) In situ purification of the transport system

Attempts have been made to remove from the red cell membrane all protein species except band 3 and to demonstrate that anion transport still persists. In our laboratory [5, 19] trypsin was incorporated into red cell ghosts. This led to a digestion of all membrane proteins, including the protein in band 3. The digested membrane proteins with the exception of one and sometimes two peptides which have no common binding sites for H_2DIDS, SITS and DNFB, are released from the membrane. By radioactive labeling techniques it was shown that band 3 is split into two peptides of molecular weights of 58,000 and 48,000, but that the split products remain associated with the membrane. The trypsinized red cell ghosts form vesicles which continue to transport anions at a somewhat reduced rate. As in intact ghosts, anion transport in these vesicles is susceptible to inhibition by a representative array of different inhibitors of anion transport. This shows that one or both of the two fragments derived from band 3 are likely to be involved in anion transport. Recently CABANTCHIK et al. [20] have formed vesicle preparations from triton extracted red cell ghosts which seem to be devoid of any membrane proteins except the protein in band 3. Certain characteristic features of the kinetics of anion transport (*p*H and temperature dependence) seem to be preserved in these vesicles, suggesting that band 3 protein is the transport protein.

(6) Reconstitution of the transport system by incorporation of band 3 protein into lipid bilayers

Unpublished work from this laboratory (RUDLOFF, OBERMEIR and PASSOW, 1975) has shown that it is possible to incorporate highly purified band 3 protein (>95% pure) into liposomes. This is associated with an increase of the liposomal anion retention, indicating an increase in liposomal volume. However, no facilitated anion transport could be detected. Using a triton extract, ROTHSTEIN et al. [21] found an increased anion exchange across the liposomal membrane. Incorporation of similar extracts from DIDS treated cells resulted in no such increase. The problem with these experiments consists in the low purity of the protein preparation (about 80%) and the fact that only the relatively low percentage of all reconstitution experiments were successful. Thus it is not quite clear whether the controls with the extracts from DIDS treated cells demonstrate the inability of the modified protein to perform the transport or if in these cases the reconstitution of the system was unsuccessful for other reasons.

Extensive studies on the in situ enzymatic dissection of the protein in band 3 [22] has shown that the H_2DIDS binding sites on that protein are associated with at least two different populations of peptides with equal molecular weights. The populations can be distinguished by differences of susceptibility to proteolytic cleavage. Treatment with chymotrypsin revealed, for example, that about 40% of the H_2DIDS binding sites reside on peptides which, in situ, are resistant to any major degradation by the enzyme. The rest is attached to a population of peptides which are split into fragments of 70,000 and 36,000 Daltons. The cleavage of the chymotrypsin sensitive component of band 3 does not lead to an inhibition of anion transport. Moreover, determination of the H_2DIDS binding sites on the chymotrypsin-resistant fraction and on the 70,000 Daltons split product revealed a linear relationship between H_2DIDS binding and inhibition of anion transport for the resistant peptide and the split product. It was further observed that the response of the peptide in band 3 to pronase and papain is also nonuniform. Moreover, the experiments with internally trypsinized ghosts described above also show that each of the two peptides derived from band 3 carries H_2DIDS binding sites. If one accepts the evidence for a participation of the H_2DIDS binding sites on the protein in band 3 as conclusive, one would be compelled to assume that protein molecules which differ with respect to their susceptibility or accessibility for the various enzymes perform the same function. Perhaps we are dealing with a mixture of isozymes. It must be kept in mind, however, that labeling with isothiocyanates and DNFB is rather non-specific and that, perhaps, among the many labeled molecules in band 3 only a minute number is involved in transport. This would imply that H_2DIDS binding to the various populations simply behaves like H_2DIDS binding to some minor competing constituent which is the true transport protein. Although this possibility cannot be excluded, experiments on the sidedness of action of APMB on the common binding sites for DNFB and H_2DIDS led us to favour the assumption of a participation of a majority of the H_2DIDS binding sites in transport [16, 23].

The protein in band 3 is rather large for a molecule which plays a role as a carrier. Although conformational changes in small parts of the protein molecule could possibly account for anion exchange with the observed turnover number ($2 \cdot 10^4$/s/site [13]), it would seem much more plausible if the carrier function would be exerted by a small lipid soluble molecule. Recently KADENBACH and HADVÁRY [24, 25] have shown that a proteolipid in the inner mitochondrial membrane is capable of binding inorganic phosphate and to carry it into organic solvents. The authors have suggested that this protein serves as a phosphate carrier. The existence of a proteolipid had also been demonstrated in the red blood cell membrane [26]. We have confirmed this latter finding and discovered that this proteolipid becomes dinitrophenylated under the conditions under which DNFB inhibits anion transport. Moreover, the proteolipid comigrates during elution in chloroform on a Sephadex LH 20 column together with the extracted sulfate. It is, however, unclear whether or not this observation indicates a true stoichiometrical binding as required for a carrier (LEPKE and PASSOW, unpublished).

Studies on the effects of sulfate concentration on the inhibition of sulfate transport by APMB revealed that there is little competition between the inhibitor and the ion to be transported. Thus, the common binding sites for APMB, DNFB, and H_2DIDS on the protein in band 3 are modifier sites rather than transfer sites [23]. It is still necessary, therefore, to identify the transfer sites and to demonstrate that they are located on the protein in band 3. Studies on the sidedness of action of reversibly binding stilbene disulfonic acids led us to discuss the possibility that both, a mobile carrier and the protein in band 3, are required to accomplish anion transport [15]. It should be noted, however, that this reflects a thought which is not yet adequately substantiated by experimental facts and that the protein in band 3 could possibly accomplish anion transport without the involvement of a separate carrier.

References

[1] STECK, T. L.: J. Cell Biol. **62**, 1 (1974)
[2] BRETSCHER, M. S.: J. molec. Biol. **58**, 775 (1971)
[3] JENKINS, R. E., and M. J. A. TANNER: Biochem. J. **147**, 393 (1975)
[4] CABANTCHIK, Z. I., and A. ROTHSTEIN: J. Membrane Biol. **15**, 207 (1974)
[5] PASSOW, H., H. FASOLD, L. ZAKI, B. SCHUHMANN, and S. LEPKE: Proceedings of the Ninth Meeting of the Federation of European Biochemical Societies, Budapest, 1974. **35**, 197. G. GÁRDOS and I. SZÁSZ (eds.). Publishing House of the Hungarian Academy of Sciences, Budapest 1975
[6] HO, M. K., and G. GUIDOTTI: J. biol. Chem. **250**, 675 (1975)
[7] PASSOW, H.: Prog. Biophys. molec. Biol. **19**, 424 (1969)
[8] PASSOW, H. in: The Molecular Basis of Membrane Function. D. C. TOSTESON (ed.). Prentice-Hall, Inc., Englewood Cliffs, New Jersey 1969, p. 319

[9] PASSOW, H., and K. F. SCHNELL: Experientia **25**, 460 (1969)
[10] OBAID, A. L., A. F. REGA, and P. GARRAHAN: J. Membrane Biol. **9**, 385 (1972)
[11] PASSOW, H.: J. Membrane Biol. **6**, 233 (1971)
[12] KNAUF, P. A., and A. ROTHSTEIN: J. gen. Physiol. **58**, 190 (1971)
[12a] MADDY, A. H.: Biochim. biophys. Acta **88**, 390 (1964)
[13] LEPKE, S., H. FASOLD, M. PRING, and H. PASSOW: J. Membrane Biol. (1976) in press
[14] ROTHSTEIN, A., Z. I. CABANTCHIK, and P. A. KNAUF: Fedn Proc. Fedn Am. Socs exp. Biol. **35**, 3 (1976)
[15] ZAKI, L., H. FASOLD, B. SCHUHMANN, and H. PASSOW: J. cell. Physiol. **86**, 471 (1975)
[16] PASSOW, H., L. ZAKI, M. PRING, and H. FASOLD: FEBS Symposium on the Biochemistry of Membrane Transport, Zürich, 1976. Springer, Berlin, Heidelberg, New York 1976, in press
[17] ROTHSTEIN, A., P. A. KNAUF, and Z. I. CABANTCHIK: Biochim. biophys. Acta (1976), submitted for publication
[18] CABANTCHIK, Z. I., P. A. KNAUF, T. OSTWALD, H. MARKUS, L. DAVIDSON, W. BREUER, and A. ROTHSTEIN: Biochim. biophys. Acta (1976) in press
[19] LEPKE, S., and H. PASSOW: Biochim. biophys. Acta (1976) in press
[20] WOLOSIN, J. M., H. GINSBURG, and Z. I. CABANTCHIK: FEBS Symposium on the Biochemistry of Membrane Transport, Zürich, 1976. Springer, Berlin, Heidelberg, New York 1976, in press — see also Abstracts of communications L 39, p. 20
[21] ROTHSTEIN, A., Z. I. CABANTCHIK, M. BALSHIN, and R. JULIANO: Biochem. biophys. Res. Commun. **64**, 144 (1975)
[22] PASSOW, H., H. FASOLD, S. LEPKE, M. PRING, and B. SCHUHMANN: Proceedings of the Ninth Rochester International Conference on Environmental Toxicity "Membrane Toxicity". M. W. MILLER and A. E. SHAMOO (eds.). Plenum Press, New York 1976, in press
[23] PASSOW, H., and L. ZAKI: 30th Anniversary Symposium of the Biophysical Laboratory "Molecular Specialization and Symmetry in Membrane Function", Boston 1976. Harvard University Press, Boston, in press
[24] HADVÁRY, P., and B. KADENBACH: Eur. J. Biochem. **39**, 11 (1973)
[25] KADENBACH, B., and P. HADVÁRY: Eur. J. Biochem. **39**, 21 (1973)
[26] REDMAN, C. M.: Biochim. biophys. Acta **282**, 123 (1972)

Acta biol. med. germ., Band 36, Seite 823—829 (1977)

Department of Cell Metabolism, National Institute of Haematology and Blood Transfusion, Budapest, Hungary

Effect of intracellular calcium on the cation transport processes in human red cells

G. Gárdos, I. Szász, and B. Sarkadi

Summary

Fresh or ATP-depleted human red cells were loaded with calcium by a short treatment of the erythrocytes with the Ca-ionophore A23187 in isotonic NaCl medium. The ionophore was eliminated by thorough washing in isotonic KCl solution containing 0.5% albumin. Passive K-transport increased up to 60fold in fresh erythrocytes loaded with 0.5—5 mM calcium. In ATP-depleted red cells much lower (0.1 mM) intracellular calcium levels had maximum effect on K-permeability. In both, Ca-loaded fresh and ATP-depleted red cells the addition of 0.5—1 mM propranolol further enhanced K-permeability.
Active Na-transport was completely inhibited by 1 mM internal calcium concentration, whereas passive Na-transport increased to about 1.5 fold.
The calcium pump (efflux) was greatly activated by intracellular calcium levels between 10 and 1000 μM. Then the Ca-pumping rate reached its maximum (85 μM/l cell/min) and levelled off. Passive Ca-permeability of Ca-loaded fresh cells corresponded to the physiological low value up to intracellular Ca-concentrations of 3.0 mM; above this level it increased rapidly (up to 25fold at 5 mM internal calcium). Passive calcium permeability of ATP-depleted cells definitely enhanced even at low intracellular calcium levels.
In energized erythrocytes the morphology of the cells during Ca-loading and Ca-pumping changed according to the intracellular Ca-level. The morphological changes showed no correlation with the altered K- and Na-transport processes.

The physiological intracellular Ca-concentration of human red cells amounts to about 5—15 μmoles/l cell, whereas the Ca-concentration of the plasma approaches 2.5 mM. The large concentration gradient is maintained by the low passive Ca-permeability of the membranes (Ca-influx at 2.5 mM $[Ca]_e$ is ≈ 0.5 μmoles/l cell/min) and the high Ca-pump capacity of the red cell [1—4]. Due to these facts Ca-ions could be introduced so far only into resealed ghosts or drug treated cells [4]. The metabolic state of these cell preparations and the characteristics of their membranes were significantly altered by the treatments. Our technique described recently enabled us to introduce varying amounts of Ca into fresh, intact erythrocytes with the Ca-ionophore A 23 187 [5]. The strategy of Ca-loading was elaborated on the basis of the following facts: Ca-ions are rapidly transported into the erythrocytes by A 23 187. The calcium taken up by the cells increases the K-permeability of the membrane (up to 60fold). In NaCl medium this manifests itself in a rapid K-efflux. The increased K-permeability exceeds the net Cl permeability of the cell and hence the membrane becomes hyperpolarized. The hyperpolarization, in turn, enhances the inward movement of the positive Ca-ions. The role of hyperpolarization is justified by the fact that in KCl medium or in the presence of chlorobutanol — inhibiting K-transport — Ca-influx is slower and Ca is not accumulated. Taking these findings into consideration red cells were incubated

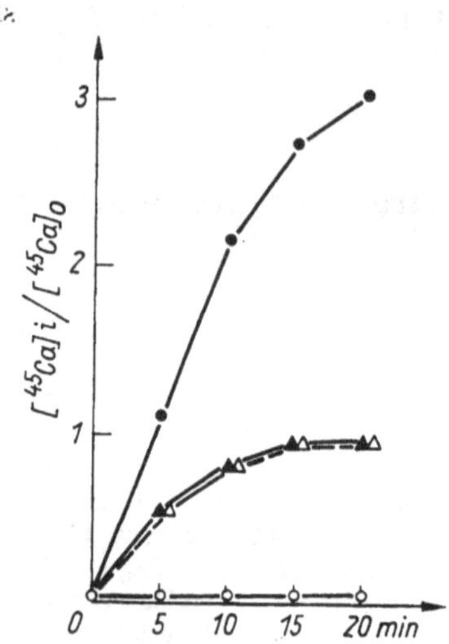

Fig. 2

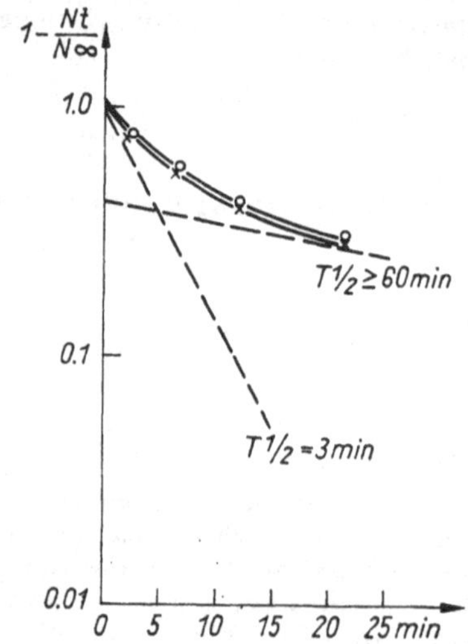

Fig. 1

Fig. 1. ^{45}Ca uptake of human erythrocytes in the presence of A23187.
○ 0.16 M NaCl; ● 0.16 M NaCl + 5 μM A23187; ▲ 0.16 M NaCl + 5 μM A23187 + 3 mM chlorobutanol; △ 0.16 M KCl + 5 μM A23187;
$t = 37$ °C; pH 7.4; hct = 20%; $[Ca^{2+}] = 0.5$ mM

Fig. 2. ^{42}K-efflux from calcium + ^{42}K-loaded fresh red cells.
Intracellular $[Ca^{2+}]$: 3.2 mM; Hct = 5%; $t = 37$ °C; Medium: 120 mM KCl, 30 mM Tris-HCl, pH 7.4.
× Control; ○ 2 mM EGTA

Table 1

Effect of Ca-loading on the cation composition and the metabolic state of red cells.

Intracellular concentration [mmole/l cells]	Normal fresh red cells ($n = 12$)	Ca-loaded red cells* ($n = 6$)	Difference
K	107.5 ± 6.8**	110.0 ± 8.5	not significant
Na	8.7 ± 1.5	9.1 ± 1.3	not significant
Mg	2.52 ± 0.46	2.62 ± 0.42	not significant
ATP	1.15 ± 0.18	1.21 ± 0.14	not significant
2,3-DPG	3.95 ± 0.56	4.01 ± 0.71	not significant

* A23187 Ca-loading; $[Ca]_i = 0.5-2.0$ mM
** Mean ± SD values

with A 23187 and varying amounts of calcium in NaCl medium. During 1—2 min at 37 °C the required amount of calcium could be introduced to the cells (Fig. 1). The ionophore could be eliminated by thorough washing in 0.16 M KCl containing 0.5% albumin. After the washing procedure the K, Na, Mg, ATP and 2,3-DPG contents of the cells were within the physiological range (Table 1).

The effect of intracellular calcium on the cation transport processes could be well investigated in this preparation. Studying the K-transport in high K-medium by following the ^{42}K-efflux from cells preloaded with ^{42}K and calcium, the loss of intracellular K and cell shrinkage could be prevented. Using this sensitive technique under well defined conditions the K-efflux as a function of time did not give a straight line in the semilogarithmic plot. The curve seemed to be composed of two straight lines representing a rapid and a slow efflux component, respectively (Fig. 2). Increasing intracellular Ca-concentration to a few mM was still insufficient to get a linear K-efflux plot. This is most probably due to the inhomogeneous distribution of calcium within the cell population. The average intracellular calcium concentration needed for evoking rapid K-transport in fresh erythrocytes is relatively high (>0.5 mM).

Much lower $[Ca]_i$ is needed if propranolol is present in the system. On addition of 0.5—1 mM propranolol to fresh, ^{42}K-loaded red cells in the presence of external calcium, rapid Ca-influx and K-efflux take place at as low as 10—20 μM intracellular calcium concentrations. K-efflux against time gives a straight line on the semilogarithmic plot; it is, however, dependent on the presence of external calcium (Fig. 3).

If red cells loaded with 0.5—5 mM Ca are treated with 0.5—1 mM propranolol, K-efflux becomes also linear in the semilogarithmic plot reflecting a homogeneously enhanced K-permeability. At this high intracellular calcium concentration,

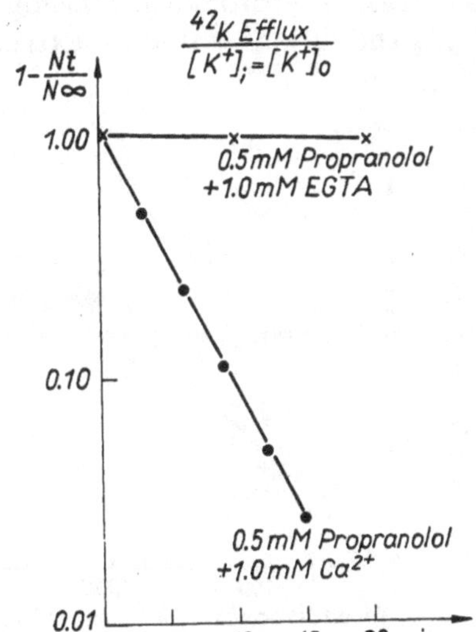

Fig. 3

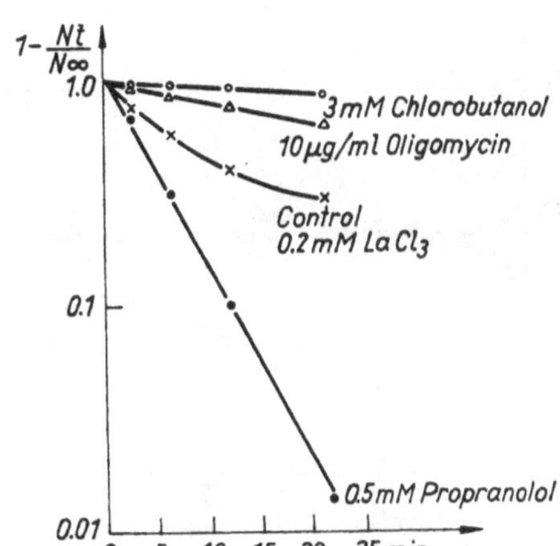

Fig. 4

Fig. 3. ^{42}K-efflux from ^{42}K-loaded fresh red cells in the presence of 0.5 mM propranolol ± 1 mM calcium.

Hct = 5%; t = 37 °C; Medium: 120 mM KCl, 30 mM Tris-HCl, pH 7.4.

Fig. 4. ^{42}K-efflux from calcium + ^{42}K-loaded fresh red cells in the presence of 0.5 mM propranolol.

Intracellular $[Ca^{2+}]$: 2.5 mM; Hct = 5%; t = 37 °C; Medium: 120 mM KCl, 30 mM Tris-HCl, pH 7.4.

× Control; ● 0.5 mM propranolol; ○ 3 mM chlorobutanol; △ 10 μg/ml oligomycin

K-efflux becomes independent of external calcium, indicating that propranolol increases not only the Ca-uptake, but has also a direct effect on the K-permeability of the membrane. Chlorobutanol and oligomycin are very potent inhibitors of the K-efflux in this system (Fig. 4).

In ATP-depleted red cells ≈ 20 μM intracellular calcium starts to evoke the rapid K-transport. This K-flux, however, becomes independent of external calcium only at inner Ca-concentrations exceeding 80—100 μM. The permeability increasing effect of 0.5—1 mM propranolol was demonstrated even in ATP-depleted red cells [6]. The varying relationships between intracellular calcium and membrane K-permeability obtained with fresh or ATP-depleted red cells ± propranolol call the attention to the paramount importance of well-defined conditions. This statement is even more obvious if we consider the exceedingly low free calcium levels evoking rapid K-transport in ATP-depleted ghosts loaded with Ca-EGTA buffers as reported by SIMONS [7].

Owing to the very high passive K-permeability the effect of intracellular calcium on the K accumulation via the K^+-Na^+-pump cannot be measured. The active Na-extrusion, however, can be well studied. On increasing the intracellular [Ca] the rate of the passive (ouabain-insensitive) Na-transport is slightly increased (to about 1.5 fold). The active Na-pump (ouabain-sensitive Na-transport) gets completely inhibited by 1 mM inner [Ca] (Fig. 5).

Next we studied the Ca-transport of Ca-loaded fresh erythrocytes. Raising the intracellular Ca-concentration from 10 to 1000 μM the rate of active Ca-extrusion

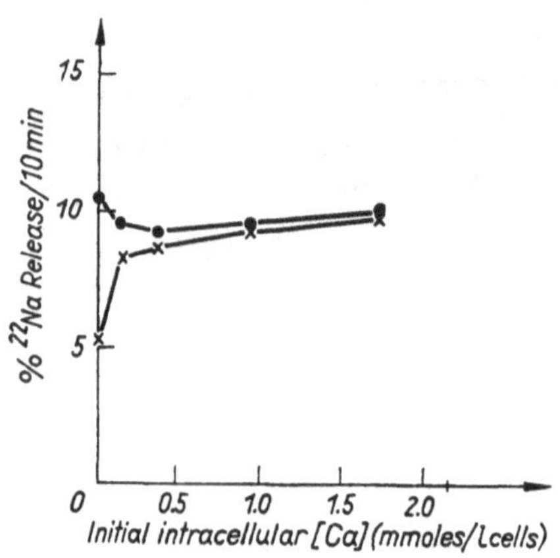

Fig. 5

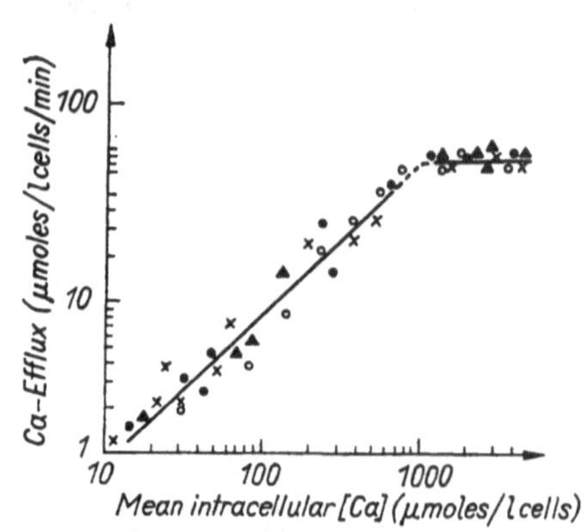

Fig. 6

Fig. 5. Effect of intracellular Ca-concentration on ^{22}Na-efflux from intact human red cells. Medium: 120 mM KCl, 15 mM NaCl, 2.0 mM $CaCl_2$, 15 mM Tris-HCl, *p*H 7.4, Hct = 5%; $t = 37$ °C.

● Control; × 10^{-5} M ouabain

Fig. 6. Effect of intracellular Ca-concentration on the rate of Ca-efflux from intact human red cells.

For incubation medium and conditions see the legend of Fig. 5. Data of four separate experiments (×, ●, ▲, ○)

continuously augmented (Fig. 6). The maximum value: 85 μM/l cell/min was reached at 1.0 mM $[Ca]_i$, thereafter the Ca-efflux rate reached a plateau. This type of Ca-activation corresponds to the so called "low Ca-affinity" Ca-Mg-ATPase activity of the red cell membrane preparations [8]. The rate of active Ca-pump was unaffected until ATP level did not fall below 100 μM. Under this level the Ca-pumping rate declined. These findings are in good agreement with the results obtained on isolated red cell membranes, showing that the Michaelis constant of Ca-Mg-activated ATPase for ATP is about 40 μM [9].
The rate of Ca-influx at intracellular calcium concentrations lower than 3.0 mM corresponded to the physiological low value. At higher $[Ca]_i$, however, it increased rapidly. At 5.0 mM $[Ca]_i$ a 25fold increase in the passive Ca-transport was observed, indicating the damage of the membrane. In ATP-depleted cells the passive Ca-permeability increased: without raising the intracellular calcium level the Ca-influx rate was doubled and at increased intracellular calcium levels the passive Ca-transport rate enhanced rapidly, at 5.0 mM $[Ca]_i$ being around 40—50 μmoles/l cell/min (Fig. 7).

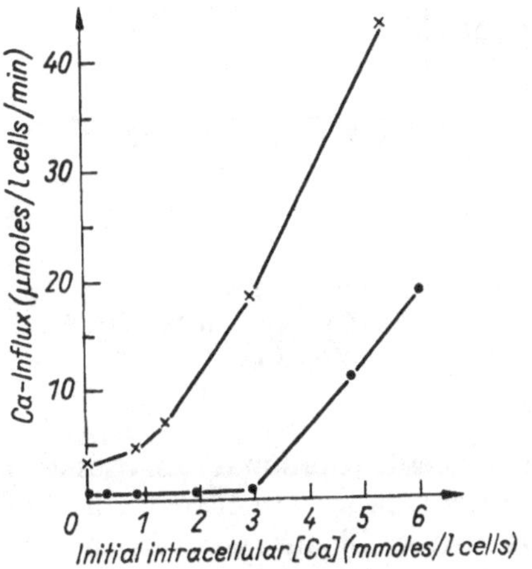

Fig. 7

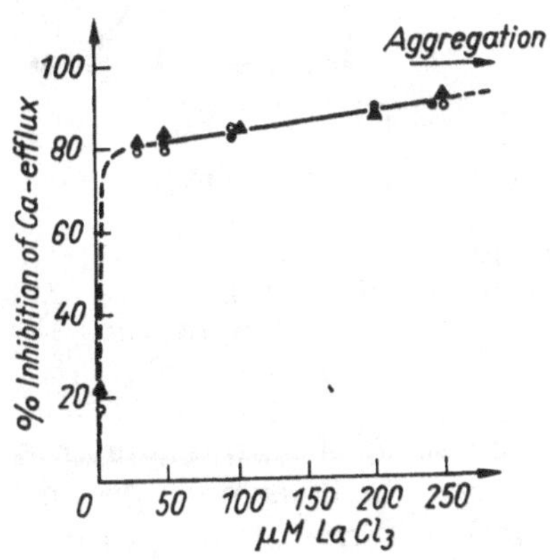

Fig. 8

Fig. 7. Effect of intracellular Ca-concentration on the rate of Ca-influx in fresh (●) and in exhaustively ATP-depleted (×) red cells.
The incubation medium (see the legend of Fig. 5) was supplemented with 5 μCi of ^{45}Ca. Hct = 20%; t = 37 °C

Fig. 8. Inhibition of Ca-efflux from Ca-loaded intact red cells by lanthanum.
For incubation medium and condition see legend of Fig. 5. Data of three separate experiments (●, ○, ▲) 10 min incubation periods. Initial intracellular [Ca] = 1.5—2.5 mM

The active Ca-pump was not affected by Ca or other divalent cations (Mg, Sr, Ba, Mn) in the medium. The presence of EDTA and EGTA as well as the substitution of Na or K by choline ions were of no consequence. These facts yielded evidence against a cation countertransport coupled to the Ca-pump. Ouabain and oligomycin did not inhibit the Ca-pump, whereas SH-reagents (NEM, PCMB, ethacrynate), propranolol and ruthenium red caused a partial inhibition of it. The most potent inhibitor of the Ca-pump was found to be the lanthanum ion applied in a concentration range from 50 to 250 μM (Fig. 8). Lanthanum induced inhibition of the Ca-pump could be suspended instantaneously by the addition of the impermeable chelator EGTA. Using ^{140}La the elimination of lanthanum from the cells could also be a well demonstrated indication that lanthanum did not penetrate the cells, but "immobilized" the Ca-pump from outside (Fig. 9).

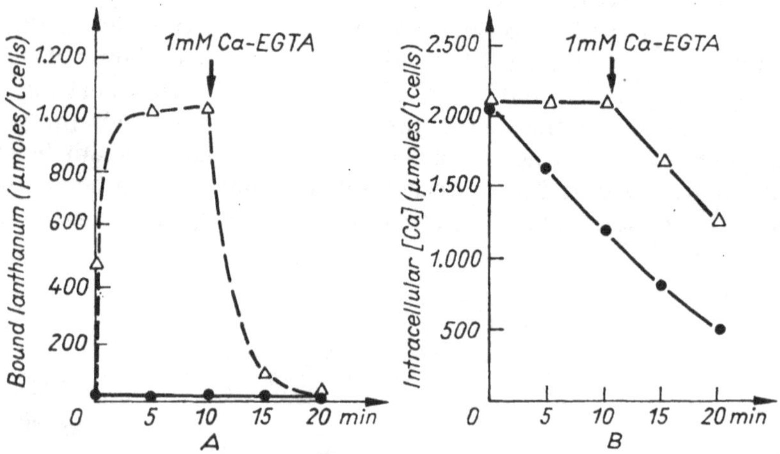

Fig. 9. Binding of lanthanum to the red cells (A) and inhibition of Ca-efflux by lanthanum (B).

For incubation medium see legend of Fig. 5. Hct = 12%; t = 37 °C. Initial intracellular [Ca] = 2.1 mM. ● 0.2 mM $LaCl_3$* + 1 mM Ca-EGTA; △ 0.2 mM $LaCl_3$*.
* In (A) supplemented with 5 μCi of $^{140}LaCl_3$.

The increase of the intracellular calcium concentration markedly distorts the cell's shape; the cells undergo an echinocytic disk-sphere transformation. The characteristic forms of this process are: the normal biconcave disk, the crenated disk, the crenated discoid, the crenated spheroid, the crenated sphere and the smooth sphere or microsphaerocyte [10].

The morphological transformation during Ca-loading and pumping is also well visible under the phase-contrast microscope. The normal biconcave disks (as shown in Fig. 10a) above 500 μM $[Ca]_i$ are rapidly converted to microsphaerocytes (Fig. 10b). The process is reversible. After pumping out Ca the cells gradually regain their biconcave disk shape (Figs. 10c, d). The same morphological changes were observed in finer details by scanning electronmicroscopy.

In energized cells the echinocytic disk-sphere transformation and its reversal is correlated only to the Ca-transport processes affecting the intracellular calcium level, and is independent of the concomitant changes in the K and Na-transport. Not only the relationship between $[Ca]_i$ and K-permeability is affected by propranolol but also the $[Ca]_i$-morphology relation is altered, so far as propranolol — being a stomatocytic agent — promotes the transitions towards the disk(oid) shape.

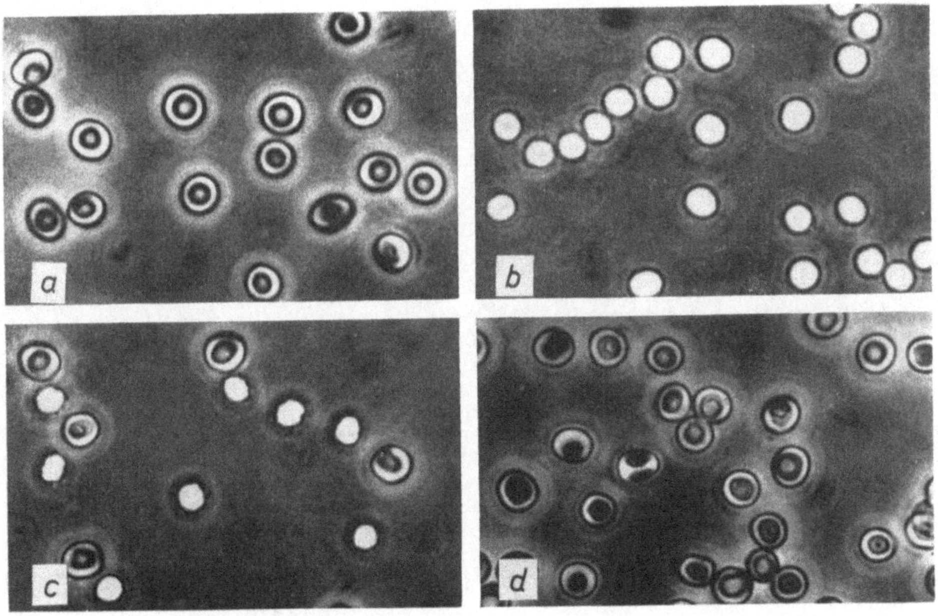

Fig. 10. The shape of red cells under phase-contrast microscope.
(a) Before ^{45}Ca loading; (b) after ^{45}Ca loading ($[Ca^{2+}]$ = 440 mM); (c) during Ca^{2+} pumping ($[Ca^{2+}]_i$ = between 10 and 50 μM); (d) after restoration $[(Ca^{2+}]_i$ = below 5 μM).

References

[1] Schatzmann, H. J.: Experientia **22**, 364 (1966)
[2] Lee, K. S., and B. C. Shin: J. gen. Physiol. **54**, 713 (1969)
[3] Olson, E. J., and R. J. Cazort: J. gen. Physiol. **53**, 311 (1969)
[4] Schatzmann, H. J.: Curr. Top. Membranes Transp. **6**, 125, (1975)
[5] Sarkadi, B., I. Szász, and G. Gárdos: J. Membrane Biol. **26**, 357 (1976)
[6] Gárdos, G., I. Szász, and B. Sarkadi: FEBS Proceedings **35**, 167 (1975)
[7] Simons, T. J. B.: J. Physiol., Lond. **256**, 227 (1976)
[8] Quist, E. E., and B. D. Roufogalis: Archs Biochem. Biophys. **168**, 240 (1975)
[9] Schatzmann, H. J.: J. Physiol., Lond. **235**, 551 (1973)
[10] Bessis, M. in: Red Cell Shape. Physiology, Pathology, Ultrastructure. M. Bessis, R. I. Weed and P. F. Leblond (eds). Springer, Berlin 1973

Acta biol. med. germ., Band 36, Seite 831—836 (1977)

Institut für Physiologische und Biologische Chemie der Humboldt-Universität Berlin, 104 Berlin, DDR

Mechanism of senescence of red blood cells

CH. TANNERT, G. SCHMIDT, D. KLATT, and S. M. RAPOPORT

Summary

An integral hypothesis is submitted on the interaction of endogenous and exogenous factors in the mechanism of senescence of erythrocytes and on their selective phagocytosis by autologous macrophages.
A method that allows quantitative determination and preparative yield of senescent erythrocytes is proposed.

A multitude of data has been accumulated on the problem of aging of red blood cells and the selective elimination of senescent cells. Recently it could be shown by KAY [1] that erythrocytes of high density, which were obtained by centrifugation according to the method of MURPHY [2], contain receptors on their surface on which autologous IgG is bound. These were old cells. If incubated with autologous macrophages about 40% of the cells of this density fraction were ingested. This was in contrast to young cells of low density, which did not bind γ-globulin.
Charging with IgG seems to be a predisposition for the selective elimination of old erythrocytes from the circulation. Model studies of KAY [1] with in vitro aged red blood cells revealed that these cells are phagocytosed by cultures of autologous macrophages only if pretreated with IgG.
Thus the mechanism of IgG-receptor exposure becomes a central question with respect to the understanding of the process of erythrocyte aging both in vivo and in vitro.
In our own investigations we used another technique to detect globulin-charged erythrocytes. We could show that erythrocytes aged in vivo can be agglutinated by an antiglobulin serum. In brief our method consists in the following: A 3%-suspension of human Od-erythrocytes was prepared in which the cells of high density are enriched by centrifugation in an isopyknic dextran gradient [3]. To 15 μl of this suspension 100 μl of an IgG enriched fraction of a goat antihuman-globulin serum was added. Agglutination macroscopically observable occurs after 45 min at room temperature. A quantification of the agglutinated cells was carried out by means of a normal light microscope.
In experiments with bloods of various volunteers agglutination could be observed almost only in the fraction preceeding the one with the highest density, that is fraction 8. In this fraction about 40% of the cells are agglutinable.
In all fractions a certain increase in the aggregation rates in the presence of anti-serum as compared with the controls was observed, only in fraction 8, however, this increase was drastic and highly significant.
Assuming that senescent cells were present only in fraction 8 a percentage of about 1% of senescent cells in the whole blood cell population is estimated. This

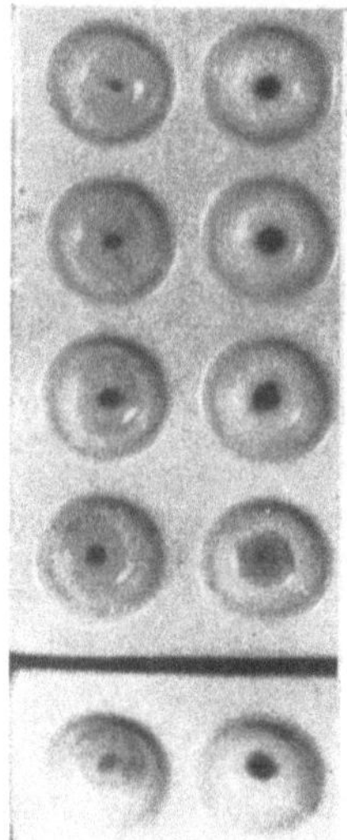

Fig. 1. Immunoagglutination of freshly drawn erythrocytes of increasing density.
Agglutination procedure see text.
Fraction No. 1, 6, 7, 8, 9: Human Od-erythrocytes of increasing density (see Table 1) prepared according to SCHMIDT et al. [3]. C: Controls with saline.
AHG: Test probes with goat antihumanglobulin serum. IgG enriched according to the following procedure: To antiserum in original concentration containing x mg protein, estimated by the Biuret method, x ml n-caprylic acid were added. After 30 min of precipitation at room temperature – with stirring – the supernatant fluid was separated from the precipitate by centrifugation and dialyzed against saline for at least 3 days.
Notice that the agglutination can be observed only in fraction 8 with AHG!

Table 1

Immunoagglutination of freshly drawn erythrocytes of increasing density.
Results given as mean microscopically estimated agglutination rates ± S.E.M. In brackets: number of counts (the number of counted cells was at least 300). Antiserum: IgG enriched fraction of goat antihumanglobulin serum (AHG); preparation procedure see Fig. 1. Fraction 1 to 9: Human Od-erythrocytes of increasing density were prepared according to SCHMIDT et al. [3].

Fr.	Density [g/ml]	Agglutination [% of agglutinated cells]		Significance of the difference (t-test)	
		with AHG (test probes)	with saline (controls)	controls vs. test probes	fr. 8. vs. other fractions (test probes
1	1.0857	12.1 ± 2.4 (9)	3.7 ± 2.4 (7)	$p < 0.05$	$p < 0.001$
2	1.0872				
3	1.0890				
4	1.0905				
5	1.9017				
6	1.0932	15.4 ± 3.3 (11)	4.6 ± 1.6 (13)	$p < 0.01$	$p < 0.001$
7	1.0960	7.2 ± 2.7 (4)	3.1 ± 1.2 (7)	nonsignif.	$p < 0.001$
8	1.0965	39.8 ± 4.7 (5)	8.1 ± 2.0 (12)	$p < 0.001$	
9	1.1000	21.1 ± 4.5 (4)	13.7 ± 1.4 (4)	$p < 0.02$	$p < 0.02$

figure is in good agreement with the daily degradation rate obtained by labeling techniques [4]. Experiments with antibody coupled to a solid phase are in progress now in order to estimate the true proportion of senescent cells.

In addition to the experiments of KAY [1] we should like to remark that the IgG-receptors of in vivo aged red blood cells and in vitro aged cells seem to be different, because in a first agglutination experiment with absorbed antisera in vivo aged cells agglutinated with antiglobulin serum absorbed by in vitro aged cells and vice versa.

Table 2

Identity of the IgG-receptors of red blood cells aged in vivo (senescent cells = SC) and in vitro (aged cells = AC).
Results given as mean agglutination rates ± S.E.M.
In brackets: number of counts (the number of counted cells was about 300).
In vitro aged cells (AC): Cells 3 times washed with 0.15 M phosphate-buffered saline, *p*H 6.9, incubated in the same medium for 14 hrs at 37 °C, washed 3 times and incubated in autologous serum with EDTA for 90 min at 37 °C, washed 4 times and resuspended in 0.15 M phosphate buffered saline at *p*H 7.4.
In vivo aged cells (SC): Prepared by means of density gradient centrifugation according to SCHMIDT et al. [3] as fraction 8.

Cells	Antiserum	Agglutination		Significance of the the differences (t-test): probes with AHG vs. controls
		macroscopic. visible	% of agglutinated cells	
AC	AHG	yes	55.8 ± 3.6 (4)	$p < 0.001$
AC	AHG_{as}	yes	78.0 ± 4.9 (3)	$p < 0.001$
AC	saline (control)	no	12.8 ± 2.0 (5)	
SC	AHG	yes	36.2 ± 3.2 (3)	$p < 0.01$
SC	AHG_{aa}	yes	61.1 ± 1.9 (3)	$p < 0.001$
SC	saline (control)	no	19.6 ± 2.2 (4)	

AHG: IgG enriched fraction of goat antihumanglobulin serum (preparation procedure see Fig. 1.)
AHG_{as}: AHG absorbed with SC; AHG_{aa}: AHG absorbed with AC

Based on the mentioned experimental data and on others, which are already known in part for a long time, we should like to propose now the following summarizing hypothesis on the mechanism of senescence and the selective phagocytosis of red blood cells aged in vivo.

The initial event in the aging of red blood cells consists most likely in intracellular changes, which are caused by increasing damage of the cell, probably by the formation of superoxide radicals.

Recently the age dependent changes of metabolic and physical properties of red blood cells have been summarized by various authors [5—7].

Of special interest in the described changes of the erythrocytes during aging is the decrease in the activities of a number of enzymes of the glycolysis and of the oxidative pentose pathway as well as the inactivation of Met-Hb-reductase. As a consequence of these alterations the concentrations of NADH and NADPH, ATP and 2,3-DPG decrease. Furthermore the Mg^{2+}-concentration and the K^{+}-concentration are diminished, while the concentrations of Na^{+}- and Ca^{2+}-ions increase in the cells [6, 8]. These changes in the ion concentrations are probably caused by damage of the membrane.

WEED et al. [8] conclude from the changes of ATP, DPG, Mg^{2+} and Ca^{2+} that during aging interaction between calcium and spectrin is enhanced and leads to a more rigid membrane. This should be regarded as an interesting hypothesis which has not yet been proved, since the concentrations of the free uncomplexed ATP and free Ca^{2+}-ions are rather uncertain. Furthermore experiments by Syllm-Rapoport et al. [9, 10] carried out with anaemic patients with total arrest of erythrocyte maturation showed that transfused cells which developed a high level of adenine nucleotides in vivo, especially of ATP, aged in a normal time span. The increase in rigidity of aging erythrocytes measured by means of filtration or aspiration into microcapillaries may be explained in different ways.

a) Evidence exists for age-dependent changes in the lipid content of the cell membrane [6, 11]. These could be caused by lipid peroxidation, e.g. by formation of superoxide radicals or to an increase in the activity of phospholipases possibly caused by the higher intracellular calcium [12].

b) A hypothesis proposed by Lessin [13] ascribes the increase of rigidity to the denaturation of hemoglobin, the amount of which increases by the deficiency of reduced pyridin nucleotides and glutathion. These so-called "Micro-Heinz-Bodies" are supposed to have a high affinity for the thiol groups at the inner membrane surface and may cause a permanent deformation and reduced deformability of the membrane (as well as a deterioration in the Na-K-ATPase activity). The possibility remains, however, that the observed deformation of the cell membrane may be due to artefacts.

Whatever the exact cause of the increase in rigidity may be, the result is the often confirmed impairment of the rheologic properties of the cells by reduced deformability. The smallest capillaries which the erythrocytes have to pass have a diameter of 2—3 μm, whereas the mean diameter of the red cell is 7 μm. Thus high flexibility of the membrane and the special shape of the red cell is required for the normal passage in certain regions of the microcirculation, e.g. in the cordsinus of the spleen [14]. This may have two remarkable consequences with regard to the more rigid old cells: a further increase in rigidity due to cell fragmentation [6, 8, 15—18], and, even more important, an accelerated partial degradation of the glycocalyx by extracellular hydrolytic enzymes either localized permanently at the places of slowed down cell passage or set free by enhanced mechanical influence of the rigid cells on the surrounding tissue.

Experiments of various authors point to the neuraminidases. The diminished surface charge of senescent erythrocytes caused by degradation of sialic acid has repeatedly been reported [19—24]. Most interesting in this connection are experiments of Bocci et al. [24] showing an accelerated loss of sialic acid from the red cell membrane by means of a brief period of stagnation of the splenic and renal circulation in vivo. The much shorter survival of extensively desialylated erythrocytes is well established [25]. In a recently published article Bocci [20] reviews the relevant literature on the problem in question and gives support to the hypothesis of an uneven body distribution of effective sialidase activity. Again the spleen seems to be the most remarkable organ. Finally experiments of Kay [1] gave evidence that the action of neuraminidases on erythrocytes exposes hidden IgG-receptors. Desialylated red cells are phagocytized by autologous macrophages in vitro if charged with autologous IgG after neuraminidase treatment.

It should be stressed, however, that not only exogenous factors contribute to the exposal of the senescence-receptors, but endogenous ones as well. Young and old erythrocytes which have been brought to the same content of neuraminic acid are degraded in different times, i.e. the old ones disappear much faster from the circulating blood [26].

The action of endogenous factors may be explained in the following way: if one imagines the cell membrane as a pyramide of glycolipoproteids it is conceivable that the conformational change of protein located on the basis of this pyramide at the inside of the membrane may alter the whole structure. The metabolic changes inside the cell during aging may be the cause for such events.

Thus the often expressed opinion must be rejected of the so-called "metabolic death" of the cells only by age-dependent loss of ATP. It is rather necessary to regard the mechanism of the senescence and the destruction of the senescent red cell as a complex process, affected by a great number of intracellular and extracellular events.

In conclusion we should like to summarize our hypothesis in a graphical form (Fig. 2).

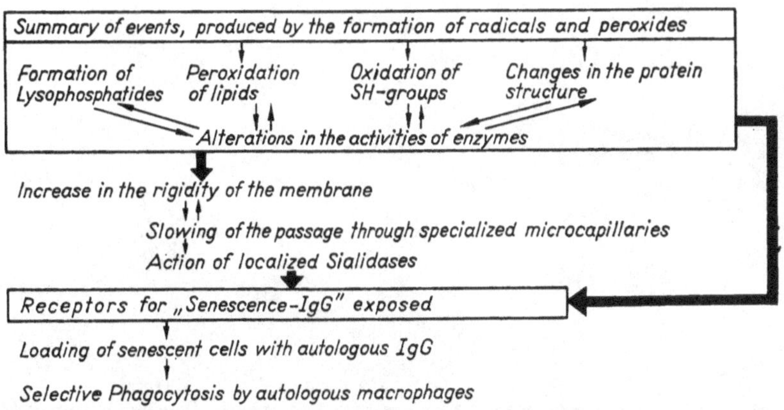

Fig. 2. Mechanism of senescence of erythrocytes and their selective phagocytosis

References

[1] Kay, M. M. B.: Proc. natn. Acad. Sci. U.S.A. **72**, 3521 (1975)
[2] Murphy, J.: J. Lab. clin. Med. **82**, 334 (1973)
[3] Schmidt, G., J. Gross, R. Moller, and R. Staak: Acta biol. med. germ. **34**, 1621 (1975)
[4] London, I. M., R. West, D. Shemin, and D. Rittenberg: J. biol. Chem. **184**, 351 (1950)
[5] Greenwalt, T. J., E. A. Steane, and N. E. Pine in: Glycoproteins of blood cells and plasma. G. A. Jamieson and T. J. Greenwalt (eds.). Philadelphia, Toronto 1971, p. 235
[6] Cohen, N. S., J. E. Ekholm, M. G. Luthra, and D. J. Hanahan: Biochim. biophys. Acta **419**, 229 (1976)
[7] Ganzoni, A. M., J. P. Barras, and H. R. Marti: Vox Sang. **30**, 161 (1976)
[8] Weed, R. J., P. L. La Celle, and E. W. Merill: J. clin. Invest. **48**, 795 (1969)
[9] Syllm-Rapoport, I., E. C. G. Hoffmann, and D. Schmidt: Acta biol. med. germ. **7**, 461 (1961)
[10] Syllm-Rapoport, I., G. Jacobasch, S. Prehn, and S. Rapoport: Blood **33**, 617 (1969)
[11] Shohet, S. B., and J. E. Haley in: Red Cell Shape. M. Bessis, R. I. Weed, and P. F. LeBlond (eds.). Springer, Berlin, Heidelberg, New York 1973, p. 41
[12] Allan, D., and R. H. Michell: Biochem. Soc. Transact. **3**, 751 (1975)

[13] LESSIN, L. S. in: Red Cell Shape. M. BESSIS, R. I. WEED, and P. F. LEBLOND (eds.). Springer, Berlin, Heidelberg, New York 1973, p. 151
[14] La CELLE, P. L.: Blood Cells **1**, 269 (1975)
[15] LA CELLE, P. L.: Biorheology. **9**, 51 (1972)
[16] WEED, R. J., and C. P. REED: Am. J. Med. **41**, 681 (1966)
[17] COOPER, R. A., and S. J. SHATTIL: New Engl. J. Med. **285**, 1514 (1971)
[18] GLASER, R., and A. LEITMANNOVÁ: Studia biophys. **48**, 219 (1975)
[19] DUROCHER, J. R., R. C. PAYNE, and M. E. CONRAD: Blood **45**, 11 (1975)
[20] BOCCI, V.: Experientia **32**, 135 (1976)
[21] DANON, D., and Y. MARIKOVSKY: C. r. Acad. Sci., Paris **253**, 1271 (1961)
[22] YAAEI, A.: Blood **33**, 159 (1969)
[23] GATTEGNO, L., D. BLADIER, and P. CORNILLOT: Hoppe-Seyler's Z. physiol. Chem. **356**, 391 (1975)
[24] BOCCI, V., G. P. PESSINA, and T. CONTI: Protides of the Biological Fluids. 21st Colloquium. H. PEETERS (ed.). Pergamon Press, Oxford 1973, p. 49
[25] JANCIK, J., R. SCHAUER, and H.-J. STREICHER: Hoppe-Seyler's Z. physiol. Chem. **356**, 1329 (1975)
[26] STOFFEL, W., and W. DÄRR: Hoppe-Seyler's Z. physiol. Chem. **356**, 385 (1975)

Acta biol. med. germ., Band 36, Seite 837—842 (1977)

Institut für Organische Chemie und Biochemie der Universität Hamburg, BRD

Transphosphatidylierungs-Reaktionen mit Phospholipase D in Erythrozytenmembranen

G. Gercken und W. Witt

Zusammenfassung

In Erythrozytenmembranen wurden durch Transphosphatidylierung mit Phospholipase D und Methanol oder Äthanol über 50 Mol-% der natürlich nicht vorkommenden Phosphoglyzerolipide Phosphatidylmethanol und Phosphatidyläthanol gebildet. Daneben entstanden 10—15 Mol-% Phosphatidsäure. Mit Propanol und Butanol nahm die Akzeptoreigenschaft für den Phosphatidylrest ab. Äthanolamin und Glyzerol wurden in der Phospholipase D-Reaktion mit Erythrozytenmembranen nicht umgesetzt.
Methanol und Äthanol erhöhten die Aktivitäten der (Na^+K^+)-ATPase, Mg^{++}-ATPase und Ca^{++}-ATPase wahrscheinlich durch Freilegung kryptischer Aktivität. Die stark veränderte Phospholipidzusammensetzung der Erythrozytenmembran nach Inkubation mit Phospholipase D und Methanol oder Äthanol hatte dagegen keinen signifikanten Einfluß auf die Aktivitäten der ATPasen.

Einleitung

Zur gezielten Modifikation der Membranlipide sind Phospholipasen wegen ihres spezifischen Angriffs auf die verschiedenen Esterbindungen geeignet. Sie wurden deshalb zur Untersuchung der molekularen Struktur von Biomembranen und funktioneller Membraneigenschaften benutzt. Durch den selektiven Abbau der Phospholipide lassen sich Aussagen über die Lokalisation der Phospholipide in der Membran, über die Art der Lipid-Protein-Wechselwirkung und über den Einfluß der verschiedenen Lipidmoleküle auf die Konformation der Membranproteine und die Aktivität membrangebundener Enzyme machen.
So sprach das Hydrolysemuster bei kombiniertem Einsatz von Phospholipase A_2 (Naja naja) und Sphingomyelinase (Staphylococcus aureus) an intakten Erythrozyten für einen Abbau nur der äußeren Lipidschicht und eine asymmetrische Verteilung der Phospholipide in der Erythrozytenmembran. Die cholinhaltigen Lipide Phosphatidylcholin und Sphingomyelin befinden sich hauptsächlich in der Außenschicht und Phosphatidyläthanolamin und -serin in der Innenschicht der Lipiddoppelmembran [1—3]. Durch den Angriff von Phospholipase A_2 und Phospholipase C an „sealed inside-out" Erythrozytenmembranen wurde die asymmetrische Verteilung der Phospholipide bestätigt [4]. Die Hydrolyse der Phospholipide in Erythrozytenmembranen durch Phospholipase A_2 und Phospholipase C verminderte die Aktivitäten der (Na^+K^+)-ATPase und Mg^{++}-ATPase [5] sowie der Ca^{++}-ATPase [6] bis zur vollständigen Inaktivität. Die (Na^+K^+)-ATPase ließ sich durch Zusatz von Phosphatidylserin reaktivieren, nicht dagegen durch Phosphatidylcholin oder -äthanolamin [5]. Dies ist ein Hinweis für die spezifische Funktion des polaren Restes der Phospholipide. Phospholipase D-Hydrolyse der Phosphoglycerolipide zu Phosphatidsäure erhöhte die (Na^+K^+)-ATPase-Aktivität [5].

Phospholipase D (Phosphatidylcholin-phosphatidohydrolase, EC 3.1.4.4) katalysiert jedoch nicht nur die Hydrolyse der Phosphat-Basen-Bindung in Phosphoglyzerolipiden, sondern kann auch den Phosphatidylrest nach folgendem Schema auf einen anderen primären Alkohol übertragen [7—9]:

Phosphatidyl-X + Enzym-SH → Phosphatidyl-S-Enzym + X-OH

Hydrolyse

Phosphatidyl-S-Enzym + HOH → Phosphatidsäure + Enzym-SH

Transphosphatidylierung

Phosphatidyl-S-Enzym + ROH → Phosphatidyl-R + Enzym-SH

In der Phospholipase D-Reaktion wird der Phosphatidylrest zunächst an eine SH-Gruppe des Enzyms gebunden und kann dann entweder in der Hydrolysereaktion auf Wasser oder in der Transphosphatidylierungsreaktion auf einen Akzeptoralkohol übertragen werden. Während in der ersten Reaktion Phosphatidsäure entsteht, wird in der zweiten ein entsprechend dem eingesetzten Alkohol nicht natürlich vorkommendes Phosphoglyzerolipid gebildet. Diese Reaktion eignet sich deshalb besonders gut, um die Bedeutung der polaren Reste der Phospholipide in Biomembranen zu untersuchen.

Material und Methodik

Erythrozyten vom Menschen wurden aus Konservenblut durch Zentrifugation und 3maliges Waschen in isotonem Phosphatpuffer isoliert und gereinigt. Nach der Methode von Dodge et al. [10] wurden „ghosts" durch hypotone Hämolyse hergestellt. Für die Umsetzung mit Phospholipase D wurden Erythrozyten bzw. „ghosts" in isotonem Tris-HCl-Puffer (50 mM Tris-HCl, 110 mM NaCl, 40 mM $CaCl_2$, *p*H 7,2) für 1 Std. bei 37 °C unter Zusatz von 1,25 Einheiten Phospholipase D (Boehringer, Mannheim, BRD) inkubiert. Für die Transphosphatidylierungsversuche enthielten die Ansätze 10 Vol.% Alkohol (Methanol, Äthanol, Propanol oder Butanol). Die $CaCl_2$-Konzentration wurde auf 10 mM gesenkt und die Phospholipase D-Aktivität auf 8 Einheiten erhöht. Das Endvolumen betrug 8 ml und die zugesetzte Erythrozytenmembran-Menge 12 mg, bezogen auf Protein. Phospholipase D wurde aus Wirsingkohl nach der Methode von Davidson und Long [11] bis zur Azetonfällung isoliert und anschließend durch Säulenchromatographie an DEAE-Zellulose bis zu einer spezifischen Aktivität von 2 Einheiten weiter gereinigt. Die Phospholipase D-Reaktion mit Erythrozyten oder „ghosts" wurde nach 1 Std. durch Abkühlen gestoppt und der Ansatz mit Mg^{++}-Sorbit-Imidazol-HCl-Puffer (68 mM Imidazol/HCl, 2,7 mM $MgCl_2$, 166 mM Sorbit, *p*H 7,2) auf 30 ml aufgefüllt und zentrifugiert. Das Sediment wurde 4mal in 30 ml dieses Puffers gewaschen.
Für die Analyse der Lipidzusammensetzung wurden die Erythrozyten bzw. „ghosts" nach Ways und Hanahan [12] (Methode II) extrahiert. Die Trennung der Lipide erfolgte durch 1- bzw. 2-dimensionale Dünnschichtchromatographie auf Kieselgel HR (Merck, Darmstadt, BRD); Fließmittel der 1. Dimension: Chloroform; Methanol-25% Ammoniak 60:30:10 (V/V); Fließmittel der 2. Dimension: Chloroform-Methanol-Eisessig-Wasser 60:30:4:2 (V/V). Nach Anfärbung mit Joddampf wurden die Lipide durch die modifizierte Phosphatbestimmung nach Bartlett [13] (Brockmann und Gercken, [14]) quantifiziert.
Die ATPasen-Aktivitäten wurden im optischen Fermenttest durch Kopplung mit der Pyruvatkinase- und Laktatdehydrogenase-Reaktion bestimmt. Der Testansatz hatte ein Endvolumen von 1,1 ml und enthielt 2 mM ATP, 2 mM Phosphoenolpyruvat, 0,2 mM NADH, 2 mM $MgCl_2$, 20 mM KCl, 94,4 mM (Gesamt-ATPase) bzw. 4,4 mM NaCl (Mg^{++}-ATPase), 0,1 mM EGTA, 44 mM Imidazol-HCl, *p*H 7,2, 4 Einheiten Pyruvatkinase und 7,7 Einheiten Laktatdehydrogenase (Boehringer, Mannheim). Den Ansätzen zur Bestimmung der Mg^{++}-ATPase wurde 0,2 mM Strophanthin zugesetzt. Zur Messung der Ca^{++}-ATPase-Aktivität wurden die Ionenkonzentrationen des Mg^{++}-ATPase-Bestimmungsansatzes wie folgt geändert: 0,6 mM $CaCl_2$, 5 mM $MgCl_2$, 100 mM KCl. Von dem gemessenen Umsatz wurde die Mg^{++}-ATPase-Aktivität abgezogen, um die Ca^{++}-ATPase-Aktivität zu erhalten. Die Aktivität

der Azetylcholinesterase wurde mit Azetylthiocholin und Ellman-Reagenz (5,5'-Dithiobis(2-nitrobenzoesäure)) bestimmt. Um die spezifischen Enzymaktivitäten zu berechnen, wurde der Proteingehalt der Erythrozytenmembranen nach Lowry et al. [15] gemessen.

Ergebnisse und Diskussion

Die Lipide der Plasmamembran intakter Erythrozyten werden von Phospholipase D nicht angegriffen, jedoch die von „ghosts", die demnach bereits eine veränderte Membran besitzen. Alkoholzusatz zu der Reaktion von „ghosts" mit Phospholipase D hatte nicht nur einen entscheidenden Einfluß auf das entstandene Endprodukt (Phosphatidylalkohol statt Phosphatidsäure), sondern auch auf die Umsetzung der verschiedenen Phosphoglyzerolipide und die Größe des Umsatzes. Während bei der Reaktion mit Phospholipase D im alkoholfreien Medium nur Phosphatidylcholin angegriffen wurde und etwa 20 Mol-% Phosphatidsäure entstanden, bewirkte Methanol eine weitgehende Umsetzung auch von Phosphatidyläthanolamin und -serin, wobei über 50 Mol-% Phosphatidylmethanol neben etwa 15 Mol-% Phosphatidsäure gebildet wurden. Hieraus wird deutlich, daß im wäßrigen Medium mit relativ geringem Alkoholzusatz die Transphosphatidylierung die bevorzugte Reaktion gegenüber der Hydrolyse ist, und daß der Alkohol wahrscheinlich durch eine Membranauflockerung Phosphatidyläthanolamin und -Serin für das Enzym zugänglich macht.

Abb. 1 a und b zeigen die 2-dimensionale Dünnschichtchromatographie der Erythrozytenmembran-Lipide, mit der eine vollständige Trennung der Phospholipide erreicht wird. Die Verteilung auf die 4 Hauptfraktionen Phosphatidylcholin, -äthanolamin, -serin und Sphingomyelin der natürlichen Lipidzusammensetzung (Abb. 1 a) wurde durch die Reaktion mit Phospholipase D und Methanol stark

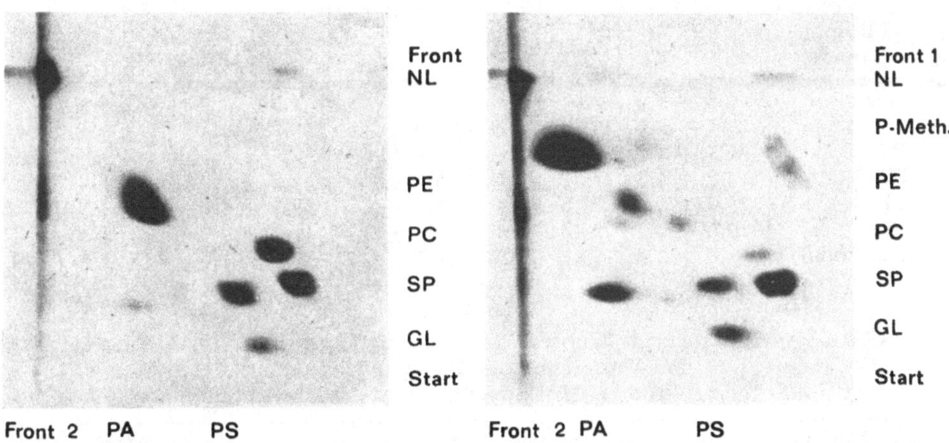

Abb. 1 a) 2-Dimensionale DC der Erythrozytenmembran-Lipide auf Kieselgel HR. Fließmittel 1: Chloroform-Methanol-25% Ammoniak 60:30:10; Fließmittel 2: Chloroform-Methanol-Eisessig-Wasser 60:30:4:2.

b) 2-Dimensionale DC der Erythrozytenmembran-Lipide nach Umsetzung mit Phospholipase D und Methanol auf Kieselgel HR.

Fließmittel 1: Chloroform-Methanol-25% Ammoniak 60:30:1:; Fließmittel 2: Chloroform-Methanol-Eisessig-Wasser 60:30:4:2

verändert: Die 3 Phosphoglyzerolipide, vor allem Phosphatidylcholin, waren deutlich vermindert; dafür war neben einer Zunahme von Phosphatidsäure eine neue Hauptfraktion aufgetreten, die als Phosphatidylmethanol identifiziert wurde (Abb. 1 b). Sphingomyelin wurde nicht umgesetzt.

Die Identifizierung des neu entstandenen, natürlich nicht vorkommenden Phosphatidylmethanols wurde abgesichert 1. durch die chemische Synthese von Phosphatidylmethanol einmal durch Kondensation von Phosphatidsäure und Methanol in Gegenwart von N,N-Dizyklohexylkarbodiimid in Pyridin, und zum andern durch Methylierung von Phosphatidsäure mit Diazomethan und 2. biochemisch durch Umsetzung von Phosphatidylcholin mit Phospholipase D und Methanol. Spektroskopisch und chromatographisch erwiesen sich die auf verschiedenen Wegen hergestellten Verbindungen als gleich mit dem bei der Erythrozytenmembran-Inkubation entstandenen Phospholipid.

Tab. 1 faßt die Werte der Lipidzusammensetzung nach Inkubation mit den verschiedenen Alkoholen und Phospholipase D zusammen. Die Wirkung von Äthanol war ähnlich der von Methanol. Es wurde eine Ausbeute von über 50 Mol-% Phosphatidyläthanol erzielt. Die Akzeptoreigenschaft des Alkohols nahm über Propanol und Butanol ab, wie die stärkere Bildung von Phosphatidsäure und die geringere Bildung des entsprechenden Phosphatidylalkohols zeigten. Mit Äthanolamin und Glyzerol, die für die Reaktion von Phosphatidylcholin mit Phospholipase D geeignete Akzeptoren sind, ließ sich keine Transphosphatidylierungsreaktion bei der Inkubation von ghosts mit Phospholipase D nachweisen.

Tabelle 1

Phospholipidzusammensetzung von „ghosts" nach Inkubation mit Phospholipase D und 10 Vol.% Alkohol.
(Mol-%; Mittelwerte $\pm$ Standardabweichung; $n = 3$)

Phospho-lipide	Kontrolle	Methanol	Äthanol	Propanol	Butanol
Start	2,8 ± 1,4	0,3 ± 0,4	—	1,6 ± 1,4	0,8 ± 1,4
SP	24,1 ± 2,9	21,0 ± 2,3	21,9 ± 2,6	19,2 ± 2,4	17,9 ± 2,6
PC	30,5 ± 1,9	0,9 ± 0,5	1,1 ± 1,3	2,4 ± 1,5	11,2 ± 4,0
PS	13,8 ± 1,3	5,9 ± 2,2	8,4 ± 1,3	11,2 ± 1,7	11,4 ± 1,3
PE	25,0 ± 2,1	3,8 ± 1,0	4,9 ± 2,2	13,6 ± 1,0	18,1 ± 4,0
PA	2,0 ± 0,7	14,8 ± 7,2	11,8 ± 3,7	12,2 ± 4,2	19,8 ± 1,4
P-Alkohol	1,8 ± 2,5	53,3 ± 8,2	51,3 ± 9,7	39,9 ± 1,6	22,5 ± 6,3

Tabelle 2

ATPase-Aktivitäten von "ghosts" nach Inkubation mit 10 Vol. % Alkohol und Umsetzung mit Phospholipase D.
(μMol/Std. · mg Protein; Mittelwerte $\pm$ Standardabweichung; $n = 3$)

Inkubation	(Na^+K^+)-ATPase	Mg^{++}-ATPase	Ca^{++}-ATPase
Kontrollen	0,08 ± 0,06	0,06 ± 0,02	0,24 ± 0,17
Ca^{++} (10 mM)	0,16 ± 0,09	0,06 ± 0,03	0,25 ± 0,14
Methanol	0,26 ± 0,11	0,15 ± 0,05	0,89 ± 0,55
Äthanol	0,23 ± 0,13	0,15 ± 0,06	0,71 ± 0,15
Meth. + PL D	0,38 ± 0,09	0,34 ± 0,15	0,81 ± 0,09
Äth. + PL D	0,28 ± 0,04	0,22 ± 0,08	0,54 ± 0,20

Tab. 2 zeigt die Veränderungen der ATPasen-Aktivitäten durch die Vorinkubation mit Zusatz von 10 mM $CaCl_2$ allein, Methanol oder Äthanol bzw. Methanol oder Äthanol zusammen mit Phospholipase D. Die Versuche mit Erhöhung der Ca^{++}-Ionenkonzentration auf 10 mM dienten dazu, den Ca^{++}-Ionen-Effekt von der Alkohol- und Phospholipase D-Wirkung abgrenzen zu können. Die Inkubation der „ghosts" mit Methanol und Äthanol führte zu einer Steigerung der Aktivitäten der (Na^+K^+)-ATPase, Mg^{++}-ATPase und Ca^{++}-ATPase. Durch Zusatz von Phospholipase D wurden diese Aktivitätssteigerungen nicht signifikant verändert. Propanol und Butanol ohne und mit Phospholipase D-Zusatz verminderten dagegen die ATPase-Aktivitäten stark. Diese Alkohole hatten in der zugesetzten Menge bereits einen denaturierenden Einfluß auf die Membranstruktur und erniedrigten das Lipid/Protein-Verhältnis. Die Azetylcholinesterase-Aktivität wurde durch die Reaktion von „ghosts" mit Phospholipase D und Methanol nicht verändert. Die Aktivitätssteigerung der ATPasen durch Methanol und Äthanol läßt sich dadurch deuten, daß diese Alkohole die Permeabilität der Membran verändern, so daß kryptische Aktivität freigelegt wird.

Um diese Annahme zu prüfen, wurden die Versuche wiederholt und die Erythrozytenmembranen nach Inkubation zur Freilegung der kryptischen Aktivität 2mal in Trockeneis-Azeton gefroren und bei 37 °C aufgetaut [16]. Gegenüber den Kontrollen der nicht mit der Gefrier-Tau-Technik behandelten „ghosts" wurden die Aktivitäten der (Na^+K^+)-ATPase und der Mg^{++}-ATPase verdoppelt (Tab. 3). Zusatz von Methanol oder Äthanol und Umsetzung mit Phospholipase D führten zu keiner weiteren signifikanten Aktivitätssteigerung. Das Ergebnis spricht für die Deutung, daß Methanol und Äthanol die ATPasen-Aktivität durch Freilegung der kryptischen Aktivität steigern. Bemerkenswert ist, daß die stark veränderte Phospholipidzusammensetzung der mit Phospholipase D behandelten Erythrozytenmembranen offensichtlich keinen Einfluß auf die Aktivität der membrangebundenen Enzyme hat.

Tabelle 3

ATPase-Aktivitäten von "ghosts", behandelt mittels Gefrier-Tau-Technik, nach Inkubation mit 10 Vol. % Alkohol und Umsetzung mit Phospholipase D (µMol/Std. · mg Protein; Mittelwerte $\pm$ Standardabweichung; $n = 3$)

Inkubation	(Na^+K^+)-ATPase	Mg^{++}-ATPase	Ca^{++}-ATPase
Kontrollen	0,15 ± 0,06	0,12 ± 0,04	0,25 ± 0,08
Ca^{++}	0,21 ± 0,15	0,14 ± 0,04	0,38 ± 0,23
Methanol	0,21 ± 0,05	0,21 ± 0,07	0,57 ± 0,34
Äthanol	0,18 ± 0,09	0,20 ± 0,08	0,38 ± 0,20
Meth. + PL D	0,37 ± 0,15	0,25 ± 0,13	0,48 ± 0,29
Äth. + PL D	0,23 ± 0,07	0,19 ± 0,05	0,39 ± 0,22

Die Möglichkeit, die polaren Reste der Phosphoglyzerolipide innerhalb der Membran verändern zu können, wobei die Vielfalt der verschiedenen Phospholipide durch ihre Umwandlung in die einheitliche Fraktion des entsprechenden Phosphatidylalkohols aufgehoben wird, erscheint uns als ein interessantes Modell für weitere Untersuchungen über die strukturelle und funktionelle Bedeutung der polaren Reste der Phospholipide.

Literatur

[1] Verkleij, A. J., R. F. A. Zwaal, B. Roelofsen, P. Comfurius, D. Kastelijn u. L. L. M. van Deenen: Biochim. biophys. Acta **323**, 178 (1973)
[2] Zwaal, R. F. A., B. Roelofsen, P. Comfurius u. L. L. M. van Deenen: Biochim. biophys. Acta **406**, 83 (1975)
[3] Renooij, W., L. M. G. van Golde, R. F. A. Zwaal u. L. L. M. van Deenen: Eur. J. Biochem. **61**, 53 (1975)
[4] Kahlenberg, A., C. Walker u. R. Rohrlick: Can. J. Biochem. **52**, 803 (1974)
[5] Roelofsen, B., u. L. L. M. van Deenen: Eur. J. Biochem. **40**, 245 (1973)
[6] Coleman, R., u. T. A. Bramley: Biochim. biophys. Acta **382**, 565 (1975)
[7] Dawson, R. M. C.: Biochem. J. **102**, 205 (1967)
[8] Yang, S. F., S. Freer u. A. A. Benson: J. biol. Chem. **242**, 477 (1967)
[9] Saito, M., E. Bourque u. J. Kanfer: Archs Biochem. Biophys. **164**, 420 (1974)
[10] Dodge, J. T., C. Mitchell u. D. J. Hanahan: Archs Biochem. Biophys. **100**, 119 (1963)
[11] Davidson, F. M., u. C. Long: Biochem. J. **69**, 458 (1958)
[12] Ways, P., u. D. J. Hanahan: J. Lipid Res. **5**, 318 (1964)
[13] Bartlett, G. R.: J. biol. Chem. **234**, 466 (1959)
[14] Brockmann, U., u. G. Gercken: Clinica chim. Acta **23**, 489 (1969)
[15] Lowry, O. H., N. J. Rosebrough, A. C. Farr u. R. J. Randall: J. biol. Chem. **193**, 265 (1951)
[16] Hanahan, D. J., J. Ekholm u. G. Hildenbrandt: Biochemistry **12**, 1374 (1973)

Adresse: Prof. Dr. Günther Gercken und Wolfgang Witt,
Institut für Organische Chemie und Biochemie der Universität Hamburg, Martin-Luther King-Platz 6, 2000 Hamburg 13

Acta biol. med. germ., Band 36, Seite 843–844 (1977)

Department of Biochemistry, Tokyo Medical and Dental University, Yushima, Bunkyo-ku, Tokyo, Japan

New technique for the separation of membrane surface proteins

M. Nakao, K. Sano, and H. Ohta

The concept that there is a large difference in the structure of the red cell membrane between outer surface and the inner surface has been well established: Various methods have been devised to label the outer surface proteins from outside. The isolation of labelled proteins, however, still poses many problems. In the 6th Erythrocyte Symposium in 1970 I reported that a PCMB derivative of aminoethyl dextran with an average molecular weight of 250,000 inhibited both (Na^+, K^+) transport and (Na^+, K^+)ATPase from the outside of the membrane.
Since the mercury compound is a reversible modifier, we synthesized bromoacetylamidoethyl dextran ($BrCH_2CONH \cdot CH_2 \cdot CH_2$-R), which behaved quite similarly to the PCMB derivative of dextran. Then we tried to utilize similar macromolecular compounds which could be split to release the original peptides after isolation.
We newly synthesized bromoacetyldextran esters with average molecular weights (M.W.) of 250,000 and 500,000, and diazotized dextran with a M.W. of 500,000. Peptides bound to these dextrans could be easily separated from almost all proteins in SDS and could be easily split after separation.
In a preliminary experiment, bromoacetyldextran (abbreviated BAD T-250, average M.W. 250,000) with 37 bromoacetyl groups/molecule was reacted with red cell membranes for 1 hr at 37 °C. After washing, the membrane fraction was solubilized and electrophoresed in 5% SDS polyacrylamide gel; As seen in fig. 1 several broad bands appeared. Intact cells reacted previously with BAD T-250 for 1 hr gave a diffuse band above the spectrins. Cells and membranes which were reacted with the reagents previously treated with cysteine, showed no change as compared with controls.

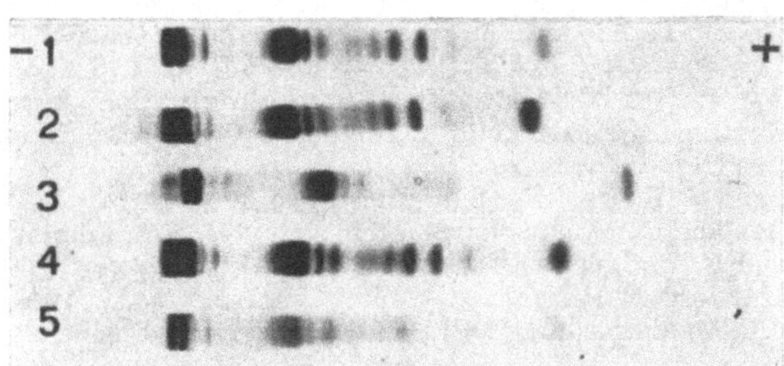

Fig. 1. SDS polyacrylamide electropherogram of BAD T-Treated red cell membranes. 1. Control without BAD T-250; 2. intact cells with BAD T-250; 3. membranes with BAD T-250; 4. intact cells with BAD T-250; previously treated with cysteine; 5. membranes with BAD T-250 previously treated with cysteine

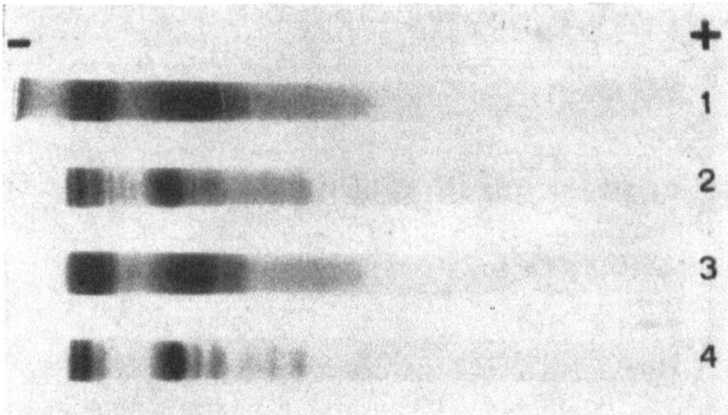

Fig. 2. Red cell membranes were treated with (1, 2) and without (4) diazotized dextran T500 and with $Na_2S_2O_4$ for 30 min at 37°. In the third column the membranes were reacted with diazo dextran, the excess of diazo compound was eliminated with resorcine, thereafter $Na_2S_2O_4$ was applied

In Fig. 2 the membranes were treated with diazotized dextran (T 500). The pattern in the control which had been split after diazotation did not differ from the original membranes without diazodextran.

The intact cell was reacted with the diazodextran and electrophorised. The gels corresponding above the molecular weight 250,000 were cut off, treated with $Na_2S_2O_4$ and extracted electrophoretically. The second SDS PAGE showed two main bands corresponding to band III. These results suggest strongly that a part of band (III) and/or PAS (I) have peptide moieties which protrude to the outside, besides the sugar moiety.

References

[1] Maddy, A. H.: Biochim. biophys. Acta **88**, 390 (1968)
[2] Berg, H. G.: Biochim. biophys. Acta **183**, 65 (1969)
[3] Bretscher, M. S.: J. molec. Biol. **58**, 775; **59**, 351 (1971)
[4] Bretscher, M. S.: J. molec. Biol. **71**, 523 (1972)
[5] Ohta, H., J. Matsumoto, K. Nagano, M. Fujita, and M. Nakao: Biochem. biophys. Res. Commun. **42**, 1127 (1971)
[6] Nicolson, G. L., V. T. Marchesi, and S. J. Singer: J. Cell Biol. **51**, 265 (1971)
[7] Nicolson, G. L., S. P. Masouridis, and S. J. Singer: Proc. natn. Acad. Sci. U.S.A. **68**, 1416 (1971)
[8] Furusawa, M., and H. Adachi: Expl Cell Res. **50**, 497 (1968)
[9] Nicolson, G. L., and S. J. Singer: Proc. natn. Acad. Sci. U.S.A. **68**, 942 (1971)
[10] Bender, W. W., H. Garran, and H. C. Berg: J. molec. Biol. **58**, 783 (1971)
[11] Carraway, K. L., D. Kobylka, and R. B. Triplett: Biochim. biophys. Acta **241**, 934 (1971)
[12] Steck, T. L., and G. Dawson: J. biol. Chem. **249**, 2135 (1974)
[13] Phillips, D. R., and M. Morrison: Biochem. biophys. Res. Commun. **40**, 284 (1971)
[14] Shin, B. C., and K. L. Carraway: Biochim. biophys. Acta **345**, 141 (1974)
[15] Gahmberg, C. G., and S. Hakamori: J. biol. Chem. **248**, 4311 (1973)
[16] Dodge, J. T., C. Mitchell, and D. J. Hanahan: Archs Biochem. Biophys. **100**, 119 (1963)
[17] Fairbanks, G., T. L. Steck, and D. F. H. Wallach: Biochemistry **10**, 2606 (1971)

Acta biol. med. germ., Band 36, Seite 845—846 (1977)

Biophysics Department, St. Mary's Hospital Medical School, London, W2 1PG, U.K.

Observations on some factors affecting the flexibility of erythrocyte membranes

M. W. Rampling and J. A. Sirs

An automatic recording centrifuge has been developed for measuring the flexibility of the erythrocyte [1], and has been used for a prolonged and continuing study of factors which affect this flexibility. The technique is based on the fact that at normal haematocrits the red cells in whole blood are all in contact, so when they are packed down by low centrifugal forces (200× *g* is used) the rate of packing is determined by the average flexibility of the cell population. It is this rate of packing that is measured with the automated centrifuge. The advantages of the technique are that it can be used with anticoagulated whole blood, the measurement takes only a few minutes to perform and it gives an average value for the whole erythrocyte population.

In a preliminary study the effects of the anticoagulants heparin, ACD and EDTA on flexibility were investigated. Of these, heparin had the least detrimental effect at normal concentration, while ACD and EDTA both reduced erythrocyte flexibility to a significant and similar extent [2]. This is of functional significance since the rate of oxygen exchange of cells from heparinised blood is known to be optimal, being determined only by the chemical reaction rate of the haemoglobin [3], while by comparison the rate for cells from ACD blood is reduced [4].

Perhaps the most interesting result to emerge from these investigations of anticoagulant techniques was the extreme loss of flexibility induced by defibrination of blood [2]. It was not at first clear whether this was due to the loss of fibrinogen, or to activation of one or more of the clotting factors which then interacted with the red-cell membrane. The action of fibrinogen was demonstrated by measurements on blood which had been anticoagulated with heparin and subsequently defibrinated with Ancrod [5], and on suspensions of washed cells to which various concentrations of fibrinogen had been added [6]. This showed that the flexibility of the erythrocyte is directly related to the concentration of plasma fibrinogen, i.e. increasing the concentration of fibrinogen increases the flexibility of the erythrocytes.

This phenomenon has recently been investigated further. It has been found that a good linear correlation exists between the flexibility measurement made with the centrifuge technique and the fibrinogen concentration. The slope of this relationship appears to be independent of subject from whom the cells are taken, but the residual flexibility at zero fibrinogen concentration varies from subject to subject. The effect of plasmic degradation products of fibrinogen [7] on flexibility was also investigated. The early digestion products, X and Y, were found to be almost as efficient as whole fibrinogen in affecting the flexibility but the late fragments, D and E, had no effect. These results show an interesting similarity to the results of binding studies of fibrinogen to staphylococci [8].

Dextran, a long chain polysaccharide molecule which has been used clinically as a volume expander and for prophylaxis against post-operative thrombosis, was also studied. It was found to have an opposite effect on flexibility to that of fibrinogen, that is dextran reduces red-cell flexibility [6]. Since it is known that in high concentration dextran can cause precipitation of fibrinogen [9], it was originally suggested that the effect of dextran was a secondary one, due to the loss of fibrinogen that it caused from plasma. It is now clear however that this is not the case. Dextran has been shown to bind directly to the erythrocyte membrane and to reduce membrane flexibility at concentrations which are too low to affect the concentration of fibrinogen in plasma [10].

A survey has been made of the variation of erythrocyte flexibility within a healthy population [11]. A comparison of the results for the two sexes showed no difference in the mean value of the flexibility for the two groups, however the female values were significantly more variable than those of the males. This increased variability presumably reflects changes in the concentration of fibrinogen, and of other circulating factors which affect flexibility, related to the menstrual cycle. Within the population as a whole there was a small (3%) but highly significant group (of both sexes) whose red cells had flexibility values more than three standard deviations from the mean. Measurements on these subjects of a wide variety of soluble plasma components showed their concentrations to be in the normal range. It was also found that the flexibility of these cells did not change significantly when they were suspended in plasma from subjects with erythrocyte flexibility in the normal range. These results suggest that the differences are due to inherent structural factors in the erythrocyte membrane.

A survey is currently taking place of various pathological conditions in order to discover those that are characterized by abnormal flexibility.

References

[1] SIRS, J. A.: Phys. Med. Biol. **15**, 9 (1970)
[2] RAMPLING, M. W., and J. A. SIRS: Phys. Med. Biol. **15**, 15 (1970)
[3] SIRS, J. A.: J. Physiol., Lond. **189**, 461 (1967)
[4] SIRS, J. A.: J. Physiol., Lond. **203**, 93 (1969)
[5] MYERS, P., M. W. RAMPLING, and J. A. SIRS: J. Physiol., Lond. **230**, 51P (1972)
[6] RAMPLING, M. W., and J. A. SIRS: J. Physiol., Lond. **223** 199, (1972)
[7] MARDER, V. J., and A. Z. BUDZYNSKI: Thromb. Diath. Haemorrhag. **33**, 199 (1975)
[8] HAWIGER, J., D. K. HAMMOND, and S. TIMMONS: Nature, Lond. **258**, 643 (1975)
[9] RAMPLING, M. W.: Biochem. J. **143**, 767 (1974)
[10] RAMPLING, M. W.: Biochem. Pharmac. **25**, 751 (1976)
[11] RAMPLING, M. W., and J. A. SIRS: Biorheology **13**, 101 (1976)

Acta biol. med. germ., Band 36, Seite 847—858 (1977)

Pharmakologisches Institut der Universität Ulm, D-79 Ulm/Donau, Oberer Eselsberg, GFR

Purification and properties of high-affinity Ca^{2+}-ATPase of human erythrocyte membranes

H. U. Wolf, G. Dieckvoss, and R. Lichtner

Summary

1. The high-affinity Ca^{2+}-ATPase of human erythrocyte membranes, an intrinsic membrane protein, was purified by gel chromatography on Sepharose CL-6B. During the chromatographic procedure the enzyme was protected against denaturation and aggregation by combining it with mixed micelles consisting of the nonionic detergent Tween 20 and crude phosphatidylcholine.
2. The highest value of specific activity of the isolated enzyme amounts to 3.1 U/mg protein. This represents a purification ratio of 155 as related to the membrane protein content and of 25,900 as related to the protein content of the entire cells.
3. The high-affinity Ca^{2+}-ATPase has been characterized as a protein complex consisting of three different subunits with molecular weights of 145,000 ± 5,000 (α-subunit), 115,000 ± 5,000 (β-subunit), and 105,000 ± 5,000 (γ-subunit) Dalton. The α-subunit can be labelled by γ-^{32}P-ATP indicating that the active site of the high-affinity Ca^{2+}-ATPase is located on this subunit.
4. The purified enzyme combined with mixed micelles consisting of Tween 20 and crude phosphatidylcholine shows kinetic properties, which do not differ significantly from those of the membrane-bound enzyme: K_m = 36—65 μM (depending on the Ca^{2+}-concentration), enzyme-Ca^{2+} dissociation constant = 1.9 μM, *p*H-optimum = 7.0—7.1, slight activation by monovalent cations, such as Na^+, K^+, NH_4^+, and Rb^+, activation by divalent cations, such as Zn^{2+}, Mn^{2+}, and Co^{2+}, and inhibition by chemical modifiers, such as Hg^{2+}, mersalyl, 5,5'-dithio-bis-(2-nitrobenzoate), 2-methoxy-5-nitrotropone, 2,4,6-trinitrobenzene-sulfonate, and diazotized sulfanilate.
5. The isolated high-affinity Ca^{2+}-ATPase does not exhibit any K^+-activated p-nitrophenylphosphatase activity.

Introduction

Erythrocyte membranes show two Ca^{2+}-ATPase activities, which differ with respect to their affinities for Ca^{2+}-ions. The so-called low-affinity Ca^{2+}-ATPase, which shows a Ca^{2+}-enzyme dissociation constant in the mM range, is often considered to be responsible for the shape and deformability of the erythrocyte membrane [1]. The so-called high-affinity Ca^{2+}-ATPase shows a Ca^{2+}-enzyme dissociation constant of about 1 μM. According to Schatzmann [2] it is responsible for the active Ca^{2+}-outward transport across the erythrocyte membrane. It thus maintains the very steep gradient between a free Ca^{2+} concentration in the 0.1 to 1.0 μM range in the cytoplasm and 2 mM Ca^{2+} outside the cell. Since Schatzmann

Abbreviations: Ca^{2+}-ATPase = Ca^{2+}-activated adenosinetriphosphate phosphohydrolase (EC 3.6.1.3); CMA = chemical modifying agent; DFP = Diisopropylfluorophosphate; MOPS = Morpholinopropane-sulfonic acid; γ-^{32}P-ATP = γ-position isotopically labelled ATP; PMSF = Phenylmethylsulfonylfluoride; SDS = Sodium dedocylsulfate; TLCK = N^α-Tosyl-L-lysyl-chloromethane-hydrochloride

[2] gave the experimental evidence for the biological significance of this enzyme, we were interested in the purification and the characterization of this enzyme in its isolated form. The final aim is to recombine this enzyme with artificial lipid bilayers and to show its biological function in the isolated form.

By now a series of very effective protein purification and separation methods is well-known. However, most of these methods do not allow to isolate a protein, especially an intrinsic membrane protein, in its original state, i.e. without any significant denaturation or even alteration of its properties. The reason is that a considerable part of the surface of the intrinsic proteins is strongly hydrophobic, and this part has to be protected against any contact with the aqueous phase during all steps of the solubilization and isolation procedures. In the following we want to present a method, by which the isolation of undenatured high-affinity Ca^{2+}-ATPase of human erythrocyte membranes is possible. In addition, this method might be useful in future for the isolation of other intrinsic membrane proteins.

Materials and methods

Acrylamide was from Roth (Karlsruhe); 2-methoxy-5-nitro-tropone from Sankyo (Japan); γ-^{32}P-ATP (15.3 Ci/mMole) from Amersham-Buchler; 2,4,6-trinitro-benzenesulfonic acid from EGA, Heidenheim; trypsin-inhibitor (soy beans) from Boehringer, Mannheim; ammoniumpersulfate, Comassie Brilliant Blue G 250, diisopropylfluorophosphate, sodium dodecylsulfate, mercaptoethanol, morpholinopropanesulfonic acid, phenylmethylsulfonylfluoride, Triton X-100, and Tween 20 were from Serva, Heidelberg; all other chemicals were from Merck, Darmstadt.

The water was twice redistilled in a quartz apparatus.

Protein determinations were made according to Lowry et al. [16].

Erythrocyte membranes were prepared essentially as described before [4], but with the addition of 1 mM DFP and 0.1 mM TLCK to all hemolyzing and washing solutions.

The Ca^{2+}-ATPase activity was measured as described before [4], the concentration of the ascorbic acid being increased to 20 g/l. 1% SDS was added to the ascorbic acid solution [17]. SDS-gelelectrophoresis was carried out according to Weber and Osborn [18] in 5% acrylamide gels; the gels were stained according to Fairbanks et al. [19]. In analytical disc electrophoresis the acrylamide concentration was 3.5%.

The SDS-gel electrophoresis for demonstration of the phosphorylated intermediate was run at 2—4 °C with the following buffer system: 93.8 mM citrate, 12.4 mM H_3PO_4, 12 mM Tris, 0.2% SDS, *p*H 2.4 (Lichtner, R. and H. U. Wolf, submitted to press). Tris does not serve as a buffer substance, but as counterion, since Na^+ and K^+ cause precipitation of dodecylsulfate at low temperatures.

Ca^{2+}-concentrations below 100 µM and the concentrations of all other divalent metal ions were adjusted by 0.4 mM Mg^{2+}-Me^{2+}-EDTA-buffers [20].

Solubilization of the membrane bound high-affinity Ca^{2+}-ATPase was achieved by Triton X-100 [3]. The erythrocyte membranes did not contain low-affinity Ca^{2+}-ATPase or (Na^+, K^+)-ATPase [4]. All kinetic parameters such as the K_m-value or the Ca^{2+}-enzyme dissociation constant were tested after solubilization in order to avoid any partial denaturation occurring during the step of solubilization.

Since the high-affinity Ca^{2+}-ATPase is not solubilized selectively, but together with a great number of other membrane proteins including certain amounts of spectrin, it has to be separated from these impurities. However, all separation methods we used — including gel chromatography, affinity chromatography and ion exchange chromatography — did not work. In all cases the enzyme was destroyed almost completely.

In the case of gel chromatography the whole column was equilibrated with stabilizing effectors of the high-affinity Ca^{2+}-ATPase, especially K^+, Ca^{2+}, and Mg^{2+}. The media for elution of the enzyme were always identical with those of the equilibration. However, in spite of the addition of these essential effectors, the enzyme was inactivated after one chromatographic run.

In order to avoid this inactivation, detergents such as Triton X-100 or Tween 20 were added to the equilibration and elution media in analogy to the purification experiments presented by WALTER and HASSELBACH [5] in the case of the Ca^{2+}-ATPase of the sarcoplasmic reticulum. But the addition of these detergents did not increase either the yield of active material to more than 3—5% of the starting activity.

Only when we added vesicles generated from crude phosphatidylcholine (or phospholipids extracted from erythrocyte membranes) by ultrasonication, we obtained a considerable amount of activity from the column. However, the activity appeared in the void volume of the Sepharose CL-6B column and a separation from other proteins including the spectrin was not possible. The molecular weight of the active material was about $2-4 \cdot 10^6$ Dalton, which is nearly identical with the particle weight of the phospholipid vesicle [6].

In order to obtain a separation from these proteins, we tried to decrease the size of the particles, by which the column had been equilibrated. This was possible by substituting the phospholipid vesicles by mixed micelles made from Triton X-100 or Tween 20 and phospholipids in a given molar ratio by ultrasonication. The scheme of the entire procedure of solubilization by Triton X-100 and purification on the basis of mixed micelles is presented in Fig. 1. The first step represents the solubilization of the membrane-bound high-affinity Ca^{2+}-ATPase by 0.2% Triton X-100 in the presence of 200 mM K^+, 1—10 mM Mg^{2+}, 0.5—5 mM Ca^{2+}, 10 mM MOPS, *p*H 7.0, 50 mM cysteine, 200 µg phosphatidylcholine/ml, 10 mM DFP, 1 mM TLCK, and 0.01 mM PMSF at 0 °C. The solubilized Ca^{2+}-ATPase forms a complex with a mixed micelle consisting of phospholipids and Triton X-100.

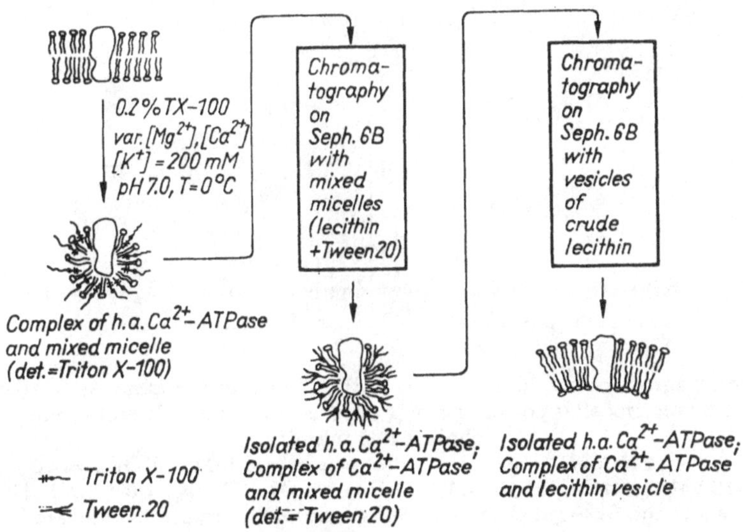

Fig. 1. Scheme of solubilization, purification and recombination with phospholipid vesicles of high-affinity Ca^{2+}-ATPase.
For details see text

For separation from other solubilized membrane proteins, the solubilized material is subjected to gel chromatography on Sepharose CL-6B under the following conditions, which have been found to be optimal for the isolation of the high-affinity Ca^{2+}-ATPase:

The column (90 cm length, 2.6 cm inner diameter) was equilibrated thoroughly with a medium consisting of 200 mM K^+, 10 mM Mg^{2+}, 1 mM Ca^{2+}, 50 mM cysteine, 10 mM MOPS, *p*H 7.7, 1 mM DFP, 0.1 mM TLCK, 0.01 mM PMSF, 0.01 mM Ca-EDTA, and mixed micelles, which had been generated from 0.5 mg phosphatidylcholine/ml + 6.85 mg Tween 20/ml by 60 min ultrasonication (Branson sonifier B 12) at 0 °C under purified N_2.

The elution of the Ca^{2+}-ATPase was performed using a medium, which was identical with the equilibration medium. The ATPase was eluted from the column combined with mixed micelles consisting of Tween 20 and phosphatidylcholine, possibly with minor amounts of Triton X-100. In this chromatographic step Triton X-100 may be substituted for Tween 20 without a significant loss of the separation effect. However, the stability of the high-affinity Ca^{2+}-ATPase is higher in a medium containing mixed Tween 20 micelles than it is in a medium containing Triton X-100 micelles.

Results and discussion

The result of the chromatographic purification is shown in Fig. 2. The Ca^{2+}-ATPase activity appears after an elution volume corresponding to about 600—800000 Dalton. As seen by the inserted SDS-gelelectrophoretograms, the high-affinity Ca^{2+}-ATPase appears as a protein complex consisting of more than one different subunits. The activity is separated from all other proteins, especially from spectrin, which still appears in the void volume. The yield of this chromatographic step is about 80% of the starting activity, and about 45—50% of the starting membrane-bound activity.

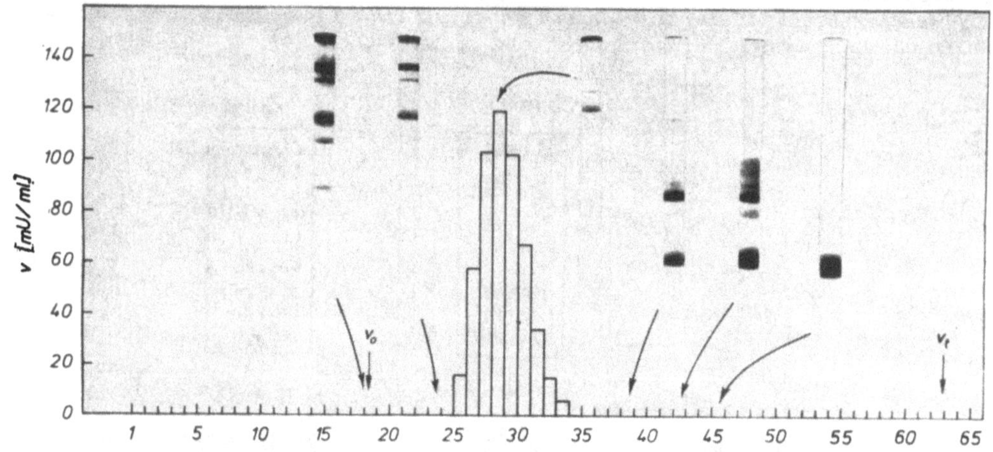

Fig. 2. Elution diagram of the high-affinity Ca^{2+}-ATPase in a Sepharose CL-6B column by application of mixed micelles consisting of Tween 20 (6.85 mg/ml) and crude phosphatidylcholine (0.5 mg/ml).
Conditions: K^+ = 200 mM, Mg^{2+} = 10 mM, Ca^{2+} = 1 mM, MOPS = 10 mM, pH 7.0, CaEDTA = 10 μM, cysteine = 50 mM, DFP = 1 mM, TLCK = 0.1 mM. The insertions show the SDS-gel-electrophoretograms of the fractions indicated

Fig. 3 shows the comparison between the SDS-gel electrophoretograms of the entire membrane (left gel), the solubilized material (middle gel), and the isolated high-affinity Ca^{2+}-ATPase (right gel). It consists of three different bands, which show molecular weights of 145000 ± 5000, 115000 ± 5000, and 105000 ± 5000 Dalton. In analogy to other ATPases we denoted them as α (145000), β (115000), and γ (105000) subunits.
The increase in specific activity is from 0.02 U/mg protein of the membrane-bound enzyme to about 2.3 U/mg protein of the most purified fractions. This value can be increased to 3.1 U/mg protein after activation by 0.1—1.0 mg/ml phosphatidyl-

choline. This corresponds to a purification ratio of 155 as related to the membrane proteins and of 25 900 as related to the protein content of the whole cell. Assuming an α, β, γ-structure for the Ca^{2+}-ATPase, 2400 copies/erythrocyte can be calculated on the basis of the specific activity.

In order to identify the active subunit of the high-affinity Ca^{2+}-ATPase, phosphorylation experiments with γ-^{32}P-ATP were undertaken. The results are shown in Fig. 4. The phosphorylation experiments were run in the presence of 2 μM ATP,

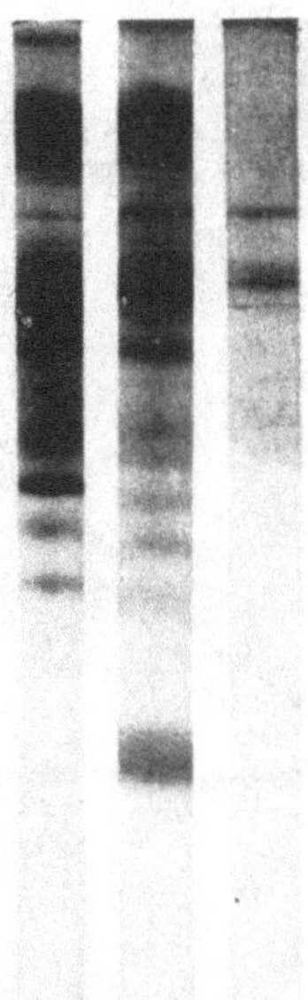

Fig. 3

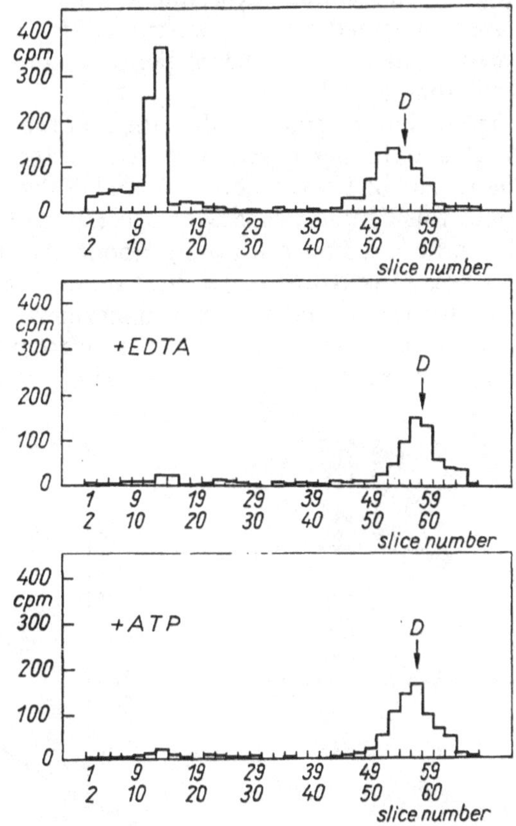

Fig. 4

Fig. 3. SDS-gelelectropherograms of erythrocyte membrane proteins (left gel), of solubilized material (middle gel), and of the purified high-affinity Ca^{2+}-ATPase (right gel). The gels were stained with Coomassie Blue

Fig. 4. Demonstration of the phosphorylated intermediate of the isolated high-affinity Ca^{2+}-ATPase by labelling with γ-^{32}P-ATP.

Upper line: Distribution of radioactivity of ^{32}P in a SDS-gelelectropherogram after formation of the phosphorylated intermediate by 2 μM γ-^{32}P-ATP in the presence of 50 mM Na^+, 10 μM Ca^{2+}, 10 μM Mg^{2+}, 100 μM CaEGTA, 50 mM Tris-HCl, pH 7.5, at $T = 0$ °C. The diagram in the middle line represents a similar experiment, in which the phosphorylated intermediate was formed under identical conditions as before, but destroyed by a 30 s incubation with 10 mM EDTA. The lower line shows again a similar experiment, in which the phosphorylated intermediate was destroyed by a 30 s incubation with 1 mM cold ATP. D denotes the position of the dye marker pyronin Y. The SDS-gelelectrophoresis was carried out at pH 2.4 and 2–4 °C

10 μM Ca^{2+}, 50 mM Na^+, 10 μM Mg^{2+}, 100 μM CaEGTA, and 50 mM Tris-HCl, *p*H 7.5, at 0 °C. The upper line represents the distribution of the radioactivity in an SDS-gelelectropherogram. The radioactive peak shows a relative mobility of 0.22 as related to the dye marker pyronin *Y*. This value is identical with the mobility of the α-subunit shown in Fig. 3. The middle line represents the SDS gel of the same experiment, in which the phosphorylated intermediate was destroyed by excess EDTA to chelate the Ca^{2+}, which is essential for the stability of the intermediate. In a similar experiment, which is shown in the lower line, the phosphorylated intermediate was destroyed by 1 mM cold ATP. In an analogous experiment, which is not shown here, the intermediate was destroyed by 0.2 M hydroxylamine at *p*H 5.5.

Thus the α-subunit could be identified as the active subunit of the high-affinity Ca^{2+}-ATPase of human erythrocyte membranes. This finding is in good agreement with the results of Knauf et al. [7], and Katz and Blostein [8]. In addition, the identity of this Ca^{2+}-ATPase with spectrin can be excluded.

As yet, we do not know anything about the functions of the β- and γ-subunits of the Ca^{2+}-ATPase. However, if these subunits are lost during gel chromatography, we also loose the enzyme activity. Furthermore, if an analytical disk electrophoresis is run in the presence of the essential effectors of the Ca^{2+}-ATPase and mixed Tween 20 micelles at *p*H 8.5 (Fig. 5), a single very slowly moving band is obtained

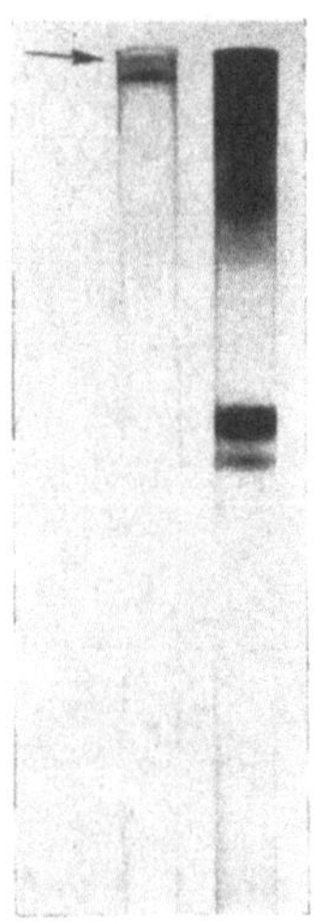

Fig. 5

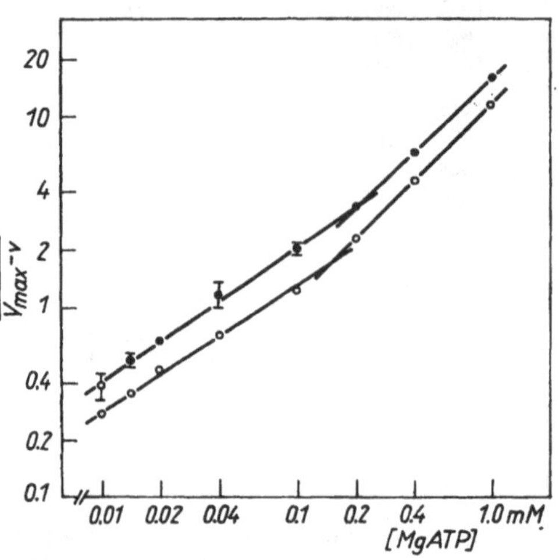

Fig. 6

Fig. 5. Analytical disc electrophoresis of high-affinity Ca^{2+}-ATPase (left gel), and of the material solubilized by Triton X-100 (right gel).
The electrophoresis was carried out at *p*H 8.5 in the presence of 200 mM K^+, 10 mM Mg^{2+}, 1 mM Ca^{2+} and mixed Tween 20-phosphatidylcholine micelles. → marks the retention of aggregated protein on top of the gel. Its position is independent of the time of electrophoresis

Fig. 6. Dependence of the reaction rate of isolated Ca^{2+}-ATPase on the concentration of substrate MgATP presented in a Hill-plot.
○ 40 μM Ca^{2+}, ● 150 μM Ca^{2+}. General conditions: 100 mM K^+, 2 mM Mg^{2+}, *p*H 7.0, $T = 30$ °C

in the case of the isolated Ca^{2+}-ATPase (left gel). For comparison, the right gel represents the analytical disk electrophoresis of the solubilized material under the same conditions: A considerable part of the proteins apparently aggregates due to their hydrophobic nature.

Thus, the two catalytically inactive β- and γ-bands seem to be essential parts of the high-affinity Ca^{2+}-ATPase complex, possibly with regulatory functions. They seem to be analogous to the 40000 Dalton subunit of the (Na^+, K^+)-ATPase [9]: Probably they are identical with the protein activator of the Ca^{2+}-ATPase demonstrated by BOND and CLOUGH [10], QUIST and ROUFOGALIS [11], and by LUTHRA et al. [12].

In order to show that the unmodified high-affinity Ca^{2+}-ATPase, but not a preparation artefact has been isolated by use of the mixed micelle gel chromatography, the enzyme was tested with respect to all common kinetic parameters.

The dependence of the reaction rate on the substrate (Mg^{2+}-ATP) concentration is shown in Fig. 6 (the two curves represent different Ca^{2+}-concentrations). The K_m-values, represented as the substrate concentration at $v/(V_{max}-v) = 1$, are nearly identical with that of the membrane-bound enzyme (cf. Table 1). The Hill coefficient is equal to 1.00 ± 0.05 at [Mg-ATP] > 0.2 mM, but it is decreased to 0.68 ± 0.06 at [Mg-ATP] < 0.2 mM. This represents a negative cooperativity, which might indicate that at least two α-subunits are combined in the native complex of the high-affinity Ca^{2+}-ATPase.

Fig. 7 shows the reaction rate of the enzyme as a function of the Ca^{2+}-concentration in a $v/v_{opt.}$—[Ca^{2+}] plot. Slight activation occurs at 0.1 μM, half maximum activation at about 1 μM, and optimum activation at about 30 μM Ca^{2+}. This is identical with the behaviour of the membrane-bound enzyme. However, if these data are plotted in a Hill plot (not shown), we find a Hill coefficient of 1.25 ± 0.11. This represents a slightly positive cooperativity, which again indicates a possible dimer structure of the enzyme.

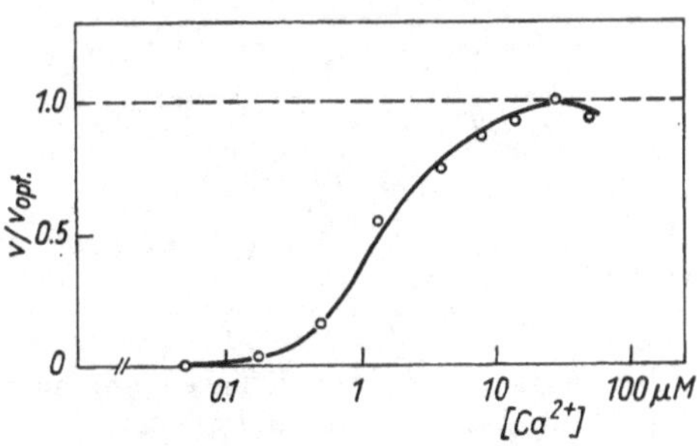

Fig. 7

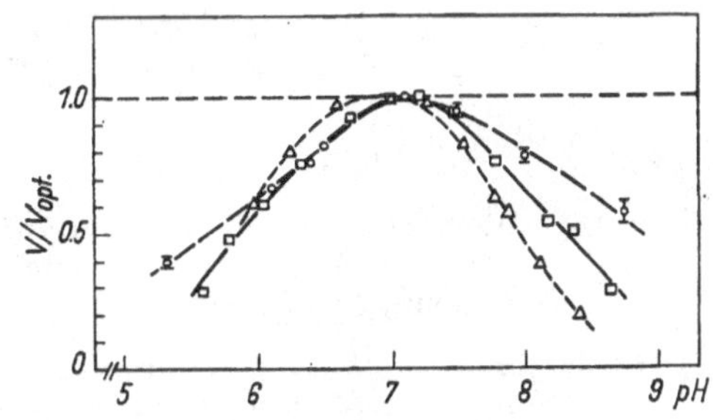

Fig. 8

Fig. 7. Reaction rate of the isolated high-affinity Ca^{2+}-ATPase as a function of the Ca^{2+}-concentration (the data being normalized to optimum reaction rate).
Conditions: 100 mM K^+, 2 mM Mg^{2+}, 1 mM MgATP, pH 7.0, $T = 30$ °C

Fig. 8. Dependence of the reation rates of membrane-bound (□), solubilized (△), and isolated (○) high-affinity Ca^{2+}-ATPase on the pH-value.
The data are normalized to optimum reaction rate.
Conditions: 100 mM Na^+ (□), or 100 mM K^+ (△, ○), 2 mM Mg^{2+}, 1 mM MgATP, 40 μM Ca^{2+}, 20 mM Tris-maleate, $T = 30$ °C

The dependence of the reaction rate on the *p*H-value is shown in Fig. 8. The curves are normalized to identical optimum reaction rates. In the case of membrane-bound Ca^{2+}-ATPase, of solubilized material and of the isolated enzyme, identical *p*H-optima between 7.0 and 7.1 were obtained.

The membrane-bound high-affinity Ca^{2+}-ATPase is not only activated by Ca^{2+} and other divalent cations, such as Zn^{2+}, Co^{2+} and Mn^{2+} [13], but also (to a smaller extent) by monovalent cations, such as Na^+, K^+, NH_4^+, and Rb^+, but not Li^+. We tested the isolated enzyme, whether it could be also activated by these cations. Some of the results are given in Fig. 9, in which the ratio (reaction rate at a given concentration of monovalent cation)/(reaction rate at 4 mM Na^+) is plotted versus the concentration of monovalent cations. The filled circles represent Li^+, which does not activate, but only inhibit. A significant activating effect is shown by NH_4^+ and K^+.

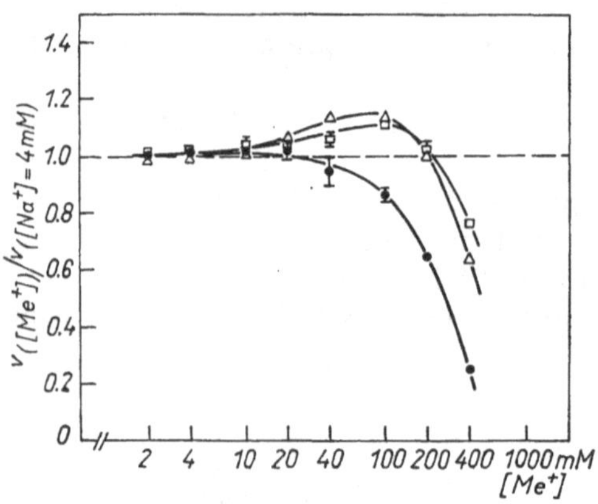

Fig. 9

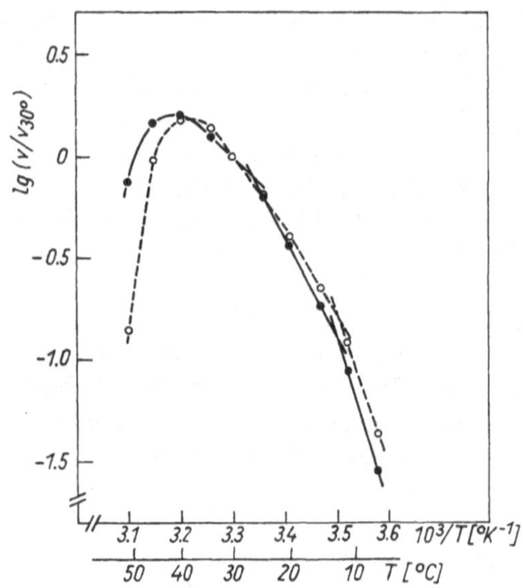

Fig. 10

Fig. 9. Activation of isolated high-affinity Ca^{2+}-ATPase by K^+ (□) and NH_4^+ (△). ● represents data obtained with Li^+.
Conditions: 2 mM Mg^{2+}, 1 mM MgATP, 40 μM Ca^{2+}; *p*H 7.0, $T = 30$ °C

Fig. 10. Dependence of the reaction rates of membrane-bound (●) and isolated (○) high-affinity Ca^{2+}-ATPase on the temperature, presented in an Arrhenius-plot.
The data are normalized to 30 °C.
Conditions: 100 mM K^+, 2 mM Mg^{2+}, 40 μM Ca^{2+}, 1 mM Mg-ATP, *p*H 7.0

Fig. 10 shows the dependence of the reaction rate on the temperature. The data are normalized to the temperature of 30 °C by plotting log $v/v_{30°}$ versus $1/T$. The curves for the membrane-bound and the isolated enzyme are very similar with respect to bends in the range of 12 °C, and of 25—27 °C, respectively, and to the temperature optimum of 37—41 °C. However, the isolated enzyme is inactivated faster than the membrane-bound one at high temperatures.

The membrane-bound enzyme is inhibited at rather low concentrations (μM and mM range) by a series of chemical modifying agents (CMA), including SH-group reagents [14]. Fig. 11 shows two examples of a strong inhibition of the isolated high-affinity Ca^{2+}-ATPase by the Hg-compound mersalyl and DTNB (Ellman's reagent). The data are presented in a normalized form, in which the ratio (reaction rate in the presence of CMA)/(reaction rate in the absence of CMA) is plotted versus [CMA].

The data presented so far are compared in Table 1 with those obtained for the membrane-bound enzyme. The Table includes some additional data, which are not shown in Figures, e.g. concerning the activation of the isolated Ca^{2+}-ATPase by Zn^{2+}, Co^{2+}, and Mn^{2+}.

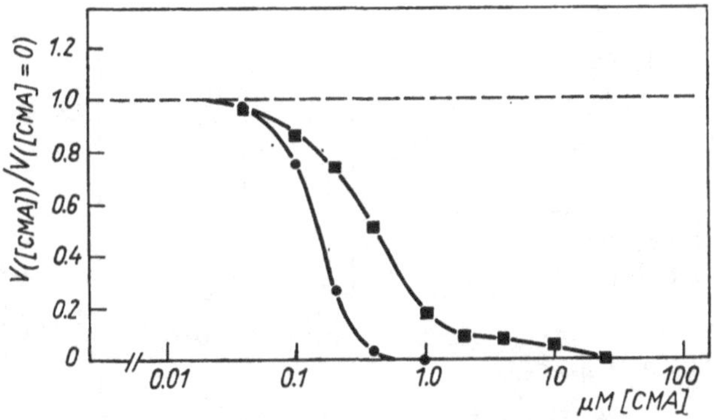

Fig. 11. Inhibition of isolated high-affinity Ca^{2+}-ATPase by mersalyl (●) and DTNB (■) (the data being normalized to [CMA] = 0).
The enzyme was incubated under the following conditions: 100 mM K^+, 2 mM Mg^{2+}, 40 μM Ca^{2+}, *p*H 7.0 (mersalyl), *p*H 8.0 (DTNB), $T = 30$ °C, $\Delta t = 2$ hrs. ATPase activity was measured at *p*H 7.0 in the presence of 1 mM MgATP

In addition, the membrane-bound and isolated enzymes have been tested for K^+-activated p-nitrophenylphosphatase activity. In both cases, this activity could not be detected under conditions which were applied by Rega et al. [15]. So, these findings do not support the hypothesis of these authors that the p-nitrophenylphosphatase may be a part of, or identical with the high-affinity Ca^{2+}-ATPase of human erythrocyte membranes.

All isolation and separation experiments presented here have been performed in the presence of protease inhibitors such as TLCK and DFP. The effects of these inhibitors are most impressive, as shown in Fig. 12. When the membranes are isolated and the enzyme is solubilized and purified only in the presence of DFP, we obtain the pattern shown in the middle SDS-gel. If DFP is omitted also, that means, if all procedures are performed in the absence of any protease inhibitors, a diffuse broad band with a molecular weight of 90—100000 Dalton is obtained, which is shown in the right gel. The pattern of the Ca^{2+}-ATPase, isolated in the presence of TLCK and DFP (left gel) is not altered, when PMSF (10 μM) and trypsin inhibitor from soy beans (1 μM) are used additionally.

Most surprisingly the kinetic parameters of the proteolytically digested Ca^{2+}-ATPases are nearly unchanged as compared to the undigested Ca^{2+}-ATPase. This

Table 1

Comparison of the properties of membrane-bound and isolated high-affinity Ca^{2+}-ATPase of human erythrocyte membranes.

The K_m-value was obtained in the presence of 40 μM Ca^{2+}. The value in parenthesis refers to experiments in the presence of 150 μM Ca^{2+}. $V_{max,Me}/V_{max,Ca}$ denotes the maximum activation potency of a given metal ion in relation to that of Ca^{2+}. The effects of the chemical modifiers (CMA) are expressed as the concentration of CMA, which produces half maximum inhibition (given in μM). The nitrophenylphosphatase activity was measured under the conditions applied by REGA et al. [15]

	Membrane-bound	Isolated
K_m [μM]	45 ± 5	65 ± 5 (36 ± 4
K_{CaE} [μM]	0.9–2.0	1.9
Activation by 100 mM Me^+:		
v_{Na^+}/v_{Li^+}	1.39	1.18
v_{K^+}/v_{Li^+}	1.42	1.28
$v_{NH_4^+}/v_{Li^+}$	1.58	1.32
v_{Rb^+}/v_{Li^+}	1.23	1.32
Activation by Me^{2+}:		
Co^{2+} K_{MeE} [M]	$1.0 \cdot 10^{-10}$	$4.6 \cdot 10^{-1}$
Zn^{2+}	$4.8 \cdot 10^{-11}$	$1.0 \cdot 10^{-9}$
Mn^{2+}	$1.2 \cdot 10^{-8}$	$2.3 \cdot 10^{-8}$
Co^{2+} $V_{max,Me}/V_{max,Ca}$	0.47	0.44
Zn^{2+}	0.51	0.77
Mn^{2+}	0.50	0.41
Optimum *p*H-value	7.0–7.1	7.0–7.1
Optimum temperature [°C]	41	37
Inhibitory effects of CMA:		
Hg^{2+} $[CMA]_{(v_i = v/2)}$ [mM]	0.00068	0.000007
5,5'-Dithio-bis-(2-nitrobenzoate)	0.001	0.0004
Diazotized sulfanilic acid	0.0042	0.0016
2-Methoxy-5-nitrotropone	0.02	0.14
2,4,6-Trinitrobenzenesulfonic acid	0.03	0.046
Diisopropylfluorophosphate	$\gg 10$	$\gg 10$
p-Nitrophenylphosphatase activity	0	0

means that unchanged kinetic properties are prerequisite, but not sufficient to prove the integrity of an isolated enzyme.

The isolation of the unmodified high-affinity Ca^{2+}-ATPase from human erythrocyte membranes shows that it is possible to isolate an intrinsic membrane protein without any severe changes in its kinetic properties by using the mixed micelle gel chromatography. Furthermore, preliminary experiments have shown that it is also possible to recombine the isolated mixed-micelle-bound Ca^{2+}-ATPase with phospholipid vesicles removing the detergent component of the protein-mixed-micelle complex on a Sepharose CL-6B column equilibrated with phospholipid vesicles. This procedure is shown on the right side of Fig. 1.

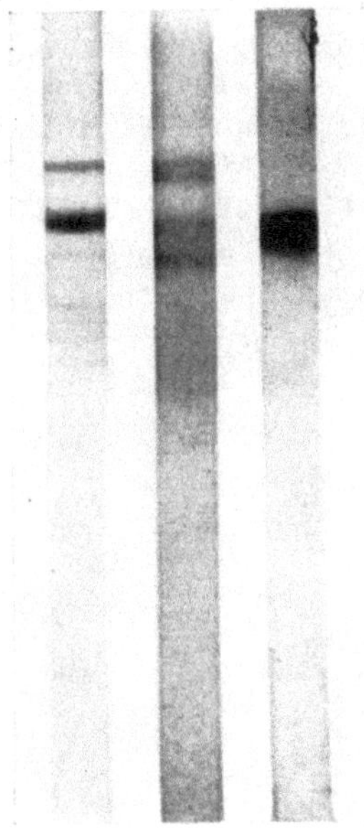

Fig. 12. SDS-gelelectrophoresis of high-affinity Ca^{2+}-ATPase, isolated under different conditions.

Left gel: Isolation of the erythrocyte membranes, solubilization and mixed micelle gel chromatography in the presence of DFP and TLCK; middle gel: all procedures were performed in the presence of DFP (TLCK being omitted); right gel: all procedures were performed in the absence of any protease inhibitor

Thus, our investigations in the future have two major aims:

1. The recombination of the purified high-affinity Ca^{2+}-ATPase with phospholipid vesicles in order to demonstrate that the active Ca^{2+}-transport across the human erythrocyte membrane is carried out by this enzyme.
2. The application of the mixed micelle gel chromatography to the purification of other erythrocyte membrane enzymes of biological significance, e.g. the low-affinity Ca^{2+}-ATPase or kinases.

References

[1] LaCelle, P. L., and F. H. Kirkpatrick in: Erythrocyte Structure and Function. Alan Riss, New York 1975, p. 535

[2] Schatzmann, H. J.: J. Physiol., Lond. **235**, 551 (1973)

[3] Wolf, H. U., and K. Gietzen: Hoppe-Seyler's Z. physiol. Chem. **355**, 1272 (1974)

[4] Wolf, H. U.: Biochim. biophys. Acta **266**, 361 (1972)

[5] Walter, H., and W. Hasselbach: Eur. J. Biochem. **36**, 110 (1973)

[6] Huang, C.-H.: Biochemistry **8**, 344 (1969)

[7] Knauf, Ph., A., F. Proverbio, and J. F. Hoffman: J. gen. Physiol. **63**, 305 (1974)

[8] Katz, S., and R. Blostein: Biochim. biophys. Acta **389**, 314 (1975)

[9] Hokin, L. E., J. L. Dahl, J. D. Deupree, J. F. Dixon, J. F. Hackney, and J. F. Perdue: J. biol. Chem. **248**, 2593 (1973)

[10] Bond, G. H., and D. L. Clough: Biochim. biophys. Acta **323**, 592 (1973)

[11] Quist, E. E., and B. D. Roufogalis: Archs Biochem. Biophys. **168**, 240 (1975)

[12] Luthra, M. G., G. R. Hildenbrandt, and D. Hanahan: Biochim. biophys. Acta **419**, 164 (1976)

[13] Pfleger, H., and H. U. Wolf: Biochem. J. **147**, 359 (1975)

[14] Wolf, H. U.: Abstracts of the 9th International Congress of Biochemistry, Stockholm 1973, p. 102

[15] Rega, A. F., D. E. Richards, and P. J. Garrahan: Biochem. J. **136**, 185 (1973)
[16] Lowry, O. H., N. J. Rosebrough, A. L. Farr, and R. J. Randall: J. biol. Chem. **193**, 265 (1951)
[17] Arnold, A., H. U. Wolf, B. P. Ackermann, and H. Bader: Anal. Biochem. **71**, 209 (1976)
[18] Weber, K., and M. Osborn: J. biol. Chem. **244**, 4406 (1969)
[19] Fairbanks, G.,.T. L. Steck, and D. F. H. Wallach: Biochemistry **10**, 2606 (1971)
[20] Wolf, H. U.: Experientia **29**, 241 (1973)

Acta biol. med. germ., Band 36, Seite 859—869 (1977)

Bereich Biophysik der Sektion Biologie der Humboldt-Universität zu Berlin
104 Berlin, DDR

Mathematical modelling of shape-transformations of human erythrocytes

R. GLASER and A. LEITMANNOVÁ

Summary

The polymorphism of human red blood cells may be explained on the basis of more recent considerations of the fluid-mosaic structure of the membrane. The origin of the pattern of surface membrane material is caused by the contraction-state of spectrins as well as by the lateral mobility of surface elements. This pattern of surface elements is characterized by the elasticity modulus (Young's modulus) E. At the present time calculations on the energetics cannot be realized because the function describing the distribution of the elasticity modulus E according to the different surface areas is unknown.
The paper presented here describes the division of the surface into the bending classes according to the principal radius of curvature R_1 for the different three cell shapes investigated (discocytes, stomatocytes, echinocytes).
Further results allow statements about the possible formation of echinocytes at the given values for the volume V and the surface A. The process of microspherulation has been investigated in a similar way, too.

Human red blood cells show a great variability of their shapes. This is of importance for their cell-physiological functions as well as for streaming conditions in the circulation. It is reasonable to assume that only cells without a nucleus may be flexible enough to pass through capillaries with a diameter, smaller than their own size. For this it should be supposed that the dimensions of the erythrocyte are given by the molecular parameter of its hemoglobin content and additionally by its osmoregulatory mechanism [1]. Not only the flexibility of the cell will be significant, but moreover the reversibility of the induced shape transformation. From the energetic point of view this means to consider the stability of the particular shape under specific conditions.

During the last few years, a number of attempts were undertaken, to solve this problem in a biophysical way [2—15]. The transformation and stability of an erythrocyte shape is to be explained on the basis of three factors: cell-volume, cell-surface and mechanical properties of the membrane (elasticity, tension, fluidity), respectively, their planar pattern distribution. In contrast to the cell volume, which can change in some degree, the membrane area should be assumed as constant. An exception is the reversible loss of the membrane material induced by various processes (microspherulation e.g.).

According to experimental observations of the physiology of the shape transformations [16—24, 32], the dynamics of membrane properties plays a significant role in this process. Especially the recently proposed concept of the fluid-mosaic-structure of the membrane gives a new basis to understand this process.

In view of this, the starting point of all discussions will be the following question: What are the possible shapes of a body with a given volume and a given surface, and secondly: can the volume be smaller than predicted by the spherical shape? Mechanical experiments with plastic balls indicate that at a volume, smaller than 0.8 relative to the sphere volume, there exist two metastable forms: a biconcave and a cup-like shape, formally comparable to these of human red cells [4, 14, 31]. It should be pointed out, however, that those experiments are able to demonstrate the shape transformations of blood cells only very roughly. In contrast to rubber or other artifical materials the cell membrane has only a very limited elasticity but a fluidity which allows a more or less plastic deformability in the plane at a constant area. Furthermore the planar inhomogenity of the membrane and the mutual influence of contractile elements on its inner surface like spectrins have to be considered. On the basis of this concept we recently proposed a hypothesis of shape transformations [15]. The membrane of red blood cells consists of planar components of different membrane components containing different kinds of lipids and/or proteins. The distribution of such components is given mainly by two factors: the contraction state of spectrins and the lateral mobility. These properties are influenced by different factors, partly independently and they work in different manner. Generally, one has to differentiate between the contraction state and the phase state of the membrane, determining the lateral fluidity. The combination of these independent factors, each of them having two distinct states (supposed for simplicity) gives 4 different cases, listed in Table 1. Calculation of the surface developments of spatial segments during the transformations: sphere-discocyte and sphere-stomatocyte indicated that the latter one predicts an assymetrical distribution of membrane material between the two hemispheres. We supposed that this is the result of clustered membrane material, induced by the spectrin contraction in combination with a high lateral mobility of them. An echinocyte will be produced, when the contraction of spectrins coincides with the low fluidity. Whether the fluidity of the membrane during the relaxation of the spectrins producing discocytes is high or low, is unknown and without relevance for this concept.

Table 1
Possible combinations of the membrane state

State of the membrane		Resulting red cell shape
Contraction state of spectrins (cluster formation)	Fluidity	
−	−	discocyte
−	+	discocyte
+	+	stomatocyte
+	−	echinocyte

Various experimental data confirming this concept have been provided earlier [15]. The nature of this hypothetical clusters is not clear up to now, however. Following this hypothesis, the present study is directed to the question of the stability of the discocyte-stomatocyte balance, the microspherulation and the form variability of the echinocytes.

The bending of discocytes and stomatocytes

The hypothesis of the shape transformation listed before demonstrates the way for plastic planar membrane deformability, leading to a discocyte or a stomatocyte (Fig. 1). Although the way to form stomatocytes is possible only by asymmetrical displacement of surface material, this fact alone is not sufficient to explain the stability of the shape formed (Fig. 2). The stability can be derived from a function of bending energy. CANHAM [5] indicated that the bending energy of a discocyte is lower than that of a flat rotational ellipsoid ("flying saucer"-form). Relations between discocyte and stomatocyte are not given there.

On the basis of the geometrical models for stomatocytes and discocytes discussed in detail earlier [15], we calculated the membrane bending for distinct small membrane areas. According to common calculations [5, 10] two principal radii of curvature in two mutual perpendicular planes are to be regarded: R_1 and R_2. Only the first one, lying in the plane of the meridional section should be significant for the question discussed here, because the other is lower, accumulates therefore less bending energy and, additionally does not vary dramatically during shape transformation. We calculated the bending (in μm^{-1}) as the reverse of the bending radius ($1/R_1$). As negative bending we nominated the case where the center of curvature lies outside the cell (e.g. invagination). For details of the calculation program see [29]).

Fig. 3 represents distribution graphs of bending in classes 0 ... 1; 1 ... 2; 2 ... 3 etc. in μm^{-1} of positive and negative sign. These graphs are given on the basis of data measured on real cells, fitted to the model curves [15]. It is clear that we have got some variations between the calculated 5 cases of each shape, given by biological variability and additionally by errors in the measurements. However, Fig. 3 indicates that the range of the distributional columns is larger in the case of stomatocytes in contrast to discocytes. This indicates that in stomatocytes small membrane areas occur, which are bent stronger than in discocytes.

It is evident that these results are not sufficient enough to prove the stability of the forms discussed. For this we need the Young's-modulus to calculate the total bending energy. The stability of discocytes and stomatocytes under particular conditions is experimentally evident, however, in spite of the difference of bending distribution demonstrated here. This underlines the hypothesis that the composition of different membrane areas in the erythrocyte induced by lateral movement, is not constant in time and we have therefore differences in the Young's modulus also. Whether the strong bending of a particular membrane area is the cause of differences in the values of Young's-modulus, or vice versa caused by it, is unknown at present. The assumption of such differences partly corresponds to the theory of FUNG and TONG [10] but differs from it in that way that this specific distribution of the stiffness of the membrane is not preformed but functionally given by the distribution of clusters in the fluid mosaic of the membrane.

Microspherulation

Human red blood cells, heated above a temperature of 48 °C became spherulated after extruding membrane material in form of microspheres [33, 34]. The same occurs under the influence of a magnetic field [27]. It is supposed that a similar process will occur during the cell's life span [35], but whether the mechanism of it is like the influence of heat or a magnetic field, is doubtful.

The spontaneous process of microspherulation contradicts all experiences of the behaviour of drops, tending to fuse, driven by the diminution of its surface energy. The microspherulation of erythrocytes, however, can be understood on the basis of a constant sum of surface and volume during this process. Nevertheless, the driving force of this phenomenon should be the minimisation of the surface energy, too, though it is not based on a diminution of surface area but rather by an energetically more favourable distribution of membrane components in the plane. It is possible that by overcoming an activation energy, there will take place a lateral separation of planar phases of membrane constituents of distinct interaction energy. This molecular process is unknown till now, but it is supposed that it has some similarities with the mosaic dynamics discussed above.

Firstly we attempt to answer the following question: What diameters of microspheres and rest-spheres are possible; and: is it explainable geometrically that in any cases the rest body is a sphere, in others — a loose sack? We solved this problem by using an iterative computer program of the following manner: Starting with an erythrocyte with a given volume and area, we separate a microsphere with a given initial radius. This process is repeated until the rest became a sphere or otherwise a volume lower than that of the microsphere itself. In such cases where the separating of a further microsphere with the initial radius was impossible, a gradual diminution of the radius was necessary until reaching a given minimum.

An example of such a calculation is demonstrated in Fig. 4. Starting with an erythrocyte with a volume of 110 μm^3 and a membrane area of 150 μm^2, it is possible, e.g., to split up 14 microspheres of a radius of 0.5 μm and a further 15th with 0.32 μm. After this, there remains a rest sphere. This means that the remaining volume is surrounded by the remaining membrane material exactly in the form of a sphere. As demonstrated in Fig. 4, the possible number of separated microspheres increases with decreasing diameter. Looking at the structure of the rest body, only for small initial microspheres it would build up a sphere. By large initial microspheres, the volume will decrease faster than the area and the rest body therefore is a sort of empty membrane sack.

In Fig. 5 the process of microspherulation is demonstrated for a shrunken erythrocyte with a volume of 80 μm^3 and an area of 150 μm^2. The curves demonstrate processes starting with different initial radii of the microspheres. The ordinate represents a logarithmic plot of a sphering index of the rest body. Until a radius of 1.68 μm is reached, the sphering index decreases during the process of microspherulation. This means: the rest will be an empty sack. For separating microspheres smaller than 1.34 μm, the curves assymptotically reached the value 1.0, i.e. there remains a rest sphere. Fig. 6 gives the critical radii as a function of the cell volume. (The extensiveness of this region is given by the chosen step of the iterative computer program.) Additionally the volume of the rest sphere is calculated, which assymptotically approaches the curve of the sphere representing the given volume of the cell.

These calculations indicate that there exist two modes of microspherulation: Production of two large spheres with radii, only slightly lower than those of the

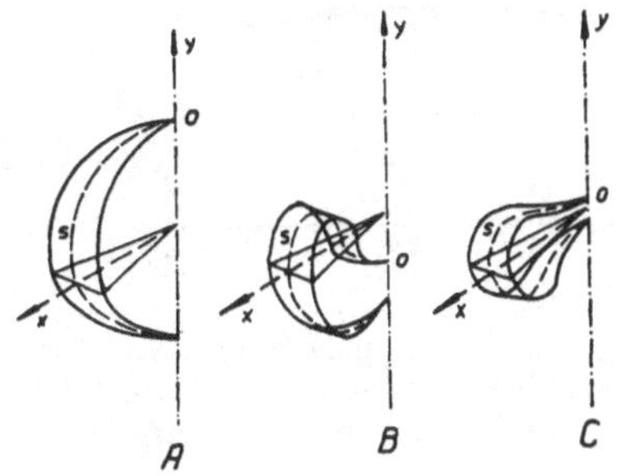

Fig. 1

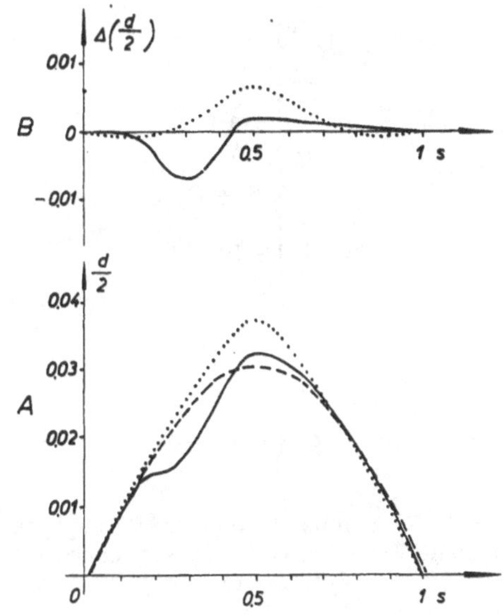

Fig. 2

Fig. 1. Spatial segments of a sphere (A), a stomatocyte (B) and a discocyte (C). The broken line (s) represents the coordinate of reference for the calculation of the breadth ($d/2$) of the segment (see Fig. 2)

Fig. 2. A: Calculated values of the breadth ($d/2$) of a segment with a space angle of 1° (ordinate in μm) versus the normalized coordinate (s) for sphere (– – –), stomatocyte (———) and discocyte (· · · · · ·).

B: Differences between curves of stomatocyte and sphere (———), and discocyte and sphere (· · · · · · ·) of the graph A

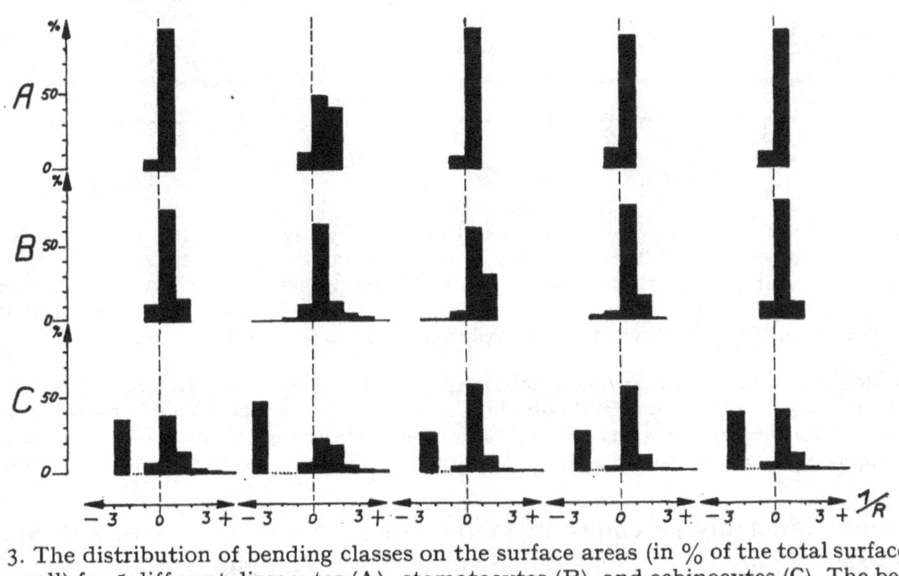

Fig. 3. The distribution of bending classes on the surface areas (in % of the total surface area of the cell) for 5 different discocytes (A), stomatocytes (B), and echinocytes (C). The bending ($1/R_1$) is given in μm^{-1}. A negative bending means the bending centre is located outside the cell and vice versa

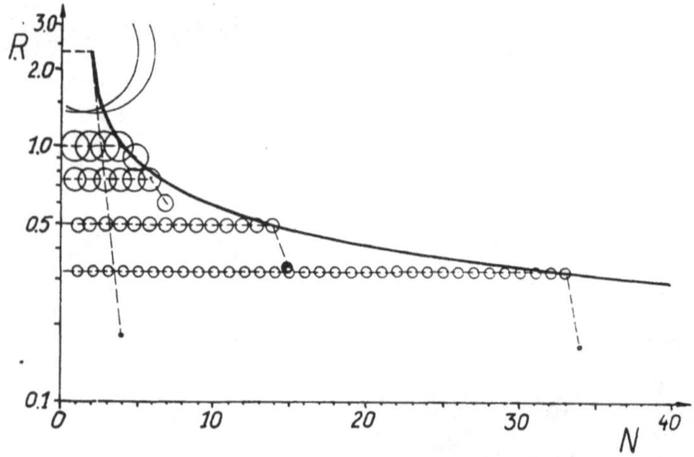

Fig. 4. The radius (R) of separated spheres (in μm) as a function of the possible number (N) of them. For some cases the sequence of microspheres is given. (The calculation based on the erythrocytes with a volume of 110 μm³ and a surface of 150 μm²).

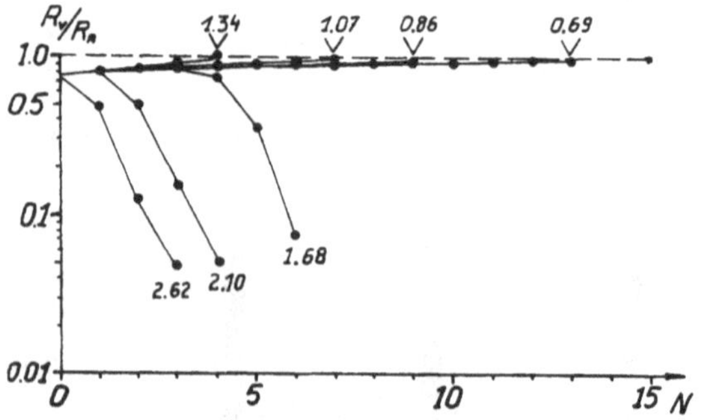

Fig. 5

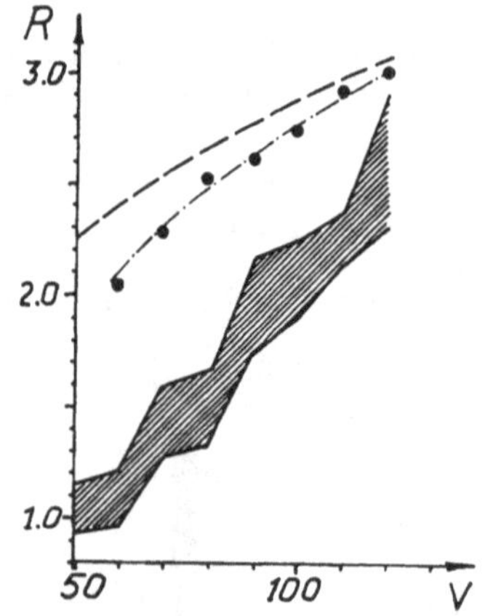

Fig. 6

Fig. 5. The sphering index (possible sphere radius given by the volume (r_v) in relation to that given by the surface (r_A) of the rest body after a maximal possible microspherulation) as a function of the number of separated spheres (N) for processes which start by the initial spherulation-radii of: 2.62, 2.10, 1.68, 1.34, 1.07, 0.86 and 0.69 μm. (The calculation are based on a shrunken erythrocyte with the volume of 80 μm³ and with the surface of 150 μm²)

Fig. 6. Region of the critical radius of initial spheres (hatched area); radius of the possible spheres (in μm) given by the initial volume (— — —); radius of the rest-sphere (in μm) after maximal microspherulation with an initial radius below the critical value (●— · —●). All these parameters are plotted against the volume (in μm³) of the cell at a constant surface of 150 μm²

cell, and additionally an empty membrane sack on the one hand, and on the other a splitting of a great number of very small spheres (partly invisible by ordinary light microscopy) producing a large rest sphere. The diameter of the large sphere in the first case, and that of the rest sphere in the second one differ only

slightly, so that by optical observations it is impossible to differentiate them exactly. The microspheres as well as the rest spheres lost their ability to swell. Little osmotic swelling therefore results in a hemolysis.
Generally this model indicates that geometrically there exist two possibilities for realising the microspherulation which indeed have been observed. The reason for the realisation of one or the other cases should be given obviously by the kinetics of this process. This is similar to the question of echinocytogenesis discussed in the following section.

Echinocytogenesis

On the basis of our general model discussed before, echinocytes are formed when spectrins contracted but the fluidity of the membrane material is too small to allow a global concentration of the clusters, leading to stomatocyte formation. Local evaginations (or invaginations, see [32]) results in echinocyte formation). Here again raises the question: what looks the distribution of the bending of membrane areas like, and: what possible types of echinocytes can be formed from the geometrical point of view?
To answer these questions, firstly a geometrical model of the echinocyte should be constructed. This is impossible to realise with simple rotating curves like stomatocyte or discocyte models [15]. Therefore we regard the echinocyte as a sphere with a number (N) of spiculae, each of them described by a curve, rotating round an axis, on which is lying a radius of the sphere. Fig. 7 shows the parameters of this curve. It is composed of the circle of the basic sphere with radius r and an additional circle with a radius ϱ, the center of which (S) is located outside

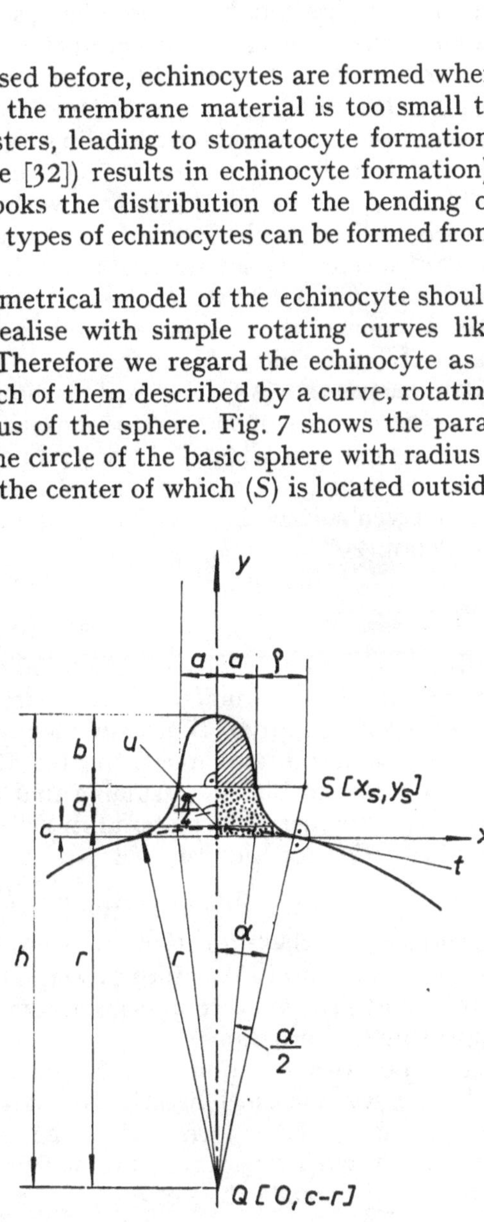

Fig. 7. Parameters characterizing the spiculum of an echinocyte

the cell, and a half ellipsoid with half axis a and b. The advantage of this kind of modelling is the similarity to the reality and additionally the possibility of its algebraic integration (for detail see [29]). For usefullness we introduced normalized parameters:

$$a^* = a/r \; ; \qquad h^* = h/r.$$

It is possible to demonstrate that by measuring the parameters a, h, N and r, all other data needed for complete modelling can be calculated.
On the basis of this concept we calculated the mean bending of the membrane in the same way as before for the cases of discocytes and stomatocytes. These calculations were based on parameters, given by scanning electron microscopy pictures published in the literature [36, 37]. The results are demonstrated in Fig. 3C. It is not surprising that given by the tops of the spiculae, we have a maximum of positive membrane bending (up to 5 μm^{-1} and partly to 7 μm^{-1}). These are very small regions, however. On the other hand we have a class containing 25% ... 45% of the membrane area with a strong negative bending of 2 ... 4 μm^{-1}. This is realised by the basis of the spiculae.
It is of interest to set this data in relation to the maximal bending (or equal: minimal radii of curvature), given by the sucking of cells into small capillaries [30]. Obviously the membrane bending realised in the echinocytogenesis can not be induced experimentally in this way. Only partly sucking producing microspherulae was possible [35]. This again indicates that particularly in echinocytogenesis dramatic changes of the surface mosaic would occur. On the basis of the mathematical model mentioned above it is possible to make some geometrical predictions concerning the real possible shapes of echinocytes with a given volume (V) and a given surface (A). We have got expressions for the following functions (see for detail [29]):

$$A = f_1(N, h^*, a^*, r) \, ,$$
$$V = f_2(N, h^*, a^*, r) \, .$$

This is a system of equations which indicates that from the four variables (N, h^*, a^*, r) only two are independently variable. Calculations of echinocytes on the basis experimentally real measurements as mentioned above indicate that about 50% of the membrane area is located at the surface of the basic sphere slightly bended (see also Fig. 3). Assuming that this value is given by the availability of the corresponding rigid material in the fluid mosaic of the membrane, we got a third equation for the rest area:

$$A_{rest} = f_3(N, a^*, r) \, .$$

Unfortunately, these functions are complicated formulas containing the variables in the second and in the third power. The solutions for different variables therefore would give very complicated equations. They are only partly soluble by approximative programs.
Fig. 8 indicates functions $h^*(a^*)$ for different volumina (continuous lines-the value are given in μm^3) directly calculated for echinocytes with a total membrane area (surface) of 152,1 μm^2. The other functions $h^*(a^*)$ for different number of spiculae (broken lines) are obtained graphically from the calculated functions $N(a^*)$ for different values of h^*. The functions $h^*(a^*)$ for different volumina were

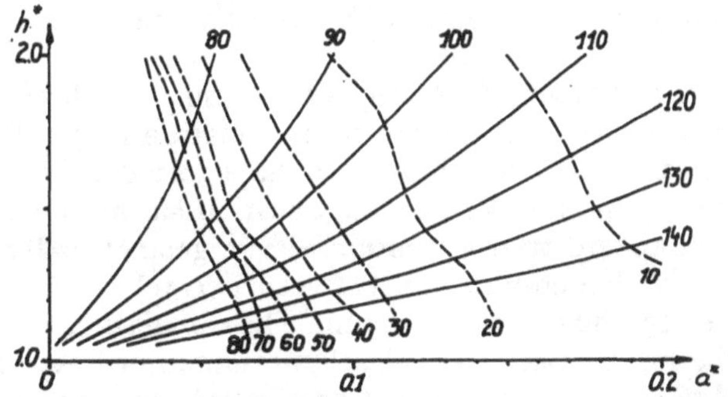

Fig. 8. Parameters of the echionocyte-shape. The relative spiculae-length (h^*) measured from the centre of the central sphere is plotted against its relative half-thickness (a^*) for echinocytes of different volumina (———), (the values are given in μm^3) and with different numbers of spiculae (– – –) for a constant membrane surface of 152.1 μm^2

obtained graphically from the calculated functions $V(a^*)$ for different values of h^*. For control the crossing points for various volume and number of spiculae N-curves were also directly calculated and found to be correct.

These calculations indicate the following conclusions:

For a given volume (V) the shape of spiculae (a^*, h^*) depends only on the number (N) of them. Therefore, we have in cells of a given volume echinocytes with a high number of thin spiculae and such with a few thick ones.

The proportion of the spiculae-shape (h^*/a^*) is determined by V and N. This relation for a given N is therefore an indicator for V.

Echinocytes with a small volume ($V < 80\ \mu m^3$) should have a high number ($N < 40$) of spiculae.

Similar to the case of microspherulation, pure geometrical evaluations allow conclusions about possible echinocyte-shapes formed. An unsolved question is the reason of a particular number of spiculae. Here again the investigation of the kinetics of this process should give an answer. On the other hand, experiments seem to show that the number of spiculae of an echinocyte once formed remains constant during subsequent changes of the volume. In this case the shape of the cell moves along the broken lines in fig. 8 and the parameters (a^*) and (h^*) are determined in this way. By subsequent shrinking of echinocytes we got the so called spherocytes or spheroechinocytes II (according to Bessis-nomenclature [38]) which have long but very thin spiculae that are usually invisible in the light microscopy.

Discussion

The calculations presented here indicate that it is possible to explain different shape transformations of human red blood cells on the basis of the fluid-mosaic behaviour of the membrane. Assuming dynamical differences in the mechanical properties of neighbouring membrane areas, controlled by the cells themselves (e.g. by internal ATP-influenced Ca^{++}-concentration) as well as by their environment (pH, temperature, drugs etc.), these transformations are explainable better than on the basis of pure mechanical calculations assuming homogeneous surface properties.

There are obviously two tpyes of questions to be answered here: The geometrical possibility of the realization of a given transformation induced by displacement of membrane material and the stability of the shape formed, given by the minimum of the bending energy. In all cases the first question has no unequivocal solution. The discocyte-stomatocyte transformation allows geometrically a mass of transition forms. The bistable equilibrium between two relatively discrete shapes is obviously predicted by the energy minimum. The calculation of it, however, is not so easy because the surface pattern of elasticity is unknown. Assuming a relatively homogeneous membrane for discocytes the calculation given by CANHAM [5] seems to be a realistic one.

In contrast to the discocyte-stomatocyte transformation the microspherulation and echinocytogenesis are geometrically stronger determined. Here, however, the kinetics of the process should be responsible for the realization of a given possibility. Here we need careful experiments and further theoretical investigations.

Our knowledge on the real molecular processes during the shape transformations is insufficient today. All determinants concerning the structure of the erythrocyte membrane we have tried to consider here. We are shure that this model should be strongly modified in the future. We look at the problem of shape transformations of human erythrocytes not only as an important medical question but additionally as a phenomenon indicating the functional regulation of mechanical properties in a cell in vivo. The erythrocyte as the simplest cell, however, again plays a role as a model object for most different problems.

References

[1] GLASER, R.: Acta biol. med. germ. **35**, 715 (1976)
[2] ADAMS, K. H.: Biophys. J. **13**, 1049 (1973)
[3] BULL, B. S., and J. D. BRAILSFORD: Blood Cells **1**, 323 (1975)
[4] BRAILSFORD, J. D., and B. S. BULL: J. theor. Biol. **39**, 325 (1973)
[5] CANHAM, P. B.: J. theor. Biol. **26**, 61 (1970)
[6] CERNY, L. C.: Biorheology **7**, 213 (1971)
[7] DANIELSON, D. A.: J. Biomechan. **4**, 611 (1971)
[8] EVANS, E. A.: Biophys. J. **14**, 923 (1974)
[9] EVANS, E. A.: Biophys. J. **13**, 926 (1973)
[10] FUNG, Y. C., and P. TONG: Biophys. J. **8**, 175 (1968)
[11] HELFRICH, W., and H. J. DEULING: J. Physique Rad. **36**, 327 (1975)
[12] JAY, A. W. L.: Biophys. J. **15**, 205 (1975)
[13] LENARD, J. G.: Bull. math. Biol. **36**, 55 (1974)
[14] PINDER, D. N.: J. theor. Biol. **34**, 407 (1972)
[15] GLASER, R., and A. LEITMANNOVÁ: Studia biophys. **48**, 219 (1975)
[16] BESSIS, M., R. I. WEED, and P. F. LEBLOND (eds): Red cell shape. Springer, Berlin 1973
[17] CHAILLEY, B., R. J. WEED, P. F. LEBLOND, and J. MAIGNE: Nouv. Rev. fr. Hématol. **13**, 71 (1973)
[18] DEUTICKE, B.: Biochim. biophys. Acta **163**, 494 (1968)
[19] FURCHGOTT, R. F., and E. PONDER: J. exp. Biol. **17**, 117 (1940)
[20] FURCHGOTT, R. F.: J. exp. Biol. **17**, 30 (1940)
[21] LÖWE, H., and F. JUNG: Pharmazie **29**, 693 (1969)
[22] SZÁCZ, J., P. TEITEL, and G. GÁRDOS: Acta Biochim. Biophys. Acad. Sci. hung. **5**, 409 (1970)

[23] Szácz, J.: Acta Biochim. Biophys. Acad. Sci. hung. **5**, 399 (1970)
[24] Weed, R. J., and M. Bessis: Blood **41**, 471 (1973)
[25] Singer, S. J. in: Structure and function of biological membranes, L. J. Rothfield (ed.). Academic Press, New York 1971
[26] Singer, S. J., and G. L. Nicolson: Science, N.Y. **175**, 720 (1972)
[27] Leitmannová, A., Stösser, and R. Glaser: Studia biophys. **60**, 73 (1976)
[28] Rakow, A. L., and R. M. Hochmuth: Biorheology **12**, 1 (1975)
[29] Leitmannová, A., and R. Glaser: Studia biophys. **64**, 123 (1977)
[30] Jay, A. D. L.: Biophys. J. **13**, 1166 (1973)
[31] Glaser, R. in: III. Winter School: Biophysics of membrane transport. School Proceedings, Post II, Wroclaw 1976 (in press)
[32] Fuji, T., T. Sato, and K. Nakanishi: Physiol. Chem. Phys. **5**, 423 (1973)
[33] Jung, F.: Folia haemat. **90**, 182 (1968)
[34] Ham, T. H., S. C. Shen, E. M. Flemming, and W. B. Castle: Blood **3**, 373 (1948)
[35] LaCelle, P. L., F. H. Kirkpatrick, M. P. Udkow, and B. Arkin in: Red cell shape. M. Bessis, R. I. Weed, and P. F. Leblond (eds). Springer, Berlin, Heidelberg, New York, 1973, p. 69
[36] Bessis, M., R. I. Weed, and P. F. Leblond (eds): Red cell shape. Springer, Berlin, Heidelberg, New York 1973
[37] Bessis, M.: Corpuscles. Atlas of red blood cell shape. Springer, Berlin, Heidelberg, New York 1974
[38] Bessis, M. in: Red cell shape. M. Bessis, R. I. Weed, and P. F. Leblond (eds). Springer, Berlin, Heidelberg, New York 1973, p. 1

Acta biol. med. germ., Band 36, Seite 871—873 (1977)

Anatomisches Institut der Friedrich-Schiller-Universität, 69 Jena, DDR

Adhäsivitätsuntersuchungen an menschlichen Erythrozyten

A. Benser

Zusammenfassung

Veränderungen an der Membran der roten Blutzellen können durch Adhäsivitätsuntersuchungen registriert werden. Nach Proteaseeinwirkung, wie Trypsinisierung oder Behandlung mit Pronase, ist eine gleichartige Minderung der Initialhaftung nach Inkubation in Plasma bzw. in Albuminlösung zu beobachten. Demgegenüber bleibt die Resuspension in Gamma-Globulin-Lösung ohne Effekt auf die Initialhaftung. Es kann angenommen werden, daß Albumin an die Zelloberfläche angelagert wird und für die Haftverringerung verantwortlich ist.

Einleitung

Die Adhäsivitätsmessung ist eine Methode, welche dazu dienen kann, bestimmte Veränderungen an der Zelloberfläche zu registrieren. In der vorliegenden Arbeit wurde das Initialhaftvermögen nach Behandlung der Erythrozytenmembran mit proteolytischen Enzymen und nachfolgender Resuspension in Serumeiweißen untersucht. Als Ausmaß der Initialhaftung wird derjenige Prozentsatz von Zellen verstanden, der in einer Kammer auf einer Haftfläche zur Sedimentation gebracht wird und nach Umdrehen der Kammer adhärent bleibt. Unbehandelte, mit PBS-gewaschene Erythrozyten zeigen eine nahezu 100%ige Initialhaftung. Sie vermindert sich nach proteolytischer Behandlung.

Material und Methodik

Es wurden Humanerythrozyten von 1—7 Tage altem Konservenblut der Blutgruppe A_1D mit 20% ACD-Zusatz verwendet. Je 1 ml des Blutes wurde zu Versuchsbeginn 3mal mit 10 ml PBS-Lösung gewaschen.

Proteasenbehandlung

Trypsin: 0,1% Trypsinlösung in PBS, *p*H 7,4; Inkubation 90 min bei 37 °C.

Pronase: 0,1% Lösung in PBS, *p*H 7,4; Inkubation 90 min bei 37 °C.

Papain: 0,5% Lösung in PBS mit 0,88% Zystein, *p*H 7,4; Inkubation 60 min bei 37 °C.

Neuraminidase: Eine 5 ml-Ampulle (Serva, Heidelberg) in 20 ml Natrium-Kakodylatpuffer, *p*H 6,5; Inkubation 18 Std. bei Zimmertemperatur.
Die Inkubation von 1 ml gewaschenem Erythrozytensediment erfolgte in jeweils 10 ml Proteasen-Lösung. Anschließend wurde das Blut 5mal in PBS-Lösung gewaschen und der Überstand abgesaugt. Die Kontrollen wurden in PBS-Lösung ohne Enzymzusatz unter sonst gleichen Versuchsbedingungen inkubiert.

Resuspension

Die Resuspension der Proben erfolgte in jeweils 0,5 ml folgender Lösungen: 1. PBS-Lösung (phosphatgepufferte NaCl-Lösung); 2. homologes Plasma; 3. 1,2% Human-Gammaglobulin-Lösung; 4. 4,8% Human-Albumin-Lösung.

Messung der Initialhaftung

Die zu untersuchenden Erythrozyten wurden mit PBS-Lösung bis zu einer mittleren Konzentration von 2250 Zellen/mm³ verdünnt. Die Suspension wurde in einer transparenten Kammer auf ein Deckglas sedimentiert. Nach Umdrehen der Kammer wurde derjenige Prozentsatz an Erythrozyten bestimmt, welcher am Deckglas adhärent war.

Ergebnisse

Erythrozyten, welche nur in PBS inkubiert wurden (Kontrollwerte), zeigten eine 100%ige Initialhaftung. Ein nahezu gleiches Ergebnis wurde auch bei nachfolgender Resuspension in Plasma und Albumin- bzw. Globulinlösung erhalten.
Erythrozyten, die nach Trypsinbehandlung in PBS-Lösung resuspendiert wurden, zeigten keine wesentliche Abnahme der Initialhaftung. Die Adhäsivität verminderte sich dagegen um 14—20%, wenn die Suspension in Plasma oder Albuminlösung erfolgte. Die Abweichungen von den Kontrollwerten waren statistisch signifikant.
Nach Resuspension in 1,2% Globulinlösung war weder bei den Kontrollerythrozyten, noch bei den trypsinierten Erythrozyten eine Verminderung des Initialhaftvermögens zu beobachten.
Ein ähnliches Verhalten zeigten die Erythrozyten nach Pronasebehandlung. Nach Resuspension in PBS-Lösung nahm die Adhäsivität um 5%, nach Resuspension in Globulinlösung um 6% ab.
Das Initialhaftvermögen veränderte sich nicht wesentlich gegenüber den Kontrollwerten nach Papainbehandlung und Neuraminidasebehandlung und nachfolgender Suspension in allen angegebenen Proteinlösungen.
Abb. 1 zeigt die Versuchsergebnisse.

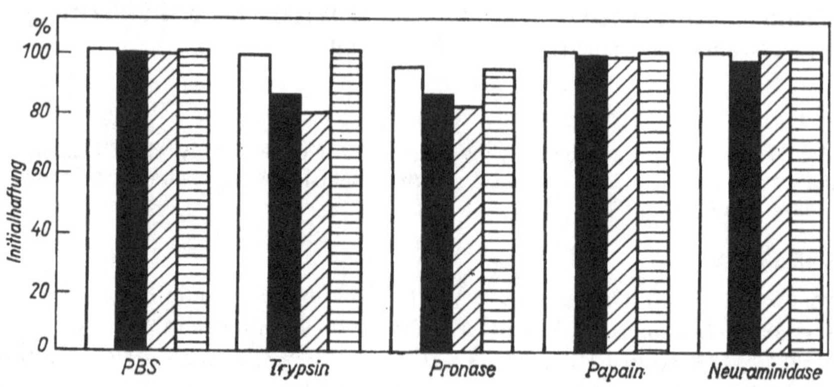

Abb. 1. Initialhaftung von Erythrozyten nach proteolytischer Behandlung. Resuspension in PBS-Lösung (1), homologem Plasma (2), Albuminlösung (3) und Gamma-Globulin-lösung (4)
Verdünnung in PBS-Lösung

Diskussion

Die Untersuchungen zeigen, daß nach proteolytischer Behandlung mit bestimmten Enzymen bei nachfolgender Resuspension in Albuminlösung und in Plasma eine Adhäsivitätsverminderung eintritt. Demgegenüber zeigen unbehandelte, PBS-gewaschene Erythrozyten keine wesentliche Abnahme der Initialhaftung nach Inkubation in diesen Medien.

Die vorliegenden Untersuchungen lassen zwar keine genaue Aussage darüber zu, inwieweit die Eryhtrozytenmembran unter den hier gegebenen Versuchsbedingungen verändert bzw. abgebaut wurde, man kann jedoch annehmen, daß durch die Einwirkung der genannten Enzyme Rezeptoren an der Zelloberfläche freigelegt werden, welche Albumin zu binden vermögen. Das würde bedeuten, daß Albumin das Haftvermögen von Erythrozyten zu Glasoberflächen herabsetzt.

Dagegen läßt das Verhalten der Erythrozyten nach Resuspension in Gamma-Globulin-Lösung vermuten, daß dieses Protein entweder nicht an die Zelloberfläche adsorbiert wird oder im Falle einer Anlagerung die Initialhaftung nicht herabsetzt.

Die Reduktion der Initialhaftung trypsinierter und pronasierter Erythrozyten in homologem Plasma ist möglicherweise vor allem auf das Serumalbumin zurückzuführen.

Welcher Art die Enzymeinwirkung ist und auf welche Weise die Anlagerung von Proteinen an die Zellmembran erfolgt, bedarf allerdings weiterer Untersuchungen, z. B. der gelelektrophoretischen Untersuchung vorliegender Zellkomponenten und der quantitativen und qualitativen Untersuchung des an die Zelloberfläche adsorbierten Proteins.

Acta biol. med. germ., Band 36, Seite 875—878 (1977)

Institut des Sciences Médicales de l'Université d'Oran
Department de Biochimie du CHU Oran
Laboratoire Central de Biochimie (Algerie)

Essais de solubilisation des antigènes B de la membrane érythrocytaire

F. SEGHIER, M. DJAFRI, A. PAGES et L. ABABEI

En 1900 LANDSTEINER [1, 2] observe l'agglutination des globules rouges par un sérum appartenant à la même espèce. Il recevait le prix NOBEL pour la découverte des groupes sanguins ABO. Bientôt on reconnut le rôle que jouaient les groupes ABO dans la transfusion sanguine, et à partir de ce moment, la transfusion devient sans danger.

En 1908, EPSTEIN et OTTENBERG [3] ont établi que les groupes sanguins ABO étaient héréditaires. Ceci a été plus tard prouvé par DUNGERN et HIRSZFELD [4] et par BERNSTEIN [5].

En dehors de l'organisme humain, des substances analogues à A, B et H sont largement distribuées dans la nature, ces substances ont été trouvées chez de nombreux animaux, certaines plantes et de nombreuses bactéries.

SPRINGER [6, 7] dans un travail de synthèse secoue un peu notre vision anthropocentrique: „le nom de substances de groupes sanguins" pour les substances ABO et MN est impropre et ne peut s'expliquer que par l'histoire de la découverte de ces structures de surface: certaines sont universellement répandues à travers le règne animal et végétal et paraissent représenter un principe de structure des surfaces que la nature a conservé depuis les micro-organismes et les plantes jusqu'à l'homme.

Ces structures de surfaces, pour employer la définition de SPRINGER ont des propriétés de récepteurs. Les anticorps, les virus, les toxines et les agents pharmacologiques ont avec elles des interactions spécifiques. En utilisant différentes méthodes de dosage on a recherché l'antigène A dans de nombreux tissus humains et on a observé que la plupart des tissus d'un sujet A possèdent l'antigène A [8]. Des substances ABO ont été trouvés dans des momies pré-colombiennes du Pérou [9], et dans des momies pré-dynastiques égyptiennes [10].

En 1967, ECONOMIDOU [11] estime que le nombre de sites antigènes A_1 sur un globule rouge d'adulte est de 810000 à 1170000, le nombre de sites antigènes A_2 est de 240000 à 290000, et le nombre de sites d'antigènes B sur une hématie d'adulte est de 610000 à 830000.

Des études chimiques détaillées sur les substances présentes dans les sécrétions et qui ont la même spécificité sérologique que les antigènes A, B, H et Lewis des globules rouges, révèlent que la spécificité de groupe sanguin est liée à des structures hydrocarbonées.

En vue de solubiliser les antigènes B de la surface érythrocytaire humaine pour une analyse ultérieurement détaillée nous avons essayé d'utiliser les différentes phospholipases dans ce but.

Matériel et méthodes

Dans nos expériences nous avons utilisé du sang humain prélevé sur héparine. Après le prélèvement nous avons centrifugé le sang à 3000 tours/minute pendant 5 minutes pour séparer les érythrocytes. Puis nous avons lavé ces érythrocytes avec de l'eau physiologique à 4 °C trois fois. Avec ces hématies lavées nous avons déterminé le groupe sanguin en utilisant des sérums humains non dilués provenant de l'Institut Pasteur de Paris.
Pour nos expériences nous avons utilisé seulement les érythrocytes du groupe *B*.
Dans les expériences dans lesquelles nous avons essayé de solubiliser les antigènes du groupe B de la membrane érythrocytaire, nous avons incubé les érythrocytes pendant une période de 30 minutes ou 1 heure en eau physiologique. A cette suspension érythrocytaire nous avons ajouté la phospholipase A_2 (Boehringer) provenant de venin de Crotalus Adamantéus, la phospholipase *D* provenant du choux (Boehringer) et la phospholipase *C* provenant des bactéries Céréus (Boehringer).
Après l'incubation à la température 37 °C, nous avons centrifugé la suspension, et nous avons utilisé le surnageant. Nous n'avons pas constaté d'hémolyse dans ces conditions.
Dans un tube à hémolyse nous avons introduit 3 gouttes de surnageant auquel nous avons ajouté 3 gouttes de sérum anti *B* dilué.
Pour nos expériences nous avons utilisé la dilution immédiatement supérieure à la dernière dilution du sérum anti *B* testé avant l'expérience et qui nous a donné une réaction positive avec les érythrocytes.
Le mélange surnageant + sérum anti *B* a été incubé pendant 20 minutes à la température 37 °C. Puis dans un deuxième temps nous avons utilisé 1 goutte du mélange avec 1 goutte d'érythrocytes du groupe *B* lavés et non incubés pour effectuer le dosage de l'agglutination.

Resultats et discussion

Après un tel traitement, des érythrocytes du groupe B avec la phopholipase C et D nous avons obtenu après l'incubation du surnageant avec le sérum anti B, en utilisant pour le dosage les érythrocytes non traités, une réaction positive. Nous constatons que les phospholipases C et D n'ont aucun effet sur les antigènes du groupe B de la membrane érythrocytaire.

Dans le schéma 1 on voit la séquence des traitements des érythrocytes pour „décoler" les antigènes B avec la phospholipase A_2.

Schéma 1

Explication sous forme de schémas de nos différentes étapes de travail

1° Etape:

Erythrocytes + phospholipase A_2 + $CaCl_2$
(Incubation 60 minutes à 37 °C)

↓ Centrifugation à 6000 tours/min, 5 min.

2° Etape:

Surnageant (3 gouttes) + Serum anti B (3 gouttes)
(agiter, incuber à 37 °C pendant 20 minutes)

3° Etape:

↓ d'agglutination

Une goutte du melange précédant
+
Une goutte d'érythrocytes B

Dans le Tableau 1 on pense que pendant l'incubation des érythrocytes avec la phospholipase A il est possible, que se soient solubilisés des antigènes B, et par la même, ont eu, la capacité de neutraliser les anticorps B de telle manière qu'ils se sont révélés quantitativement insuffisants pour donner une réaction positive avec les érythrocytes B. Il existe aussi la possibilité suivante: la phospholipase A_2 agirait sur les anticorps en modifiant leur activité. Cette hypothèse a été rejetée par nos expériences: nous avons incubé seulement du sérum anti B avec la phospholipase A_2 à la température du laboratoire 20 °C et à 37 °C. Après ce traitement des anticorps B avec la phospholipase A_2, nous avons obtenu la même réaction positive.

Tableau 1

La neutralisation des anticorps B par des facteurs solubilisés de la membrane érythrocytaire (de groupe B)

	Agglutination
Le surnageant des érythrocytes traités d'après le Schéma 1	— —
Le surnageant des érythrocytes traités dans les mêmes conditions sans phospholipase A_2	+++

Tableau 2

L'agglutination des érythrocytes par le serum anti B après le traitement avec la phospholipase A_2

	Agglutination
Erythrocytes lavés après traitement avec la phospholipase A_2	+
Erythrocytes lavés dans les mèmes conditions sans phospholipase A_2	+++

Donc l'effet de la phospholipase A_2 doit être localisé au niveau des antigènes B de la membrane érythrocytaire.

D'autre part, nous avons recherché l'activité antigénique des érythrocytes après l'incubation avec la phospholipase A_2 (Tableau 2), pour cela, dans un premier temps, nous avons lavé 3 fois les érythrocytes avec de l'eau physiologique. Puis nous avons utilisé dans un deuxième temps le sérum anti B non dilué pour tester l'activité antigénique. Nous avons constaté que les érythrocytes maintiennent à leur surface un nombre important d'antigènes, la réaction étant positive. Mais l'agglutination cette fois-ci, demande un temps beaucoup plus important pour s'établir par rapport à la réaction d'agglutination habituelle, c'est-à-dire avant le traitement.

Or on sait que sur chaque érythrocyte, il existe plus de 600000 antigènes [11].

Conclusions

1) Avec la phospholipase A_2 nous avons réussi à séparer de la surface érythrocytaire un certain nombre d'antigènes B, lesquels antigènes B ont eu la capacité de neutraliser le sérum anti B dans la phase suivante de l'expérience.

2) Le „décolage" des antigènes B n'est pas total mais important.

3) La phospholipase A_2 n'influence pas l'activité du sérum anti B donc des anticorps anti B.

4) Des études complémentaires dans la direction d'une purification chromatographique des antigènes solubilisés est en cours.

Bibliographie

[1] Landsteiner, K.: Zentbl. Bakt. Parasitkde. **27**, 357 (1900)
[2] Landsteiner, K.: Wien. klin. Wschr. **14**, 1132 (1901)
[3] Epstein, A. A. et R. Ottenberg: Proc. N.Y. Path. Soc. **8**, 117 (1908)
[4] Dungern, E. et L. Hirszfeld: Z. ImmunForsch. **6**, 284 (1900)
[5] Berstein, F.: Z. indukt. Abstamm. Vererb. Lehre **37**, 237 (1925)
[6] Springer, G. F.: Microbes and higher plants. Their relation to blood group substances. Proceedings of the 10th Congress of international Society of Blood Transfusion, Stockholm 1964 p. 465
[7] Springer, G. F.: Angew. Chem. **5**, 909 (1966)
[8] Holvorow, E. J., P. C. Brown, L. E. Glynn, M. D. Hwes, G. A. Gresham, T. F. O'Brien et R. R. A. Coombs: Br. J. exp. Path. **48**, 430 (1967)
[9] Furuhata, T., H. Nakajima, E. Ishida, S. Izumi, K. Ferada et Y. Amano: Proc. irup. Acad. Japan **35**, 305 (1959)
[10] Boyd, W. C. et G. Boyd Lyle: J. Immun. **32**, 307 (1937)
[11] Economidou, Y., G. Hughes, N. C. Jones et B. Gardner: Vox Sang. **12**, 325 (1967)

Acta biol. med. germ., Band 36, Seite 879—883 (1977)

Medizinische Klinik I der Universität, Klinikum Großhadern, D-8000 München 70, GFR

Increased number of ouabain binding sites in human erythrocyte membranes in chronic hypokalaemia

E. ERDMANN and W. KRAWIETZ with technical assistance of P. GROSSHANS

Summary

In chronic hypokalaemia the intracellular K^+ content of the heart and of erythrocytes stays almost unchanged despite a decreased potassium concentration of the serum. Under those conditions an increased (Na^+, K^+)-ATPase activity has been measured. This has been explained as a result of an adaptive induction of that enzyme system, which is responsible for the active transmembranous sodium and potassium transport. The ouabain receptor represents a subunit of the (Na^+, K^+)-ATPase.
In erythrocyte membranes from patients with chronic hypokalaemia an increased number of membrane-bound ouabain binding sites (476 ± 185, $n = 7$) has been measured. The control values (235 ± 48, $n = 15$) did not differ significantly from those from patients with acute hypokalaemia (256 ± 54, $n = 2$) lasting for less than two days. After correction of the hypokalaemia by potassium administration the number of ouabain receptors per single erythrocyte reached the normal again after 75—130 days.
It is concluded that the number of (Na^+, K^+)-ATPase molecules in the erythrocyte membrane is increased in chronic hypokalaemia. This may take place during erythroipoesis.

Introduction

The active transport of sodium and potassium across the erythrocyte membrane is regulated by the (Na^+, K^+)-activated ATPase (EC 3.6.1.2.) [1]. This membrane-bound enzyme system as well as the active cation transport are specifically inhibited by cardiac glycosides, which bind to the outer surface of the membrane whereas ATP is hydrolysed at the internal side. Recent experiments indicate that a protein being exposed to both sides of the membrane ("spanning the membrane") contains the enzymatically active site ((Na^+, K^+)-ATPase) on the inside surface and the glycoside binding site ("receptor") on the outside [2]. Thus, the quantitative measurement of membrane-bound erythrocyte ouabain (g-strophanthin) binding sites and (Na^+, K^+)-ATPase activity should allow the calculation of the turn-over number for ATP per enzyme molecule and thereby the cation transport capacity of the red cells [3].
The concentration gradient maintained by the cell membrane for potassium is considerably raised in chronic potassium deficiency when serum potassium concentrations fall below 3 mequiv./l. Nevertheless intracellular sodium and potassium concentrations stay relatively normal [4], although the active outward transport of sodium is coupled to an active inward transport of potassium.
As a possible explanation for this phenomenon former communications have shown that there is an increase in the (Na^+, K^+)-ATPase activity ("transport ATPase") in the membranes of rat erythrocytes [4] and guinea pig heart [5, 6], when the animals are kept on a low potassium diet for several weeks.

Supported by the Deutsche Forschungsgemeinschaft

The present study was designed to find out whether the number of membrane-bound ouabain binding sites being closely related to (Na^+, K^+)-ATPase is increased in severe chronic hypokalaemia in human erythrocytes, too.

Materials and methods

[^{3}H]Ouabain (= g-strophanthin) (lot-nr. 747186, spec. activity = 12 Ci/mmole) was purchased from New England Nuclear (Dreieichenhain, GFR). The liquid scintillation fluid used was either Insta-Gel or Insta-Fluor (Packard Instruments GmbH, Frankfurt, GFR). All other chemicals were of analytical grade and obtained through Merck (Darmstadt) or Boehringer (Mannheim).

Erythrocytes were counted routinely in a Coulter Counter Modell S, Coulter Electronics (Dunstable, Bed. U.S.A.).

20 ml of freshly drawn venous blood from haematologically healthy adults were used for the membrane preparation, which has been described in detail in a previous paper [3].

0.5—1.0 ml of human erythrocyte membranes were usually used for the ouabain binding assay (for details see [3]). In brief, the equilibrium [^{3}H] ouabain binding procedure is as follows: the binding sites for cardiac glycosides ("receptors") are membrane bound, it is therefore possible after incubation with [^{3}H] g-strophanthin to separate receptor-bound from unbound (free) cardiac glycosides by ultracentrifugation or rapid filtration and to determine the relative quantities.

The total amount of [^{3}H]ouabain bound to the membranes has to be corrected by subtraction of that amount that is not displaceable by high concentrations (10^{-4} M) of unlabelled ouabain (= unspecific binding). The result is defined as specific binding.

Details of the incubation conditions and times are given in the respective figures.

Results

[^{3}H] ouabain binding to isolated human erythrocyte membranes is a time dependent process, which proceeds to an equilibrium (Fig. 1). The time course of the drug binding is the same in the control as well as in the hypokalaemic erythrocyte membranes. The amount bound at identical [^{3}H] ouabain concentrations in the incubation medium is, however, greatly increased in chronic hypokalaemia.

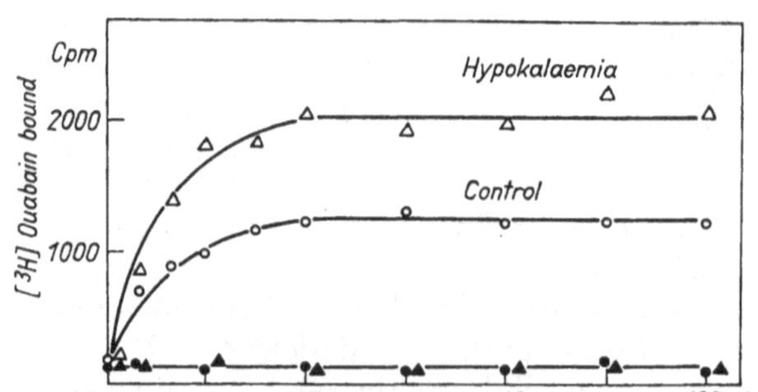

Fig. 1. Time course of [^{3}H]ouabain binding to human erythrocyte membranes.
The incubation medium contained 50 mM imidazole-HCl buffer, *p*H 7.25, 3 mM $MgCl_2$, 3 mM imidazole-phosphate, $8.3 \cdot 10^{-12}$ moles [^{3}H]ouabain and 1 ml erythrocyte membrane suspension originating from 1 ml venous blood; t = 37 °C; total volume = 2 ml.
○ $4.8 \cdot 10^9$ erythrocytes/ml, serum potassium = 4.6 mequiv/l, control person. △ $4.9 \cdot 10^9$ erythrocytes/ml, serum potassium = 2.8 mequiv/l, prolonged diuretic therapy with chronic hypokalaemia.
Filled circles and triangles: in the presence of additional unlabelled ouabain (10^{-4} M)

In order to quantitate the maximal ouabain binding capacity, equilibrium binding experiments were performed with increasing ouabain concentrations in the incubation medium. The experimental data were plotted according to SCATCHARD [7] (Fig. 2). In this plot the maximal, extrapolated number of binding sites is shown on the ordinate intercepting with the experimental line. The dissociation constant (K_D) of the ouabain-receptor-complex may be calculated from the slope of the plot. The steeper the slope, the higher the K_D (i.e. that free drug concentration at which half saturation of the receptors occurs). The straight line of the plot indicates that the affinity of the receptors for ouabain does not change with the increasing concentration (i.e. there is only one type of receptors in human erythrocytes). The two lines run parallel indicating that the affinity of the receptors for ouabain did not change in hypokalaemia but the number of receptors.

In three patients with chronic hypokalaemia (due to diuretic treatment, anus praeter naturalis associated with potassium loss and frequent use of laxatives plus diuretics), we were able to follow up the loss of receptor sites after correction of the

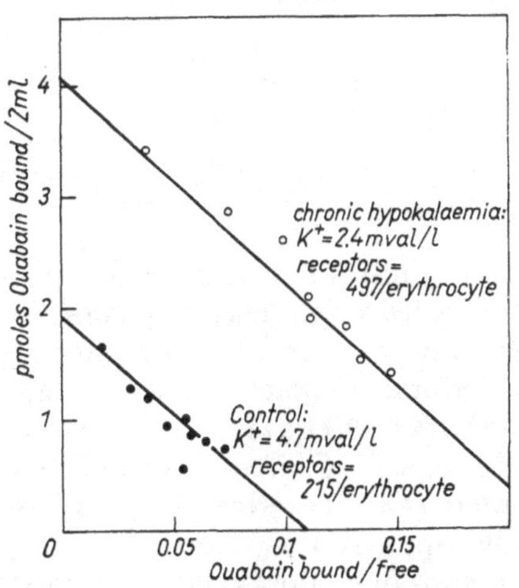

Fig. 2

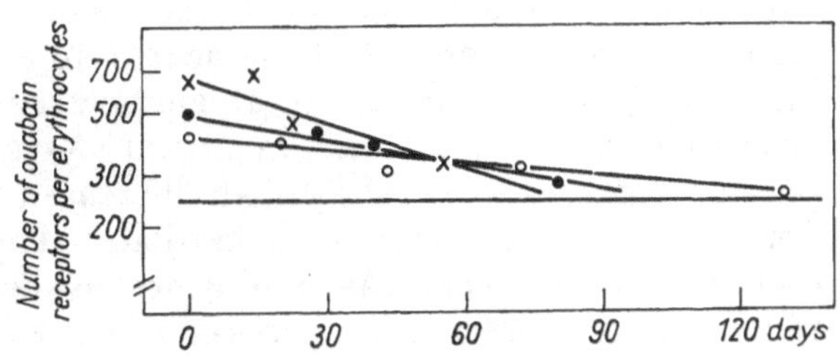

Fig. 3

Fig. 2. Receptors for ouabain on human erythrocyte membranes in chronic potassium deficiency.

Conditions of incubation as in Fig. 1 but with increasing ouabain concentrations. The experimental data are displayed according to SCATCHARD [7], where the amount "bound" is plotted versus the ratio "bound/free". In this graph a straight plot indicates an identical affinity of the receptor for the ligand at all concentrations. This means the existence of only one type of receptors. The intercept of the plot with the ordinate gives the maximal number of binding sites (receptors). This graph shows that the erythrocyte membranes from the patient with chronic potassium deficiency (○) contain more receptors than those from the control person (●), both determined at identical conditions and the same amount of erythrocytes incubated. The affinity of the receptors for ouabain is unchanged as evidenced from the parallel course of the plot

Fig. 3. Number of ouabain binding sites per single erythrocyte after treatment of hypokalaemia.

In three patients with chronic potassium deficiency the time course of ouabain receptors could be followed up after correction of the hypokalaemia by administration of potassium.

serum potassium levels (all of them had been below 2.8 mequiv/l on admission to the hospital). Although the scarce data are not conclusive yet (Fig. 3), one can clearly see that the number of ouabain receptors in erythrocyte membranes decreases after the normalisation of potassium levels and reaches the "normal" again after 75—130 days.

In two patients with acute hypokalaemia (due to diarrhoea and vomiting, and forced diuresis) the number of ouabain receptors was not changed significantly. In these patients the hypokalaemia did not last for more than two days (Table 1).

Table 1
Number of ouabain receptors per single human erythrocyte

	Ouabain receptors per erythrocyte	Serum potassium (mequiv./l)
Chronic hypokalaemia	476 ± 185 ($n = 7$)	3.0 ± 0.4 ($n = 7$)
Acute hypokalaemia	256 ± 54 ($n = 2$)	3.1 ± 0.4 ($n = 2$)
Control	235 ± 48 ($n = 15$)	4.6 ± 0.5 ($n = 15$)

The values represent the mean $\pm$ SD, n = number of patients. The number of ouabain receptors is calculated from a concentration-dependent binding experiment plotted according to SCATCHARD [7] as in Fig. 2.

Discussion

Although serum K^+ concentrations are markedly diminished in chronic hypokalaemia, the K^+ content of the heart and erythrocytes stays relatively constant [4, 6, 8]. This might be due to an enhanced active inward transport of potassium occurring in the heart and in erythrocytes in chronic hypokalaemia. In fact, previous studies have shown an increased (Na^+, K^+)-ATPase activity of rat erythrocytes [4] and guinea pig cardiac cell membranes [5, 6] after a K^+ deficient diet for several weeks. As it is generally accepted that the (Na^+, K^+)-ATPase represents the enzymatic expression of the sodium-potassium pump of the cell membrane, an increased activity of that enzyme system could explain the possibility of an enhanced active transport against a steeper intra-extracellular potassium gradient in chronic hypokalaemia, sufficient to maintain the "normal" intracellular K^+ content.

An increased (Na^+, K^+)-ATPase activity may be caused by structural changes of the cell membrane affecting the enzyme, altered enzyme properties or an augmentation of enzyme molecules in the cell membrane. Although previous studies could not demonstrate altered enzyme properties, they were not able to prove an induction of (Na^+, K^+)-ATPase molecules ("de novo synthesis") serving as adaptative mechanism in chronic hypokalaemia, either.

Recent experiments have indicated that the (Na^+, K^+)-ATPase is composed of two polypeptide chains [2]. The enzymatically active site, which is accessible from the inside surface and the binding site for cardiac glycosides, which is accessible from the outside surface of the plasma membrane are both on one, the larger polypeptide subunit of the (Na^+, K^+)-ATPase. This means that this polypeptide chain is exposed to both sides of the membrane and probably spans the membrane. Further experiments demonstrated a 1:1 relationship between the binding sites for cardiac glycosides and for nucleotides in isolated cell membranes, thus supporting this concept [9]. Assuming the measured constant relationship between ouabain binding sites and (Na^+, K^+)-ATPase activity in chronic hypokalaemia, too, an increase in cardiac glycoside receptors means an increased number of enzyme molecules. This, in fact, has been demonstrated in human erythrocytes by the above experiments.

Some further aspects seem to prove interesting, though. The disappearance, of receptors within 70—130 days after ample potassium administration might represent the appearance of newly synthetized erythrocytes and the disappearance of the old ones. Within experimental error ($n = 3$!) the time course coincides roughly with the life span of erythrocytes. In acute hypokalaemia we could not find an increased number of ouabain receptors. This might either mean that only prolonged hypokalaemia will cause an increased number of (Na^+, K^+)-ATPase molecules in the cell membranes or the formation of enzyme molecules in erythrocyte membranes occurs only during erythropoiesis.

References

[1] Schatzmann, H. J.: Helv. physiol. pharmac. Acta **11**, 346 (1953)
[2] Kyte, J.: J. biol. Chem. **247**, 7642 (1972)
[3] Erdmann, E., and W. Hasse: J. Physiol., Lond. **251**, 671 (1975)
[4] Chan, P. C., and W. R. Sanslone: Archs Biochem. Biophys. **134**, 48 (1969)
[5] Erdmann, E., H.-D. Bolte, and B. Lüderitz: Archs Biochem. Biophys. **145**, 121 (1971)
[6] Bluschke, V., R. Bonn, and K. Greeff: Eur. J. Pharmac. **37**, 189 (1976)
[7] Scatchard, G.: Ann. N.Y. Acad. Sci. **51**, 660 (1949)
[8] Bolte, H.-D., B. Lüderitz, E. Erdmann, and G. Steinbeck: Verh. dt. Ges. inn. Med. **78**, 1052 (1972)
[9] Erdmann, E., and W. Schoner: Biochim. biophys. Acta **330**, 316 (1973)

Acta biol. med. germ., Band 36, Seite 885—886 (1977)

Anatomisches Institut der Friedrich-Schiller-Universität, 69 Jena, DDR

Zur saccharoseinduzierten Agglutinabilität menschlicher Erythrozyten

K.-J. HALBHUBER, W. LINSS und G. GEYER

Zusammenfassung

Nach mehrfacher Waschung von Humanerythrozyten in wäßriger isotoner Saccharoselösung kann eine spontane Agglutination der Zellen in Gegenwart dieses Zuckers (aber auch von Glukose, Glykogen etc.) beobachtet werden. Behandlungen der Erythrozyten mit ionalen Spülmedien (PBS, NaCl etc.) führen zu keiner Zellagglutination und die zuckerinduzierte Agglutination von Erythrozyten wird durch Zugabe kritischer Mengen ionaler Flüssigkeiten zur Zuckerlösung wieder aufgehoben. Sie setzt wiederum spontan ein, wenn steigende Mengen Zuckerlösung zugeführt werden.

Um den Mechanismus des Agglutinationsverhaltens zu klären, wurden in PBS-Lösung gewaschene A_1-Erythrozyten in PBS-Lösung 60 min bei 47—52 °C inkubiert. Die anschließend erfolgende Nachbehandlung der Erythrozyten mit Saccharose führte zu keiner oder sehr stark abgeschwächter Hämagglutinationsreaktion. Wurden die in warmer PBS-Lösung vorbehandelten Erythrozyten zunächst im autologen Plasma/Saccharosegemisch (2:1), Anti-B-Serum/Saccharosegemisch, AB-Mischserum/Saccharose bzw. in 5%iger Albuminlösung/Saccharose wie ebenso in dem PBS-Überstand der PBS-Inkubation bei 47—52 °C resuspendiert und 16 Std. bei 1 °C inkubiert, so verhielten sich die Erythrozyten in Saccharoselösung unterschiedlich.

Während sich die Erythrozyten nach Inkubation im AB-Mischserum/Saccharose bzw. in der saccharosehaltigen Albuminlösung im Saccharosemilieu nichtagglutinabel zeigten, konnte eine ausgeprägte Agglutinationsreaktion nach Inkubation im autologen Plasma, Anti-B-Serum und im PBS-Überstand unter den oben genannten Bedingungen nachgewiesen werden. Dieses Agglutinationsverhalten konnte gleichermaßen mit DAB 7-Saccharose, Saccharose p. a. wie auch mit ultrafiltrierter Saccharoselösung (Filter UM 2, Amicon, Diaflo Ultrafilter, Ausschlußmolekulargewicht 1000) erzielt werden.

Aus den Befunden von NAJJAR et al. [3, 4] und HALBHUBER et al. [2] geht hervor, daß Erythrozyten in vivo γ-Globuline (erythrophile Proteine) vergleichsweise fest und spezifisch an der Blutgruppendeterminante binden können. Diese Proteine sind zum großen Teil mit dem autologen Isohämagglutinin identisch. Nach den Auffassungen von NAJJAR et al. besitzen die reaktiven Stellen der blutgruppenspezifischen Determinante und die des autologen Isohämagglutinins keine komplementäre Anpassung wie etwa ein Antikörper zu einem Antigen, doch sollen sich die Bindungsplätze beider Reaktionspartner subkomplementär zueinander verhalten und offenbar etwas geringere hydrophobe Bindungskräfte als bei der Antigen-Antikörperbindung besitzen.

Mit Hilfe der Ouchterlony-Technik läßt sich mit Anti-Humanglobulin an nicht gewaschenen und an 1mal mit PBS-Lösung bei 47 °C gewaschenen A_1-Erythrozyten die Beladung mit Globulin nachweisen. Nach 3- und mehrmaliger Waschung ergibt diese Technik ein negatives Resultat. In ähnlicher Weise ist auch in der 1. und 2. Spülflüssigkeit (Inkubationszeit 5 min im 10fachen Überschuß an PBS bezogen auf das Erythrozyten-Sediment) ein positiver Ausfall im Ouchterlony-

57*

Test zu beobachten. Die 3. und weiteren Spülflüssigkeiten enthalten keine erfaßbaren Globulinmengen. Selbst nach 6maliger Waschung mit Saccharose fällt an den Erythrozyten der Nachweis von Globulin positiv aus. Die Saccharosespülflüssigkeiten enthalten noch in der 3. Spülflüssigkeit nachweisbare Mengen an Globulin.

Mehrfache Waschung der Erythrozyten mit Saccharoselösung direkt aus dem Plasmamilieu läßt eine weitgehende Besetzung der Blutgruppendeterminanten auf der Erythrozyten-Oberfläche mit autologem Isohämagglutinin erwarten. Hier sind die interzellulären Bindungskräfte so hoch, daß bei genügend hoher Zellannäherung eine Agglutination der Erythrozyten zustandekommt (A-Erythrozyten agglutinieren also durch Anti-B im inerten Zuckermilieu). Nach Waschung im ionalen Spülmedium dissoziieren die meisten Agglutininmolekel infolge zu geringer hydrophober Attraktion von der Determinante ab und die Agglutinate werden aufgelöst. Nach Vorinkubation der Erythrozyten in warmer ionaler PBS-Lösung werden die Agglutininmolekel in besonders hohem Maße oder fast vollständig von der Erythrozyten-Determinante entfernt und mehrfache Nachspülungen mit Saccharose führen folglich zu einer schwachen oder vollständig aufgehobenen Hämagglutinationsreaktion in diesem Milieu.

Die so weitgehend hämagglutininverarmten oder -freien Erythrozyten agglutinieren im Saccharosemilieu nach prolongierter Inkubation im isohämagglutininfreien Milieu (AB-Mischserum oder Albuminlösung) ebenfalls nicht mehr. Lediglich Nachbehandlung der Erythrozyten mit Medien, die autologes Isohämagglutinin enthalten (Anti-B-Serum, autologes Plasma oder Überstand der Wärmeinkubation in PBS) verursacht eine deutliche Agglutinationsneigung der Erythrozyten nach mehrfacher Saccharosenachspülung der Zellen.

Es wird deshalb angenommen, daß die saccharoseinduzierte Hämagglutinationsreaktion eine Bindung des autologen Isohämagglutinins an der blutgruppenspezifischen Determinante der Erythrozyten voraussetzt. Unter geeigneten Bedingungen (ionales Milieu, Wärme) wird der Determinanten-Isohämagglutininkomplex weitgehend dissoziiert.

Nach Reassoziation dieses Komplexes (im ionalen und nichtionalen Milieu, Kälte) tritt die Hämagglutination in Saccharoselösung wieder auf. Die Assoziationskräfte sind offensichtlich so stark, daß eine Zellagglutination ermöglicht wird.

Eine mögliche Mitbeteiligung von Lektinen an der Agglutinationsreaktion kann ausgeschlossen werden, da alle hier eingesetzten Saccharosechargen unterschiedlichen Reinheitsgrades dasselbe Verhalten der Erythrozyten zeigten.

Literatur

[1] Dodd, B. E., u. P. J. Lincoln: Br. J. Haemat. **26**, 93 (1974)

[2] Halbhuber, K.-J., A. Benser, H. Brandt, H. Feuerstein u. G. Geyer: Folia haemat., Lpz. **101**, 220 (1974)

[3] Harshman, S., u. V. A. Najjar: Biochem. Biophys. Res. Commun. **11**, 411 (1963)

[4] Najjar, V. A., K. Nishoika, A. Constantopoulos u. P. S. Satoh: VII. Internationales Symposium über Struktur und Funktion der Erythrozyten, Berlin 1973 (= Abhandlungen der Akademie der Wissenschaften der DDR 1973). S. Rapoport u. F. Jung (Hrsg.). Akademie-Verlag, Berlin 1975

Dr. K.-J. Halbhuber, Anatomisches Institut der Friedrich-Schiller-Universität Jena, 69 Jena DDR, Teichgraben 7

Acta biol. med. germ., Band 36, Seite 887–890 (1977)

Anatomisches Institut der Friedrich-Schiller-Universität, 69 Jena, DDR

Elektronenmikroskopischer Nachweis von Bindungsorten für Kationen an der Erythrozytenmembran

W. LINSS und G. GEYER

Summary

Numerous Ca^{2+}-binding sites were localized ultrahistochemically at the inner surface of the erythrocyte membrane and small amounts at the outer surface. La^{3+}-binding sites were demonstrated at the outer surface only. The results were discussed in relationship to the binding capacity of the filamentous matrix and the glycocalyx of the human erythrocyte membrane.

Einleitung

Während der Lagerung von Erythrozyten wird bei gleichzeitigem Abfall des Adenosintriphosphats ein Anstieg des intrazellulären Kalziums beobachtet [1–5]. In Abhängigkeit vom Kalziumgehalt ändern sich auch die Gestalt und die Verformbarkeit der Erythrozyten, so daß an eine Bindung des Kalziums an Membranbestandteile zu denken ist. Mit ultrahistochemischen Techniken versuchten wir, den Ort der Kalziumbindung zu erfassen.

Am gleichen Material oder in parallelen Versuchen wurde auch die Bindungsfähigkeit der Erythrozytenmembran für Lanthanionen untersucht.

Material und Methode

Dreimal mit isotonem Phosphatpuffer, *p*H 7,4, gewaschene Erythrozyten fixierten wir in Anlehnung an OSCHMANN und WALL [6] für 60 min in 3%igem Glutaraldehyd, der mit 0,05 M Natriumkakodylatpuffer auf *p*H 7,4 eingestellt und dem 5 oder 90 mM $CaCl_2$ zugesetzt war. Gleiche $CaCl_2$-Mengen enthielt auch der zum Spülen verwendete Kakodylatpuffer und während der Nachfixierung das 1%ige OsO_4. Mittels steigender Alkoholkonzentrationen wurde entwässert und über Propylenoxid in Durcupan eingebettet. In ähnlicher Weise wurden Erythrozytenschatten, zum Teil nach Dialyse gegen 0,5 mM ÄDTA-Na-Lösung, behandelt und eingebettet. Zum Nachweis der Lanthanbindungsorte setzen wir den Fixierungs-, Spül- und Nachfixierungsflüssigkeiten außer 5 mM oder 90 mM $CaCl_2$ noch 5 mM Lanthannitrat zu. Bei einigen Chargen wurde der $CaCl_2$-Zusatz völlig weggelassen, oder es war nur im Fixans enthalten. Die Einbettung erfolgte in jedem Falle in der oben beschriebenen Weise.

Nach der Herstellung von Ultradünnschnitten untersuchten wir das Material bei Primärvergrößerungen zwischen 40000 und 50000 am JEM 100B[1].

Ergebnisse und Diskussion

An den unter $CaCl_2$-Zusatz fixierten und weiterbehandelten Erythrozyten wurden elektronendichte Ablagerungen gefunden. Diese lagen der Erythrozytenmembran innen mit glatter Fläche an, während ihre dem Zellinnern zugewandte Oberfläche unregelmäßig gestaltet war (Abb. 1). Einigen dieser Ablagerungen lag an der

[1] Für die Arbeitsmöglichkeiten in der Elektronenmikroskopischen Abteilung des Bereiches Medizin der FSU Jena möchten wir danken.

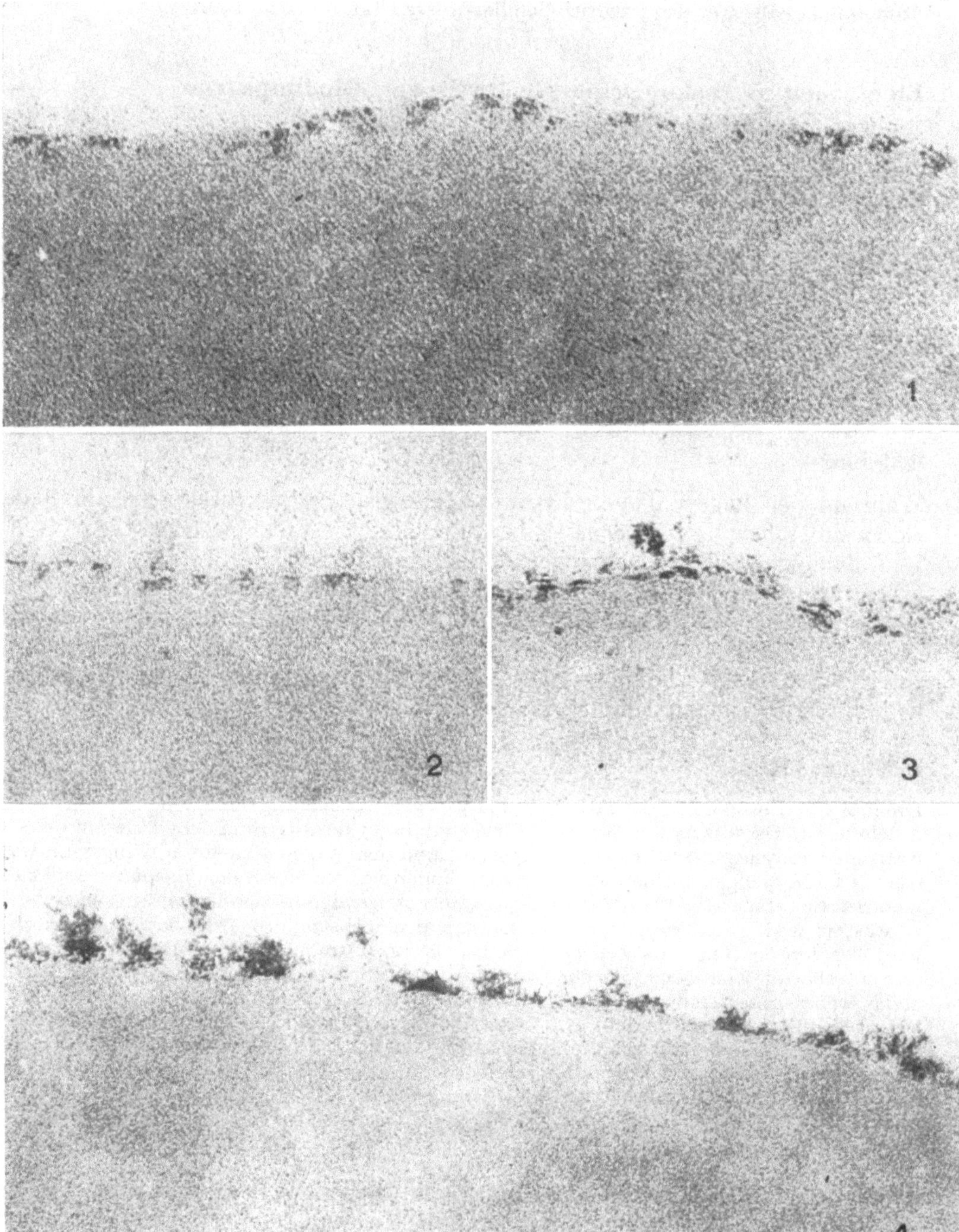

Abbildungslegenden siehe Seite 899

Außenseite der Membran eine Ablagerung von geringer Elektronendichte gegenüber (Abb. 2). Wurde während der Vorbehandlung der $CaCl_2$-Zusatz ganz oder auch nur teilweise aus den Spül- und Nachfixierungslösungen weggelassen, fanden sich im elektronenoptischen Bild derartige Ablagerungen nur vereinzelt. Es darf also angenommen werden, daß diese elektronendichten Ablagerungen Kalzium enthalten, und auch an menschlichen Erythrozyten mit dieser Methode in ähnlicher Weise wie an Muskelzellen, Axonen und Synapsen, Haar- und Stützzellen des Cortischen Organs und an Hühnererythrozyten [27, 7—10] kalziumbindende Proteine nachzuweisen sind.

Wird außer Kalzium den Fixierungs-, Spül- und Nachfixierungsflüssigkeiten noch 5 mM Lanthannitrat zugesetzt, so werden neben den Niederschlägen an der Innenseite auch an der Außenseite unterschiedlich große elektronendichte Niederschläge gefunden. Nicht selten lagen sich die elektronendichten Ablagerungen an der Innen- und Außenseite der Membran unmittelbar gegenüber, durch einen etwa 40 Å breiten elektronendurchlässigen Spalt voneinander getrennt (Abb. 3), der der Lipidschicht der Zellmembran entsprechen dürfte. Wird während der Vorbehandlung der Erythrozyten den Flüssigkeiten kein $CaCl_2$ zugesetzt, so lassen sich Niederschläge nur an der Außenseite der Membran nachweisen (Abb. 4). Die Bildung von La^{3+}-Ablagerungen erfolgt also ausschließlich an der Außenseite. Für die Bindung von La^{3+} kommen die Glykoproteine in Frage, und es ist hierbei in erster Linie an die hydrophilen Segmente der Glykophorine zu denken. Die an ihnen vorhandenen Sialinsäurereste können auch Kalzium binden [11, 12] und können für die in den $CaCl_2$-Versuchen nachgewiesene schwache Ca^{2+}-Bindung an der Membranaußenfläche verantwortlich gemacht werden.

Die starke Bindung von Ca^{2+} an der Innenseite könnte auf Spektrin und Tektin bezogen werden, welche die innere filamentäre Matrix der Erythrozytenmembran aufbauen [13—17]. Sie können durch Puffer geringer Ionenstärke aus dem System leicht entfernt werden, und die isolierten Komponenten präzipitieren mit Ca^{2+} und anderen Ionen [15, 17—25]. In unseren Untersuchungen an Erythrozyten-

Abb. 1. Nach Zusatz von $CaCl_2$ zu den Fixierungs- und Spülflüssigkeiten werden an der Innenseite der Erythrozytenmembran elektronendichte Ablagerungen gefunden, deren membrannahe Fläche glatt ist.
Vergrößerung 200000fach

Abb. 2. Vorbehandlung wie in Abb. 1. Einigen elektronendichten Ablagerungen an der Innenseite liegen an der Außenseite der Erythrozytenmembran geringe Mengen eines elektronendichten Materials gegenüber.
Vergrößerung 200000fach

Abb. 3. Wird den Fixierungs- und Spülflüssigkeiten Kalziumchlorid und Lanthannitrat zu gesetzt, so erscheinen an der Außenseite der Erythrozytenmembran stärker elektronendichte Ablagerungen. Sie liegen zum Teil den elektronendichten Komplexen an der Innenseite unmittelbar gegenüber.
Vergrößerung 200000fach

Abb. 4. Bei alleinigem Zusatz von Lanthannitrat zu den Fixierungs- und Spülflüssigkeiten werden nur an der Außenseite der Erythrozytenmembran elektronendichte Ablagerungen beobachtet.
Vergrößerung 200000fach

schatten fanden wir am dialysierten Material keine elektronendichten Ablagerungen, während wir sie am nicht dialysierten Material regelmäßig an der Innenseite der Membran nachweisen konnten. Dieser Befund spricht für eine Bindung der Ca-Ionen durch das Spektrin-Tektin-System, zumal von Stibenz [26] nach Dialyse der Erythrozytenschatten eine wesentliche Verminderung des Spektringehaltes nachgewiesen wurde.

Die Bindung von La^{3+} und Ca^{2+} an korrespondierenden Stellen der Membranaußenseite und -innenseite läßt sich in der Weise deuten, daß die Glykophorine der Erythrozytenmembran einerseits mit ihrem kohlenhydrathaltigen Anteil für die La^{3+}-Bindung verantwortlich sind, während sie an der Innenseite mit dem Spektrin-Tektin-System verknüpft sind und über dieses System mit der Ca^{2+}-Bindung in Beziehung stehen.

Literatur

[1] Nakao, M., T. Nakao, M. Tatibana, H. Yoshikawa u. T. Abe: Biochim. biophys. Acta **32**, 564 (1959)
[2] Lionetti, F. J., u. J. McKay: Vox Sang. **17**, 34 (1969)
[3] Weed, R. I., P. L. LaCelle u. E. T. Merrill: J. clin. Invest. **48**, 795 (1969)
[4] Palek, J., W. A. Curby u. F. J. Lionetti: Blood **40**, 261 (1972)
[5] Feo, C. J., u. P. F. Leblond: Blood **44**, 639 (1974)
[6] Oschmann, J. L., u. B. J. Wall: J. Cell Biol. **55**, 58 (1972)
[7] Hillmann, D. E., u. R. Llinás: J. Cell Biol. **61**, 146 (1974)
[8] Politoff, A. L., S. Rose u. G. D. Pappas: J. Cell Biol. **61**, 818 (1974)
[9] Stanka, P.: Cell. Tiss. Res. **156**, 223 (1975)
[10] Geyer, G., u. Ch. Meyer: nicht publizierte Befunde 1976
[11] Forstner, J., u. J. F. Manery: Biochem. J. **124**, 563 (1971)
[12] Seamen, G. V. F., P. S. Vassar u. M. J. Kenndall: Archs Biochem. Biophys. **135**, 356 (1969)
[13] Tillack, T. W., S. L. Marchesi, V. T. Marchesi u. E. Steers: J. Cell Biol. **29**, 135a (1968)
[14] Guidotti, G.: Archs intern. Med. **129**, 194 (1972)
[15] Bretscher, M. S.: Science, N.Y. **181**, 622 (1973)
[16] Juliano, R. L.: Biochim. biophys. Acta **300**, 341 (1973)
[17] Steck, T. L.: J. Cell Biol. **62**, 1 (1974)
[18] Palek, J., W. A. Curby u. F. J. Lionetti: Fedn Proc. Fedn Am. Socs exp. Biol. **28**, 339 (1969)
[19] Palek, J., G. Stewart u. F. J. Lionetti: Blood **44**, 583 (1974)
[20] Jacob, H., T. Amsden u. J. White: Proc. natn. Acad. Sci. U.S.A. **69**, 471 (1972)
[21] Triplett, R. B., J. M. Wingate u. K. L. Carraway: Biochem. biophys. Res. Commun. **49**, 1014 (1972)
[22] Mirkevova, L., L. Vikora u. J. Fiala: Folia haemat. **99**, 326 (1973)
[23] Reynolds, J. A.: Fedn Proc. Fedn Am. Socs exp. Biol. **32**, 2034 (1973)
[24] Lux, S. E., u. K. M. John: Blood **44**, 909 (1974)
[25] White, J. G., G. H. R. Rao u. D. D. Mundschenk: Blood **44**, 909 (1974)
[26] Stibenz, D.: Persönliche Mitteilung 1976
[27] Oschmann, J. L., T. A. Hall, P. D. Peters u. B. J. Wall: J. Cell, Biol. **61**, 156 (1974)

Acta biol. med. germ., Band 36, Seite 891—896 (1977)

Országos Munka- és Üzemegészségügyi Intézet, Budapest, Hungary

Reaction of erythrocyte membrane-bound acetylcholinesterase with reversible inhibitors: The role of apolar interactions

S. Mányai

Summary

The spasmolytic drug, bencyclan — a dimethylamino substituted cycloalcanol-ether — reversibly inhibits various cholinesterase activities in vitro. Since the protonated drug is soluble in water, whereas the uncharged molecule is highly lipophilic, bencyclan seemed to be a proper model for studies on the role of non-polar interactions in the inhibition of the erythrocyte membrane-bound acetylcholinesterase (AChE) by an amphophilic enzyme inhibitor.

AChE activity of isolated erythrocyte ghosts at different *p*H values in the presence of varying concentrations of substrate (acetylthiocholine), inhibitor (bencyclan) and a non-ionic detergent (Triton X-100) was measured. Inhibition at physiological *p*H of erythrocyte membrane-bound AChE by bencyclan revealed itself as the sum of inhibitory actions of protonated as well as uncharged forms of the drug. The reaction of bencyclan cation with AChE is competitive with the substrate, whereas that of the non-ionized drug is non-competitive.

Triton X-100 at alkaline *p*H competes with the non-ionized bencyclan for hydrophobic membrane sites and/or bound AChE resulting in a decreased AChE inhibition by the drug. Solubilization of erythrocyte ghosts with Triton X-100 abolishes the inhibition of AChE caused by non-ionized bencyclan but not that by the protonated compound.

Introduction

Acetylcholinesterase (AChE) activity of human red blood cells is localised on the outer surface of their plasma membrane [1, 2]. The enzyme is thought to be bound electrostatically since it dissociates from the membrane by high ionic strength [3—5]. According to the classification of Singer and Nicolson [6] this characteristic may correspond to the criteria of the "peripheral proteins" of biological membranes. However, the sensitivity of allosteric properties of erythrocyte AChE to modification of the composition of membrane lipids could only be explained by assuming hydrophobic forces involved in the enzyme-membrane binding [7—10]. Inhibition of AChE activity of intact erythrocytes by added lipids due to a modification of the membrane supports this idea. Determination of the AChE activity can serve as a probe for intactness of the structure of erythrocyte membrane [11].

A great number of papers dealing with the structure-activity relationships of different chemicals with cholinesterase-inhibiting capacity has been published (cf.[12—16]). Among others correlation between hydrophobicity and anticholinesterase potency of some compounds has been found [17]. However, little is known about the influence of the amphophilic character of chemicals on the mechanism of their cholinesterase-inhibiting action.

Bencyclan (N-[-3[(1-benzyl-cycloheptyl-oxy)-propyl]N,N-dimethyl-ammonium-hydrogen fumarate) — a spasmolytic drug [18] — offers a suitable model for such studies. It was found to be a reversible inhibitor of several cholinesterases tested in vitro [19]. The compound is readily soluble in water in its protonated form ($pK_a = 10.5$), whereas the nonionized molecule is of lipophilic character. It was, therefore, of interest to investigate

(1) how do changes in ionization and solubility of bencylcan influence the mechanism of its AChE-inhibitory action;

(2) to what extent are hydrophobic interactions involved in the inhibition of membrane-bound AChE by bencyclan.

Material and methods

Human erythrocyte ghosts were prepared according to Dodge et al. [20] from ACD blood conserves not older than a week. The concentrations of the membrane preparations were characterized by their sialic acid [21] and protein [22] content using bovine serum albumin as standard for the determination of the latter. A 100,000 × *g* (60 min) supernatant of ghosts suspended in a 20 mOsm phosphate buffer, *p*H 7.4, and treated with 2 per cent Triton X-100 was used as solubilized erythrocyte AChE preparation.

It was dialysed against the suspending buffer solution for four days, yet it contained Triton X-100 in a concentration of 0.5 per cent based on a spectrophotometric determination of the detergent. AChE activity was assayed by the Ellman method [23].

Acetylthiocholine iodide was purchased from Fluka, 5,5'-dithiobis-(2-nitrobenzoic acid) and Triton X-100 from Serva. Other chemicals were of reagent grade obtained from Reanal (Budapest).

Results

Effect of pH on the inhibition by bencyclan of the erythrocyte membrane-bound AChE activity

The *p*H of the reaction mixture exerts both quantitative and qualitative effects on the inhibition by bencyclan of erythrocyte membrane-bound AChE activity. In the presence of 1 mM acetylthiocholine (i.e. at saturation of the enzyme with substrate) the rise of the *p*H from 5.9 to 8.9 results in an increasing inhibition of the membrane-bound AChE by the drug. The *p*H-dependence of the inhibitory action of bencyclan is influenced by its own concentration (Fig. 1). Whereas changing the *p*H by 3 units more than triples the per cent inhibition of AChE by 3 mM bencyclan, in a 5 mM concentration it produces under the same circumstances only about 2.3fold and in 10 mM concentration as little as 1.3fold increase in the enzyme inhibition.

The Lineweaver-Burk plots of the inhibitory actions of bencyclan at *p*H 6.15 and *p*H 8.0, respectively, clearly show the profound effect of the hydrogen-ion concentration on the kinetics of the AChE inhibition (Fig. 2). At *p*H 6.15 the inhibition of membrane-bound AChE by bencyclan is competitive with the substrate. In contrast, at *p*H 8.0 the inhibitory action of the drug is either of non-competitive or of mixed type. The frequency of occurrence of either type is equal.

Effect of Triton X-100 on the erythrocyte membrane-bound AChE activity and its inhibition by bencyclan

In human erythrocyte ghosts incubated at 37 °C in phosphate buffer (100 mM, *p*H 8.0) in the presence of increasing concentrations of Triton X-100, a rise in their AChE activity can be observed at final detergent concentrations exceeding

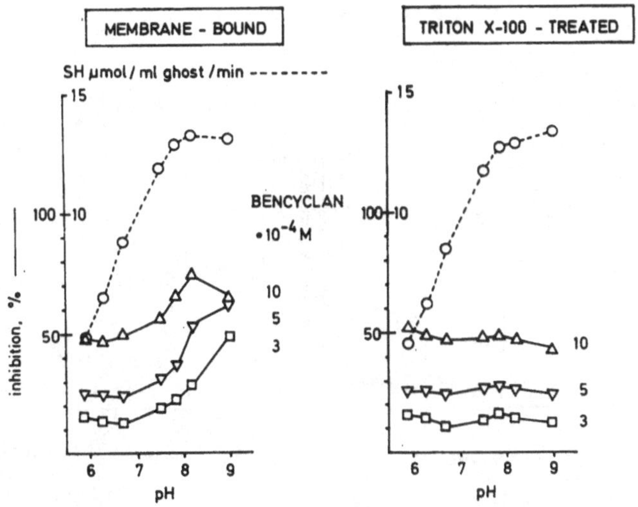

Fig. 1. The *p*H-dependence of the inhibition by bencyclan of erythrocyte membrane-bound as well as of solubilized AChE activities.

AChE activities were determined of ghosts as well as of solubilized ghosts (see Methods) incubated at 37 °C in mixtures containing 0.1 M phosphate buffers of varying *p*H; bencyclan $= 3-5-10 \cdot 10^{-4}$ M; 5,5′-ditihiobis-(2-nitrobenzoic acid) $= 3 \cdot 10^{-5}$ M; acetylthiocholine $= 1 \cdot 10^{-3}$ M.

Dotted lines: enzyme activity without bencyclan; continuous lines: inhibition of AChE activity (in per cent)

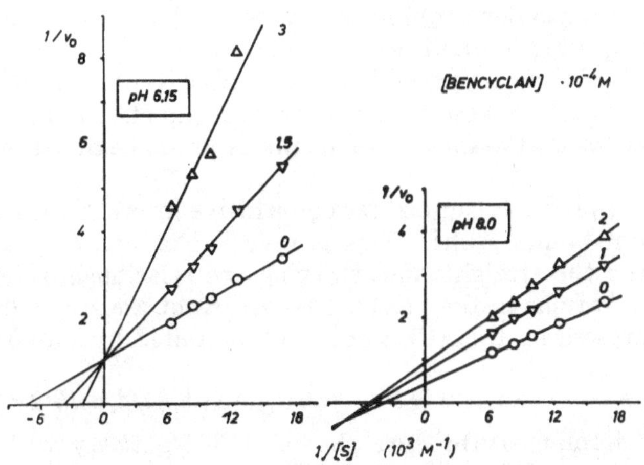

Fig. 2. Lineweaver-Burk plot of inhibition at *p*H 6.15 and *p*H 8.0, resp., of erythrocyte membrane-bound AChE by bencyclan

0.01 per cent (Fig. 3). The extent of Triton X-100-induced activation of the enzyme varies with the membrane preparations. In about 1/3 of the cases the increase in activity is preceded by a small (less than 30 per cent) initial drop in the presence of Triton X-100 concentrations ranging between 0.001 and 0.01 per cent.

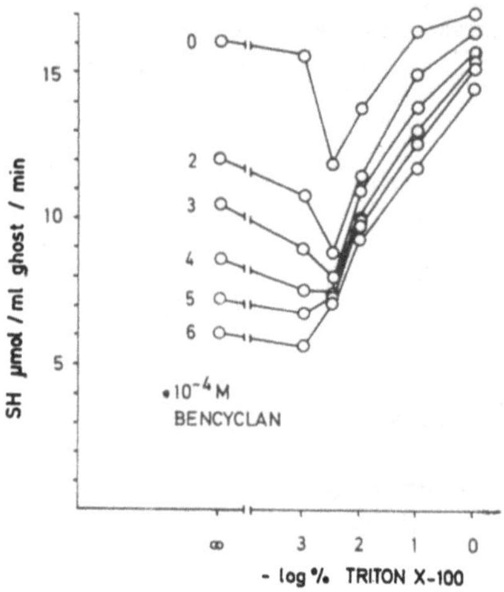

Fig. 3

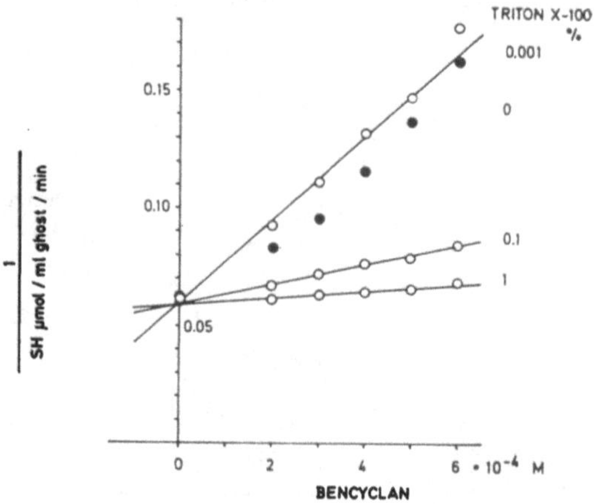

Fig. 4

Fig. 3. Effect of Triton X-100 on the erythrocyte membrane-bound AChE activity and on its inhibition by bencyclan.
Ghosts were incubated at 37 °C in mixtures containing 0.1 M phosphate buffer, *p*H 8.0, and various concentrations of bencyclan as well as Triton X-100 (both diluted with the same buffer). AChE activities were determined after 30 min in the same mixtures.

Fig. 4. Effect of Triton X-100 on the inhibition of membrane bound AChE activity by bencyclan. Data of Fig. 3

The concentration-dependent inhibition by bencyclan of the membrane-bound AChE activity is gradually abolished when ghosts are simultaneously incubated with varying concentrations of bencyclan + Triton X-100 (Fig. 3).
The semireciprocal plot of these results proves a competitive interaction between bencyclan and Triton X-100 for the AChE in the presence of saturating substrate level (Fig. 4).
Triton X-100 most probably suppresses cooperative interactions between bencyclan and membrane-bound AChE. This is revealed in Fig. 4 by the remarkable difference between the straight lines showing the relationships between $1/v$ and bencyclan concentrations in the presence of detergent and the curvilinear course of the points representing the AChE activities inhibited by the drug alone.

Effect of bencyclan on erythrocyte AChE activity solubilized with Triton X-100

Solubilization of human erythrocyte ghosts' AChE activity with Triton X-100 results in a loss of the *p*H-dependence of its inhibition by bencyclan (Fig. 1), although — in accordance with data in the literature [24] — changes in the membrane-bound as well as solubilized AChE activities with the hydrogen-ion concentration are exactly identical. In contrast to the membrane-bound AChE, there is not any increment in the inhibition by bencyclan of the solubilized enzyme at *p*H values exceeding *p*H 7.5. Therefore, a reduction in the cholinesterase-inhibiting action of bencyclan due to Triton X-100 can be observed in an alkaline medium only but not in a slightly acidic one.

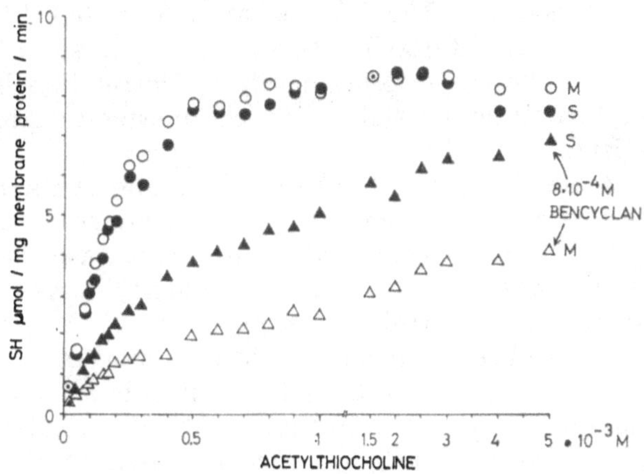

Fig. 5. Effect of solubilization of erythrocyte membranes on the inhibition of their AChE activity by 8 · 10^{-4} M bencyclan. Dependence of enzyme activity on the concentration of substrate.

M: membrane-bound AChE; S: AChE solubilized with Triton X-100 (see Methods)

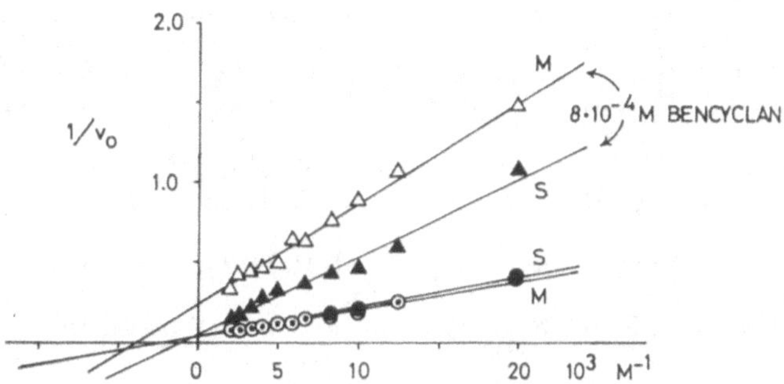

Fig. 6. Lineweaver-Burk plot of inhibition by bencyclan at *p*H 8.0 of membrane-bound as well as of solubilized AChE. Data of Fig. 5

Not only a reduction of the inhibition at *p*H 8.0 of membrane-bound AChE by bencyclan can be found after solubilization with Triton X-100 but also the type of the residual inhibition of the enzyme changes from a mixed type into a competitive one (Figs. 5 and 6).

Discussion

The results presented in the previous section show that at physiological *p*H the inhibition of the erythrocyte membrane-bound AChE by bencyclan is a sum of the inhibitory actions of the protonated as well as of the uncharged form of the drug. Circumstances favoring ionization of bencyclan (i.e. slightly acidic medium) result most probably in a direct attachment of the water-soluble bencyclan cation on the AChE molecule. On the other hand, a rise of the *p*H suppresses the protonation of bencyclan, promoting in addition cooperative interactions between hydrophobic regions of the enzyme and/or of the membrane and the highly lipophilic uncharged bencyclan molecule. The different actions at slightly acidic as well as alkaline *p*H values of Triton X-100 on the anticholinesterase effect of bencyclan may support this hypothesis; so does the strong correlation between the *p*H-dependence of the hemolytic as well as the cholinesterase inhibiting action of bencyclan (unpublished results).

The modification of the charged state of the AChE and/or its membrane-micro-environment — besides that of bencyclan — in the course of the investigations mentioned above should not be neglected. However, in case the *p*H is increased from 5.9 to 8.9, as for ionizing groups possibly playing a role in the inhibition of the AChE activity most probably the suppression of the protonation of the imidazole group of histidyl residues only, could be important. Any change in the dissociation of the anionic groups of the enzyme in this *p*H interval is unlikely. Results of investigations in progress indicate a stabilizing of the conformation of membrane-bound AChE by bencyclan, rendering thereby the enzyme insensitive against the effects of further changes of the *p*H.

References

[1] AUGUSTINSSON, K.-B.: Acta physiol. scand. **15**, Suppl. 52 (1940)
[2] ZWAAL, R. F., B. ROELOFSEN, and C. M. COLLEY: Biochim. biophys. Acta **300**, 159 (1973)
[3] HELLER, M., and D. J. HANAHAN: Biochim. biophys. Acta **255**, 251 (1962)
[4] MITCHELL, C. D., and D. J. HANAHAN: Biochemistry **5**, 51 (1966)
[5] PANIKER, N. V., A. B. ARNOLD, and R. C. HARTMANN: Proc. Soc. exp. Biol. Med. **144**, 492 (1973)
[6] SINGER, S. J., and G. L. NICOLSON: Science, N.Y. **175**, 720 (1972)
[7] MORERO, R. D., B. BLOJ, R. N. FARÍAS, and R. E. TRUCCO: Biochim. biophys. Acta **282**, 157 (1972)
[8] BLOJ, B., R. D. MORERO, R. N. FARÍAS, and R. E. TRUCCO: Biochim. biophys. Acta **311**, 67 (1973)
[9] MARTINEZ DE MELIAN, E. R., R. D. MORERO, and R. N. FARÍAS: Biochim. biophys. Acta **422**, 127 (1976)
[10] SIHOTANG, K.: Eur. J. Biochem. **63**, 519 (1976).
[11] ALONI, B., and A. LIVNE: Biochim. biophys. Acta **339**, 359 (1974)
[12] FOLDES, F. F., G. VAN HEES, D. L. DAVIS, and S. P. SHANOR: J. Pharmac. exp. Ther. **122**, 457 (1958)
[13] CHRISTIAN, S. T., C. W. GORODETZKY, and D. V. LEWIS: Biochem. Pharmac. **20**, 1167 (1971)
[14] MICHAŁEK, H.: Biochem. Pharmac. **22**, 1067 (1973)
[15] MAAYANI, S., H. WEINSTEIN, N. BEN-ZVI, S. COHEN, and M. SOKOLOVSKY: Biochem. Pharmac. **23**, 1263 (1974)
[16] COCOLAS, G. H., J. G. CRANFORD, and H. S. Y. CHOI: J. med. Chem. **17**, 938 (1974)
[17] HELLENBRECHT, D., and K.-F. MÜLLER: Experientia **29**, 1255 (1973)
[18] KOMLÓS, E., and L. E. PETŐCZ: Arzneimitt.-Forsch. **20**, 1338 (1970)
[19] CSEH, J. R., E. KOCH, É. SÜVEGES, S. VÉGH, and S. MÁNYAI: 40th Meeting of the Hungarian Physiological Society, 1974, Abstracts No. C-57
[20] DODGE, J. T., C. MITCHELL, and D. J. HANAHAN: Archs Biochem. Biophys. **100**, 119 (1963)
[21] WARREN, L.: J. biol. Chem. **234**, 1971 (1959)
[22] LOWRY, O. H., N. J. ROSEBROUGH, A. L. FARR, and R. J. RANDALL: J. biol. Chem. **193**, 265 (1951)
[23] ELLMAN, G. L., K. D. COURTNEY, V. ANDRES, and R. M. FEATHERSTONE: Biochem. Pharmac. **7**, 88 (1961)
[24] JACKSON, P., and M. WHITTAKER: Enzymologia **43**, 359 (1972)

Acta biol. med. germ., Band 36, Seite 897—902 (1977)

Anatomisches Institut der Friedrich-Schiller-Universität, 69 Jena, DDR

Streulichtmessungen an Blut: Ein Verfahren zur Beurteilung von Formänderungen der Erythrozyten in Blutkonserven

C. Scheven

Zusammenfassung

Es wird eine Versuchsanordnung zur Messung der Lichtstreuung an laminar gerührtem Blut im Auflicht beschrieben. Näher untersucht wurden die Abhängigkeit der Streulichtintensität von der Drehzahl des Rührers und von der Temperatur sowie der Einfluß einer sauerstoffhaltigen bzw. -verarmten Gasatmosphäre über der Probe.
Vorläufige Resultate von vergleichenden Messungen an Blutkonserven verschiedenen Alters weisen auf das Vorhandensein einer merklichen lagerungszeitabhängigen Veränderung der Streulichtintensität hin.

Einleitung

Erythrozyten in Blutkonserven unterliegen einer allmählichen Formveränderung. Während bei einer frischen Konserve noch alle Erythrozyten als Diskozyten vorliegen, findet sich mit zunehmender Lagerungsdauer ein immer größerer Anteil Echinozyten. In hinreichend alten Konserven sind schließlich alle Erythrozyten zu Sphärozyten umgewandelt. Parallel zur Formveränderung findet außerdem eine Änderung der Deformierbarkeit der Erythrozyten statt [1]. Die Kenntnis des Formverhaltens konservierter Blutzellen kann bei bestimmten klinischen Anwendungsfällen von Interesse sein.

Das nächstliegende Verfahren zur morphologischen Beurteilung ist die mikroskopische Auszählung der verschiedenen Formanteile in einer stark verdünnten Erythrozytensuspension. Dabei muß gewährleistet sein, daß im mikroskopischen Präparat keine zusätzlichen Formveränderungen stattgefunden haben. Das Problem der Formerhaltung beim mikroskopischen Verfahren könnte umgangen werden, wenn ein praktisch brauchbares Verfahren zur Verfügung stünde, um aus makroskopisch meßbaren Eigenschaften am unverdünnten Blut Rückschlüsse auf die darin vorherrschende Form der Erythrozyten zu ziehen. Eine der hierfür infrage kommenden Eigenschaften ist die Rückstreuung von Licht an der Oberfläche des Blutes. Meßbare Änderungen der Intensitätsverteilung des rückgestreuten Lichts in Abhängigkeit von der vorherrschenden Zellform sind insbesondere dann zu erwarten, wenn das Blut einer Strömungsbewegung unterworfen wird, durch welche die Erythrozyten — je nach ihrer Beschaffenheit stärker oder schwächer — eine räumliche Ausrichtung erfahren.

Den Ausgangspunkt für vorliegende Untersuchungen bildete ein älteres Experiment von Hoffmann et al. [2] zur Lichtstreuung an stark verdünnten Erythrozytensuspensionen im durchfallenden Licht. Diese Autoren stellten fest, daß die Stärke der bei Umrühren auftretenden Intensitätsfluktuationen des durchfallenden Lichts abhängig ist von der Form der Zellen. Beim Versuch, diesen Effekt zur

Untersuchung der morphologischen Beschaffenheit von Konservenblut heranzuziehen, ergaben sich Schwierigkeiten hinsichtlich der Erhaltung der im unverdünnten Blut vorliegenden Erythrozytenformen nach Ausführung der Verdünnung. Außerdem erzeugte der Rührvorgang infolge der niedrigen Viskosität der Suspension stets Turbulenz und damit ein meßtechnisch schlecht zu verarbeitendes fluktuierendes Signal. Mit dem Übergang zu einem Meßverfahren an unverdünntem Blut wurde bezweckt, diese beiden Nachteile der Methode von Hoffmann et al. für unsere Aufgabenstellung zu vermeiden.

In diesem Beitrag wird über Versuche berichtet, die darauf abzielen, mittels der Rückstreuung von Licht aus der der Küvetteninnenoberfläche unmittelbar angrenzenden Schicht von strömendem Blut Rückschlüsse auf dessen Formverhalten zu ziehen. Im Mittelpunkt stehen dabei methodische Fragen, die vor der Durchführung vergleichender Untersuchungen an Blutkonserven verschiedenen Alters zu klären sind. Über Resultate von Untersuchungen an letzteren soll ausführlich an anderer Stelle berichtet werden.

Methodik

Das Prinzip unserer Versuchsanordnung zeigt Abb. 1. Weißes Licht, das aus der beleuchteten Lochblende B1 austritt, wird von der Linse L1 parallel gerichtet und fällt durch die Lochblende B2 als schmales Bündel auf die Küvette K, die mittels eines Thermostaten auf konstanter Temperatur gehalten wird. Der Spiegel S ist so angeordnet, daß das von der vorderen Küvettenwand reflektierte Licht an der Linse L2 vorbeigeht, während das von dem beleuchteten Fleck in der Grenzfläche Blut-Küvettenwand rückgestreute Licht auf den Phototransistor PT abgebildet wird. Wenn das Blut durch ein in die Küvette eingeführtes Rührwerk bewegt wird, treten Helligkeitsschwankungen des Streulichts auf. Hält man die Umwälzung des Bluts im laminaren Bereich, so empfängt der Phototransistor ein mit konstanter Amplitude annähernd sinusförmig um den Ruhewert schwankendes Lichtsignal. Der periodische Anteil wird über den Verstärker V dem Röhrenvoltmeter RV zugeführt. Die Anzeige des Röhrenvoltmeters stellt das Meßresultat dar, dessen Abhängigkeit von verschiedenen Parametern bei der vorliegenden Untersuchung interessierte.

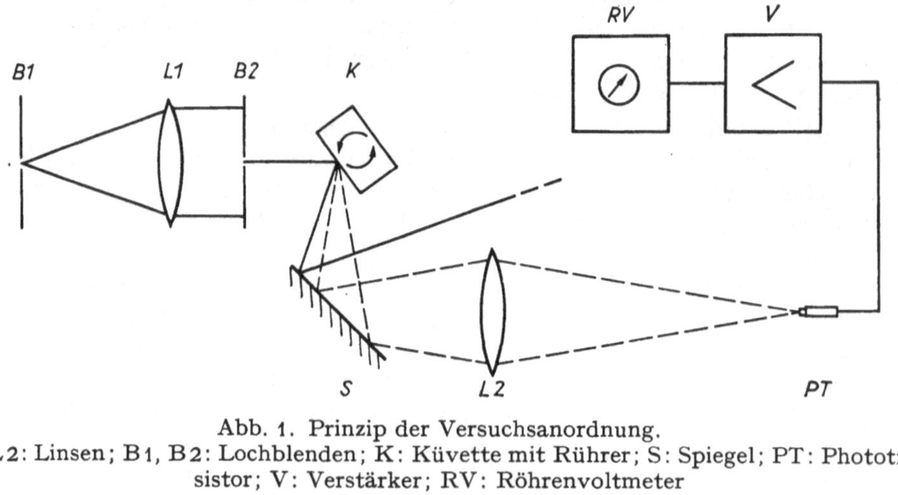

Abb. 1. Prinzip der Versuchsanordnung.
L1, L2: Linsen; B1, B2: Lochblenden; K: Küvette mit Rührer; S: Spiegel; PT: Phototransistor; V: Verstärker; RV: Röhrenvoltmeter

Resultate

Zunächst wurden die Abhängigkeit des Signals von der Drehzahl des Rührers und der Temperatur sowie der Einfluß der über der Probe befindlichen Gasatmosphäre untersucht. Für diese Messungen brauchte jeweils nur Blut aus einer einzigen Konserve zur Verfügung zu stehen. Auf die Resultate von vorläufigen vergleichenden Messungen an verschiedenen Konserven wird am Schluß kurz eingegangen. Den Einfluß der Drehzahl bei verschiedenen Temperaturen zeigt Abb. 2. Bei kleineren Drehzahlen steigt die Signalspannung annähernd drehzahlproportional an. Mit weiterer Zunahme der Drehzahl strebt das Signal zunächst einem Sättigungswert zu. Wird schließlich eine kritische Drehzahl überschritten, steigt die Signalspannung wieder an und beginnt in zunehmendem Maße zu fluktuieren, d. h. in der vom Lichtstrahl getroffenen Flüssigkeitsregion setzt Turbulenz ein. Die

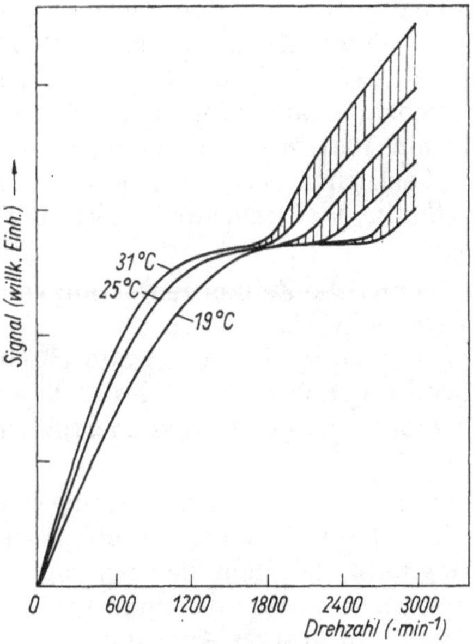

Abb. 2

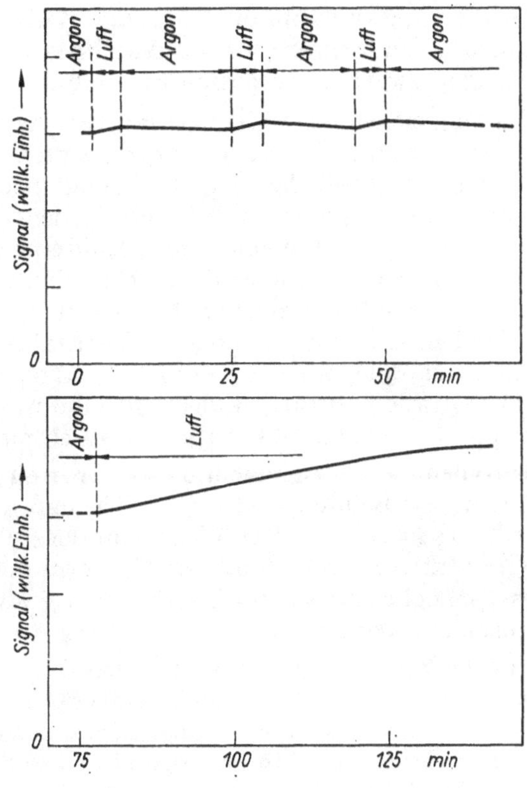

Abb. 3

Abb. 2. Abhängigkeit der Signalgröße von der Drehzahl des Rührers bei verschiedenen Temperaturen.
Schraffiert: Fluktuation des Signals infolge Turbulenz. Alter der Konserve: 3 Tage

Abb. 3. Zeitlicher Verlauf der Signalgröße bei wechselweisem Zutritt von Argon und Luft zur freien Oberfläche.
$T = 25$ °C; $n = 1850 \cdot \text{min}^{-1}$; Alter der Konserve: 4 Tage

Darstellung der Turbulenzverhältnisse in Abb. 2 hat nur qualitative Bedeutung, da das Einsetzen der Turbulenz wesentlich von der Form des Rührers bestimmt wird. Abb. 2 gibt auch ein Bild von dem Einfluß der Temperatur auf das Signal. Je niedriger die Temperatur, desto höher liegt die dem Einsatz der Turbulenz entsprechende Drehzahl und umso breiter ist das Plateau in der Drehzahlcharakteristik des Signals. Im Plateau ist der Temperatureinfluß im Rahmen unserer Meßgenauigkeit unmerklich.

Befindet sich Luft über der freien Blutoberfläche, so wird eine kontinuierliche Vergrößerung des Signals mit der Rührzeit beobachtet. Dieser Effekt wird unterbunden, wenn man unter Argon-Atmosphäre rührt. Abb. 3 zeigt den zeitlichen Verlauf der Signalspannung bei ununterbrochenem Rühren, wenn abwechselnd Argon und Luft Zutritt zur Blutoberfläche erhalten. Wurde zunächst unter Argon gerührt, so vergrößert sich unter Luft das Signal innerhalb 5 min um fast 3%. Wird dann erneut Argon zugeführt, verkleinert sich das Signal wieder, wobei mehr als 15 min nötig sind, um die 5minütige Wirkung der Luftatmosphäre rückgängig zu machen. Bei langzeitigem Luftzutritt strebt die Vergrößerung des Signals einem Sättigungswert zu, zu dessen Einstellung mehr als 1 Std. benötigt wird.

Bei vergleichenden Messungen an verschiedenen Blutkonserven muß neben dem interessierenden Effekt der Erythrozytenform auf das Signal auch der Einfluß einer unterschiedlichen Erythrozytenkonzentration berücksichtigt werden. Vorläufige Messungen an Proben verschiedener Konzentration bei gleichem Formenbestand zeigten, daß eine solche Konzentrationsabhängigkeit in der Tat besteht. So wurde bei gegenüber durchschnittlichem Konservenblut um 30% künstlich erhöhter Zelldichte ein flacherer Anstieg der Drehzahlcharakteristik des Signals gefunden, der erst bei höheren Drehzahlen in ein Plateau einmündete, dessen Höhe sich dabei nicht wesentlich änderte. Bei künstlicher Verringerung der Zellkonzentration um 30% änderte sich die Anstiegflanke der Drehzahlcharakteristik kaum, dagegen lag das Plateau deutlich niedriger.

Turbulenz setzte allgemein umso früher ein, je kleiner die Zellkonzentration war. Aus diesen Befunden ist ersichtlich, daß die Streulichtintensität in unserer Anordnung in verwickelter Weise von der Zellkonzentration abhängt, wodurch ein quantitativer Rückschluß auf die Form der Erythrozyten während des Meßvorgangs möglicherweise erschwert wird. Detailliertere Messungen zu diesem Problem stehen zur Zeit noch aus.

Für die Feststellung von lagerungsbedingten Änderungen der Lichtstreuung von Konservenblut ist die zeitliche Unveränderlichkeit der Meßanordnung über viele Wochen hinweg Voraussetzung. Dieser Forderung muß vor allem die mechanische Konstruktion des Rührers angemessen sein. Bei einem Rührerdurchmesser von maximal 8 mm und einer Küvettenweite von 10 mm in unserer Anordnung steht dem umgewälzten Blut periodisch ein Spalt von minimal 1 mm zur Verfügung. Kleinste Unregelmäßigkeiten im Lauf des Rührers wirken sich auf die Streulichtintensität aus. Sie müssen durch zweckmäßige Anordnung der Kugellager, Gestaltung der Kupplung und Auswahl des Antriebsmotors ausgeschaltet werden. Der bei den bisherigen Untersuchungen eingesetzte Rührer erwies sich in seinen Eigenschaften noch nicht als hinreichend zeitlich konstant und wird gegenwärtig durch eine verbesserte Ausführung ersetzt.

Unsere bisherigen Befunde bei vergleichenden Messungen an Blutkonserven verschiedener Lagerungszeit wurden noch nicht unter Einhaltung aller oben beschriebenen methodischen Forderungen gewonnen, so daß für jede Einzelmessung ein beträchtlicher Meßfehler anzunehmen ist. Qualitativ ist trotzdem bereits aus den vorliegenden Daten eine Veränderung der Lichtstreuung mit der Lagerungsdauer klar erkennbar. Vergleichende Messungen an bis zu 9 Konserven, die an mehreren Tagen wiederholt wurden, zeigten stets, daß zunehmende Lagerungszeit mit abnehmender Streulichtintensität korreliert ist. Während der ersten zwei Wochen verringert sich das Signal zunächst nur geringfügig, ist nach 6 Wochen dann aber bereits auf etwa 50% des Anfangswerts abgesunken. An einer 10 Wochen alten Konserve wurden nur noch etwa 10% der Signalgröße einer frischen Konserve gemessen.

Diskussion

Zur Interpretation des eigentümlichen Verlaufs der Drehzahlabhängigkeit des Signals (vgl. Abb. 2) ist es naheliegend anzunehmen, daß mit wachsender Drehzahl die flachen Erythrozytenformen zunehmend ausgerichtet werden. Durch die bei Erreichung einer gewissen Drehzahl vollständige Ausrichtung aller solcher Zellen wäre auch die Bildung des Plateaus bereits verständlich. Daß normale Erythrozyten in einer Strömung mit niedrigem Schergefälle tatsächlich ausgerichtet werden, ist durch direkte mikroskopische Beobachtungen, allerdings an mit Eigenplasma verdünnten Suspensionen, erwiesen [3]. Die Ausrichtung allein jedoch kann nicht der einzige zu einer Änderung der Streulichtintensität führende Vorgang sein, da das in der Versuchsanordnung bei Drehzahlen um $1800 \cdot \text{min}^{-1}$ maximal auftretende Schergefälle von der Größenordnung einiger $10^2 \cdot \text{s}^{-1}$ ist. Unter diesen Verhältnissen werden normale Erythrozyten zusätzlich zu länglichen Ellipsoiden deformiert [3], die sich in Strömungsrichtung einstellen. Dabei führt die Zellmembran eine Umlaufbewegung in der Art eines Treibriemens um den Zellinhalt aus, der seinerseits durch die Membranbewegung in eine zirkulierende Strömungsbewegung versetzt wird. Da einer beliebigen Verlängerung der Ellipsoide aber infolge der geringen Dehnbarkeit der Erythrozytenmembran Grenzen gesetzt sind, wird die mikroskopische Struktur des strömenden Bluts bei hinreichend hohem Schergefälle immer weniger von diesem abhängen, d. h. es sollte ein Plateau in der Drehzahlcharakteristik des Signals auftreten. Nach dieser Interpretation hängt die Lichtstreuung nicht schlechthin von der Form der Zellen ab, sondern auch von deren dynamischer Verformbarkeit in einem Strömungsfeld mit Schergefälle.

Die beobachtete Vergrößerung des Signals bei Rühren unter Luft dürfte durch die Sauerstoffaufnahme des Hämoglobins verursacht sein. Vom O_2-Gehalt des Hb abhängige Parameter, welche die Streulichtintensität des Bluts beeinflussen können, sind das Absorptionsspektrum des Hämoglobins und die Viskosität des überwiegend aus Hämoglobin bestehenden Innern der Erythrozyten. Die spektrale Intensität des vom Blut rückgestreuten Lichts ist näherungsweise umgekehrt proportional dem Extinktionskoeffizienten und erreicht größere Werte daher nur bei Wellenlängen, für die der Absorptionskoeffizient des Blutfarbstoffs klein ist, d. h. bei $\lambda > 600$ nm. Im Wellenlängenbereich 600 bis 800 nm ist der Absorptionskoeffizient des HbO_2 niedriger als der des Hb und entsprechend die Intensität des rückgestreuten Lichts für oxygeniertes Blut größer als für reduziertes. Hierauf beruhen die Verfahren der photometrischen Oxymetrie des Bluts [4, 5]. Der gefundene Effekt einer reversiblen Signalvergrößerung bei Rühren unter Luft ist demnach qualitativ durch das unterschiedliche Absorptionsspektrum von HbO_2 und Hb zu erklären. Nicht auszuschließen ist außerdem aber ein Einfluß des Sauerstoffs über die Viskosität des Hämoglobins. Da die Viskosität im Zellinnern mitbestimmend ist für das Ausmaß der Verformung der Erythrozyten in einem vorgegebenen Schergefälle, beeinflußt sie auch das Lichtstreuverhalten des strömenden Bluts. Eine extreme Erhöhung der Viskosität bei erniedrigtem O_2-Partialdruck ist vom Sichelzellenhämoglobin her bekannt [6]. Unsere Befunde wurden allerdings an nichtpathologischem Blut gewonnen. Es bleibt zu prüfen, ob die pO_2-Abhängigkeit der Viskosität des normalen Hämoglobins ausreicht, um merklich zu der gefundenen Veränderlichkeit der Streulichtintensität mit dem O_2-Gehalt der umgebenden Atmosphäre beizutragen.

Danksagung

Herrn Prof. Dr. Geyer danke ich für die zahlreichen Ratschläge, verständnisvolle Unterstützung und beständigen Ansporn zu dieser Arbeit.
Danken möchte ich außerdem Herrn Dr. Horn für eine nützliche Diskussion.

Literatur

[1] La Celle, P. L.: Transfusion **9**, 238 (1969)
[2] Hoffmann, J. F., R. H. Collier u. M. J. Deitch: J. cell. comp. Physiol. **45**, 355 (1955); J. F. Hoffmann in: Red Cell Shape. M. Bessis, R. I. Weed, P. F. Leblond (Hrsg.). Springer, New York, Heidelberg, Berlin 1973, S. 51
[3] Schmid-Schönbein, H., u. R. Wells: Science, N.Y. **165**, 288 (1969)
[4] Zijlstra, W. G.: A Manual of Reflection Oximetry. Assen 1958
[5] Polanyi, M. L., u. R. M. Hehir: Rev. scient. Instrum. **33**, 1050 (1962)
[6] Dintenfass, L. in: Theoretical and Clinical Hemorheology. H. H. Hartert u. A. L. Copley (Hrsg.) Springer, Berlin, Heidelberg, New York 1971, S. 174

Summary

C. Scheven: Tyndallimetry in blood: A method to estimate shape variations of erythrocytes in preserved blood

A device for the measurement of light scattering by laminarily streaming blood with incident light is described. The main objects of this investigation was the dependence of the scattering intensity from the rate of revolutions of the stirrer and from temperature als well as the influence of an oxygen containing resp. oxygen depleted atmosphere above the blood. Preliminary results of comparative measurements with preserved blood of different storage ages indicate that there is marked storage time-dependent change in the scattered light intensity.

Acta biol. med. germ., Band 36, Seite 903—908 (1977)

Anatomisches Institut der Friedrich-Schiller-Universität, 69 Jena, DDR

Experimentell gesteigerte Verformbarkeit und Adhäsivität von Erythrozyten

J. Unger und A. Benser

Einleitung

Die abnehmende Verformbarkeit gilt als eine Größe, von der die Elimination roter Blutkörperchen aus dem zirkulierenden Blut beeinflußt wird. Es ist deshalb von Interesse, den Einfluß solcher Agentien zu prüfen, die eine Veränderung der Erythrozytenform herbeiführen. In den hier vorgelegten Untersuchungen wurden Verformbarkeit und Adhäsivität an Humanerythrozyten nach Prokainierung bzw. Karbamylierung gemessen.

Material und Methoden

Humanblut der Gruppen A_1D und BD (aus ACD-Konserven) wurde in isotoner PBS-Lösung, *p*H 7.4, 3mal gewaschen. Nach der 3. Waschung wurde 5 min zentrifugiert und der klare Überstand weitgehend entfernt. Die Prokainierung bzw. Karbamylierung wurden wie folgt ausgeführt.

Prokainierung: 0,025 ml Erythrozytensediment wurden in 5 ml einer isotonen 0,6%igen Prokainlösung suspendiert und unter Schütteln bei Zimmertemperatur 30 min inkubiert. Anschließend erfolgte Zentrifugation und Überführen der sedimentierten Erythrozyten im Verhältnis 1:600 bzw. 1:1000 in 3 ml Prokainlösung gleicher Konzentration in die Meßkammer. Kontrollmessungen wurden an zeitlich gleichlang in PBS-Lösung inkubierten Erythrozyten ausgeführt.

Karbamylierung: 0,5 ml ungewaschenes Erythrozytensediment wurde in einer 50 mM Kaliumzyanatlösung, welcher 2 mg/ml Dextrose zugesetzt waren (*p*H 7,4) isoton im Verhältnis 1:10 suspendiert und 120 min bei 37 °C inkubiert. Anschließend erfolgte Zentrifugation und Überführen der sedimentierten Erythrozyten im Verhältnis 1:600 bzw. 1:1000 in 3 ml Kaliumzyanatlösung gleicher Konzentration in die Meßkammer. Kontrollmessungen wurden an zeitlich gleichlang in PBS-Lösung inkubierten Erythrozyten ausgeführt. Die Verformbarkeit wurde mit Hilfe einer Glaskapillare an einzelnen Erythrozyten gemessen (methodische Einzelheiten siehe [1].

Die Adhäsivität der Erythrozyten wurde als prozentualer Anteil derjenigen Zellen bestimmt, welche nach Haftung an einem Deckglas in einer Durchströmungskammer durch eine strömende Flüssigkeit (isotone Pufferlösung) bekannter Geschwindigkeit nicht von ihrer Unterlage abgelöst werden (methodische Einzelheiten siehe [2]).

Aus der Erythrozytenform wurde nach folgendem Schema ein Index für die jeweils untersuchte RBC-Population errechnet:

$$\frac{\text{D} \quad \text{E}_{\text{I}} \quad \text{E}_{\text{II}} \quad \text{E}_{\text{III}} \quad \text{Sph.-E}_{\text{I+II}}}{1{,}0 \quad 0{,}8 \quad 0{,}6 \quad 0{,}4 \quad 0{,}2} \cdot 100 \text{ Zellen}$$

Ergebnisse

Prokainierte Erythrozyten

In den Untersuchungen konnte sowohl in 1—10 Tage altem ACD-Blut als auch noch sehr deutlich an Erythrozyten aus 6—8 Wochen altem Konservenblut eine

Erhöhung der Verformbarkeit nachgewiesen werden, nämlich von 0,944 mm auf 0,882 mm des P_c-Mittelwertes von 1—10 Tage alten bzw. 0,757 mm auf 0,704 mm des P_c-Mittelwertes von 6—8 Wochen alten Diskozyten.

Die P_c-Meßwerte für Echinozyten liegen deutlich höher, wobei auch hier eine Verformbarkeitszunahme von 1,163 mm auf 1,044 mm 1—10 Tage alter Echinozyten bzw. 1,903 mm auf 1,240 mm des Mittelwertes von 6—8 Wochen alten Echinozyten nachgewiesen werden konnte. Die P_m-Mittelwerte zeigen ähnliche Relationen, liegen aber niedriger als die P_c-Mittelwerte (Tab. 1 und 2: Vertrauenswahrscheinlichkeit 95%).

Tabelle 1

1—10 Tage altes Konservenblut, Blutgruppen A, D und BD

Serie (Zellformen)	n	Inkubationszeit (Std.)	Zellformindex	P_m (mm H_2O)	P_c (mm H_2O)
Kontrolle (Echinozyten III)	41	0,5	0,507	0,508 ± 0,023	1,163 ± 0,006
Kontrolle (Diskozyten)	40	0,5	0,507	0,402 ± 0,010	0,944 ± 0,022
Prokain (Echinozyten III)	21	0,5	0,856	0,450 ± 0,088	1,044 ± 0,056
Prokain (Diskozyten)	32	0,5	0,856	0,379 ± 0,056	0,882 ± 0,021

Tabelle 2

6—8 Wochen altes Konservenblut, Blutgruppe A, D und BD

Serie (Zellformen)	n	Inkubationszeit (Std.)	Zellformindex	P_m (mm H_2O)	P_c (mm H_2O)
Kontrolle (Echinozyten III)	60	0,5	0,359	0,854 ± 0,015	1,903 ± 0,071
Kontrolle (Diskozyten)	4	0,5	0,359	0,353 ± 0,032	0,757 ± 0,027
Prokain (Echinozyten III)	40	0,5	0,698	0,579 ± 0,016	1,240 ± 0,046
Prokain (Diskozyten)	24	0,5	0,698	0,316 ± 0,006	0,704 ± 0,017

Der Zellformindex für prokainierte Erythrozyten (1—10 Tage) lag bei 0,856, für gleichaltrige Kontroll-Erythrozyten bei 0,507. Erythrozyten aus Konservenblut (6—8 Wochen alt) besaßen nach Prokainierung einen Zellformindex von 0,698, gleichaltrige Kontroll-Erythrozyten einen solchen von 0,359.

In den 4 Serien des 1—10 Tage alten Konservenblutes wurde die Verformbarkeit von 134 Einzelzellen gemessen. Die dabei gemessenen Unterdrucke P_m prokainierter Echinozyten bewegten sich in den Grenzen von 0,36—0,51 mm, die der Kontroll-Echinozyten in den Grenzen von 0,39—0,62 mm. Die Unterdrucke P_c

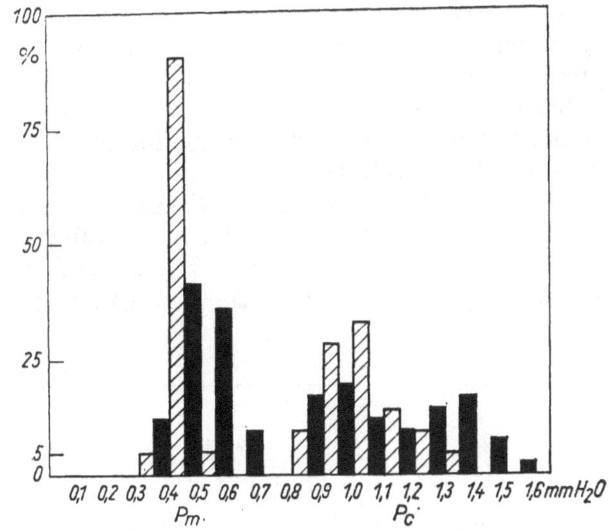

Abb. 1. Erforderlicher Unterdruck zur Verformung des ganzen Echinozyten (P_c) bzw. einer 1,8 µm langen Zellprotrusion (P_m) bei PBS-gewaschenen (schwarze Säulen) und prokainierten Erythrozyten (schraffierte Säulen) 1—10 Tage alter ACD-Konserven

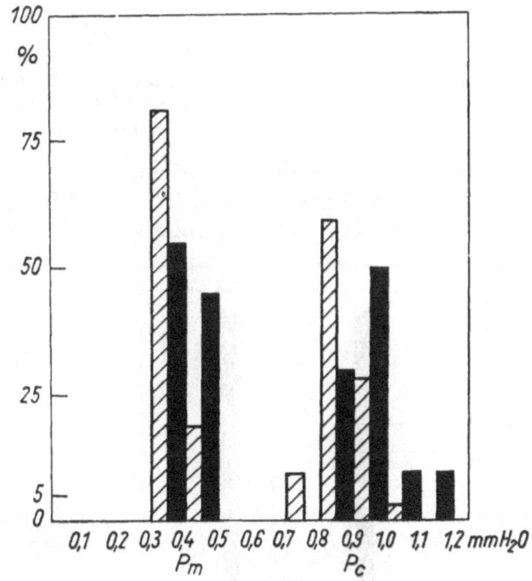

Abb. 2. Erforderlicher Unterdruck zur Verformung des ganzen Diskozyten (P_c) bzw. einer 1,8 µm langen Zellprotrusion (P_m) bei PBS-gewaschenen (schwarze Säulen) und prokainierten Erythrozyten (schraffierte Säulen) 1—10 Tage alter ACD-Konserven

befinden sich für prokainierte Echinozyten in den Grenzen von 0,81—1,13 mm und für Kontroll-Echinozyten von 0,83—1,54 mm H_2O (Abb. 1).

Die gemessenen Unterdrucke P_m prokainierter Diskozyten bewegten sich hingegen in den Grenzen von 0,33—0,46 mm, die der Kontroll-Diskozyten in den Grenzen von 0,34—0,48 mm. Die Unterdrucke P_c befinden sich für prokainierte Diskozyten in den Grenzen von 0,77—1,04 mm, die der Kontroll-Diskozyten von 0,81—1,12 mm H_2O (Abb. 2).

In den 4 Serien des 6—8 Wochen alten Konservenblutes wurde die Verformbarkeit von 128 Einzelzellen gemessen. Die dabei gemessenen Unterdrucke P_m prokainierter Echinozyten bewegten sich zwischen 0,49—0,68 mm, die der Kontroll-Echinozyten zwischen 0,74—0,97 mm. Die Unterdrucke P_c liegen für prokainierte Echinozyten zwischen 1,09—1,39 mm, die Werte für die Kontroll-Echinozyten zwischen 1,66—2,09 mm H_2O (Abb. 3).

Die gemessenen Unterdrucke P_m prokainierter Diskozyten befinden sich hingegen in den Grenzen von 0,27—0,38 mm, die der Kontroll-Diskozyten zwischen 0,33—0,37 mm. Die Unterdrucke P_c liegen für prokainierte Diskozyten zwischen 0,61—0,81 mm, die Meßwerte der Kontroll-Diskozyten zwischen 0,74—0,78 mm H_2O (Abb. 4).

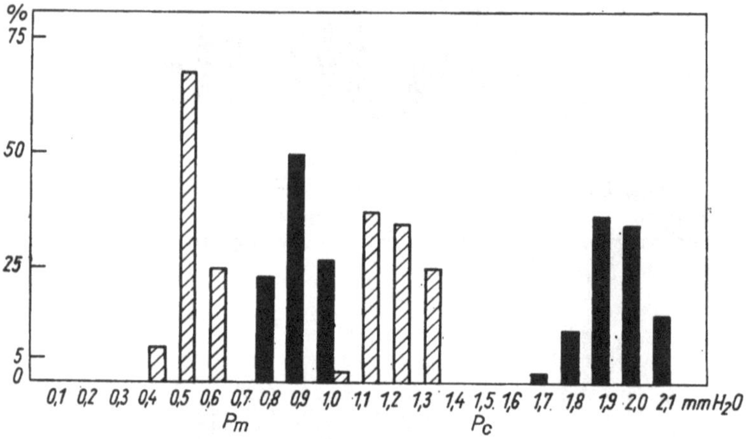

Abb. 3. Erforderlicher Unterdruck zur Verformung des ganzen Echinozyten (P_c) bzw. einer 1,8 µm langen Zellprotrusion (P'_m) bei PBS-gewaschenen (schwarze Säulen) und prokainierten Erythrozyten (schraffierte Säulen) 6—8 Wochen alter ACD-Konserven

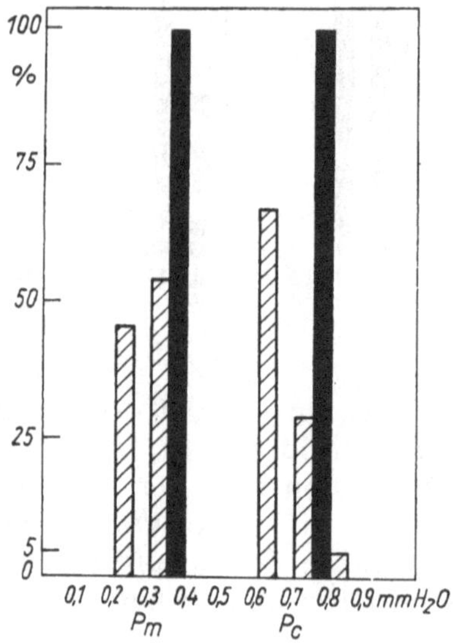

Abb. 4. Erforderlicher Unterdruck zur Verformung des ganzen Diskozyten (P_c) bzw. einer 1,8 µm langen Zellprotrusion (P_m) bei PBS-gewaschenen (schwarze Säulen) und prokainierten Erythrozyten (schraffierte Säulen) 6—8 Wochen alter ACD-Konserven

Meßresultate an Erythrozyten aus 11—12 Wochen altem Konservenblut ergaben keine deutliche Verformbarkeitsänderung nach Prokainierung gegenüber den Kontrollmeßergebnissen mehr. Der Zellformindex für prokainierte Erythrozyten lag hierfür bei 0,300 und für die Kontrollen bei 0,215.
Die nach einer Mindestinkubation von 30 min im Prokainmilieu aus der Echinozyten- in Richtung Diskozytenform transformierte Erythrozytensuspension zeigte eine mittlere Adhäsivität von 74,9% zu Glasoberflächen im Vergleich zu unbehandeltem, mit PBS gewaschenem Blut mit 55,8%. Der Zellformindex der an der Glasfläche haftenden Zellen stieg von 0,21 auf maximal 0,72 an. Der prozentuale Anteil adhärierter, prokainierter Erythrozyten nach Durchströmung in Abhängigkeit von der Zellform betrug: Diskozyten = 91,3%, Echinozyten I = 86,1%, Echinozyten II u. III = 47,4%.
Die Haftfähigkeit der Diskozyten ist etwa um das Doppelte höher als die der Echinozyten II und III.

Karbamylierte Erythrozyten

In 16 Serien konnte an 260 karbamylierten Einzelzellen 1—4 Tage alter Blutkonserven eine Verformbarkeitszunahme gegenüber den Kontrollerythrozyten gemessen werden, nämlich von 1,627 ± 0,062 mm auf 0,942 ± 0,091 mm des P_c-Mittelwertes bzw. 0,791 ± 0,014 mm auf 0,416 ± 0,033 mm des P_m-Mittelwertes (Vertrauenswahrscheinlichkeit 95%). Die Zellform-Indizes für karbamylierte Erythrozyten lagen dabei zwischen 80% und 90%. Die Indizes für Kontrollerythrozyten dagegen zwischen 50% und 70%.
Karbamylierte Erythrozyten, welche an Glasflächen zur Haftung gebracht wurden, zeigen gegenüber normalen, mit PBS-Lösung gewaschenen roten Blutzellen eine leichte Transformation aus der Echinozytenform in Richtung der Diskozytenform. Der Zellformindex der haftenden Erythrozyten stieg bis auf maximal 0,590 an. Es konnten keine Unterschiede hinsichtlich der Adhäsivität zwischen karbamylierten und unbehandelten Erythrozyten beobachtet werden. Im Vergleich zur Prokainierung ist die Erythrozytentransformation geringer. Die Umwandlung von Sphäroechinozyten ergibt vor allem eine Erhöhung der E_{II}- und E_{III}-Formen, während die Anzahl der E_I-Formen begrenzt ist und Diskozyten fast völlig fehlen. Eine Transformationsänderung dieser Art ist offenbar nicht mit einer erheblichen Kontaktflächenzunahme verbunden, welche entscheidenden Einfluß auf eine Adhäsivitätsvergrößerung besitzt.

Diskussion

Die Waschung von Erythrozyten mit PBS-Lösung führt zur weitgehenden Ablösung von adsorbierten Proteinen von ihrer Oberfläche. Gleichzeitig ist damit eine Gestaltsänderung verbunden, die praktisch alle Diskozyten in Echinozyten unterschiedlicher Formausprägung umbildet. Die in dieser Weise in vitro transformierten Zellen lassen sich durch das kationische Lokalanästhetikum Prokain nahezu vollständig in Diskozyten überführen. Da aus vorhergehenden Untersuchungen bekannt ist, daß Echinozyten eine geringere Verformbarkeit haben, ist hier untersucht worden, ob der Gestaltwandel im Prokainmilieu mit einer meßbaren Änderung der Verformbarkeit einhergeht.

Die vorliegenden Untersuchungen haben gezeigt, daß die Prokainierung von Echinozyten eine Erhöhung der Verformbarkeit verursacht.
Der Angriffsort des Prokains liegt mit hoher Wahrscheinlichkeit ausschließlich an der Erythrozytenmembran, aber es ist nicht geklärt, welche Effekte die Gestalt und Verformbarkeit des Erythrozyten hervorrufen. Als Ursache kommen a) eine Zunahme der Fluidität der Lipide und b) eine Entkopplung von kontraktilen Proteinen an der Membraninnenfläche in Frage. Wenn man davon ausgeht, daß eine höhere Membranfluidität vorliegt, sollte sich die Zelle besser an eine Unterlage anschmiegen und eine größere Zahl von Haftpunkten ausbilden können. Dem entspricht die gefundene Steigerung der Adhäsivität prokainierter Erythrozyten.
Die Karbamylierung ist bei Sichelzellen mit einer Verformbarkeitszunahme und Erhöhung der Überlebensrate verbunden.
Durch beide hier angewendete Meßverfahren konnte nachgewiesen werden, daß Karbamylierung an normalen Zellen ebenfalls eine Zunahme der Verformbarkeit ergibt. Parallele Viskositätsmessungen an Hb bzw. karbamyliertem Hb ergaben keinen Anhaltspunkt dafür, daß Karbamylierung eine größere Viskosität zur Folge hat. Weitere Untersuchungen sind erforderlich, um die Ursachen für die gesteigerte Verformbarkeit karbamylierter Erythrozyten zu klären.

Literatur

[1] UNGER, J., u. G. GEYER: CDM-Untersuchungen an Erythrozyten. II. Methodische Grundlagen. Wiss. Z. Friedrich-Schiller-Univ. Jena (1976) (im Druck)
[2] BENSER, A.: Adhäsivitätsuntersuchungen an Erythrozyten. Historischer Überblick — Methode. Wiss. Z. Friedrich-Schiller-Univ. Jena (1976) (im Druck)

Anschrift der Verfasser: Dipl.-Biol. J. UNGER und Dipl.-Biol. A. BENSER, Anatomisches Institut der Friedrich-Schiller-Universität, 69 Jena, Teichgraben 7

Acta biol. med. germ., Band 36, Seite 909—912 (1977)

Semmelweis Orvostudományi Egyetem Biofizikai Intézet Budapest, Hungary

A possible explanation of the cation selectivity in the active transport of erythrocytes

S. GYÖRGYI and I. SUGÁR

Summary

The effect of conformationally induced alteration in the charge distribution of transport ATPase and the formation of membrane binding sites with special shape and charge due to transformation of ATPase conformation on the Na^+/K^+ and the $K^+/Rb^+/Cs^+$ selectivity in the active transport of erythrocytes are discussed. As a consequence of phosphorylation (Na^+, K^+)-activated ATPase is assumed to be activated to a metastable state resulting in a more negative surface potential at the intracellular surface and/or the appearance of specific ion binding sites on both sides of the membrane. Due to these alterations a relatively high Na^+ concentration at the intracellular surface and a $Rb^+ > K^+ \gg Cs^+ \gg Na^+$ selectivity sequence of the extracellular binding sites has been formed. When the activated, metastable state of ATPase ends, the ions having bound at the surfaces are released inducing a Na^+-outflow and a K^+ (or Rb^+, Cs^+)-inflow in accordance with the local electrochemical potential gradients.

Red blood cell membrane, like other biological membranes and non-biological systems shows striking differences in selectivity towards different cations. Our understanding of the mechanism governing ionic selectivity in the exchange and transport processes is imperfect, however, one would expect to find the physical basis of selectivity sequences in the different types of interactions between fixed charges or polarizable groups of membrane surface with the ionic species. Two of the numerous possible factors resulting in the well known Na^+/K^+ and the interesting $K^+/Rb^+/Cs^+$ selectivity in the active transport of erythrocytes will be discussed: (a) the effect of conformationally induced alteration in the charge distribution of transport ATPase on the formation of Na^+ ion concentration at intracellular interface and (b) the formation of membrane binding sites with special shape and charge due to transformation of ATPase conformation.

Applying the theory of statistical ensembles BUFF and STILLINGER [1] worked out the theory of double layer structure and its properties. Equilibrium distribution and no specific binding sites have been assumed in this treatment. Using this theory PLESNER and MICHAELI [2] have shown that in the case when the surface potential ($\psi_{surf.}$) is more negative than a critical value a close-packed layer of the smallest ions is formed at the surface of the membrane because of the ion-ion short range interaction due to ion size and solvent molecular structure. Thus a concentration can be produced at the surface of membrane basically different from that in the bulk phase (Fig. 1).

Another selectivity factor can be the existence of specific binding sites. The steric structure of the binding sites, the size and charge distribution of the "cavity" formed by molecular structure plays a role not only in the case of synthetic complex forming compounds but in that of biological membranes, as well. SIMON et al.

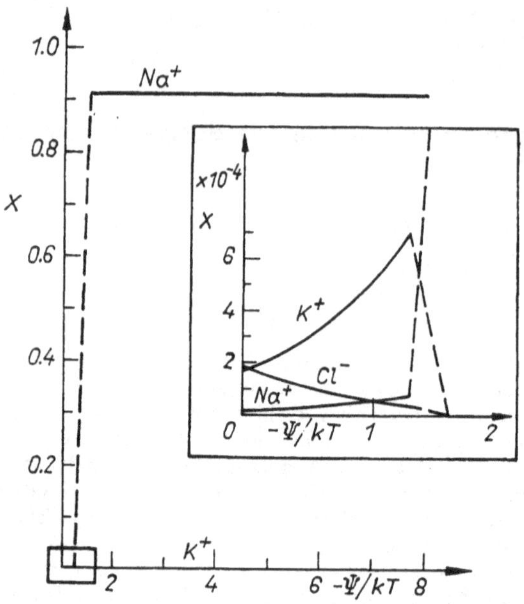

Fig. 1

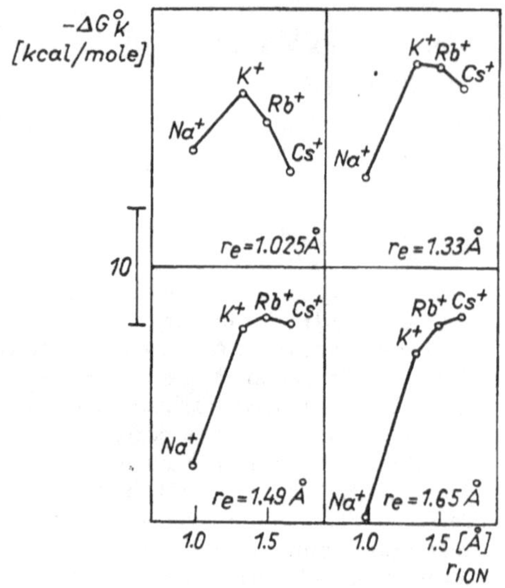

Fig. 2

Fig. 1. Computed surface concentration x vs electrostatic potential for a NaCl-KCl solution in equilibrium with a negatively charged surface. Bulk concentrations are: $c_{NaCl} = 0.01$ M, $c_{KCl} = 0.1$ M. [2]

Fig. 2. Calculated free energy of transfer (ΔG_K^0) of alkali cations from aqueous solution into a cavity with radius r_e [3]

[3] have proved that cavities with different radii can produce quite different sequences from cations. As it is shown in Fig. 2 ligands that are differentiated only by the radius of their equilibrium cavities can easily discriminate between cations of different size. Among the various conformations, equilibrium cavity is the one which corresponds to a local minimum in conformational energy. Any modification of this cavity to accomodate a cation involves an increase in conformational energy (ΔG_K^0). Of the cations belonging to one group of the periodic system, the cation that calls for the smallest modification of the cavity is preferentially bound corresponding to the free energy minimum.

The alkali cation selectivity in the active transport of erythrocytes can also be explained on the basis of the results discussed above. As a consequence of phosphorylation (Na^+, K^+)-activated ATPase is assumed to be activated to a metastable condition. Due to this conformational change the electrostatic potential on the intracellular surface of ATPase will be more negative and at the same time specific binding sites appear on both sides of this membrane region (see Fig. 3/2).

As a result of the negative potential and the existence of specific Na^+ binding sites the local concentration of Na^+ ions at the intracellular surface may increase to even a hundredfold of the average intracellular Na^+ concentration resulting in a local inversion of Na^+ concentration gradient (Fig. 3/3).

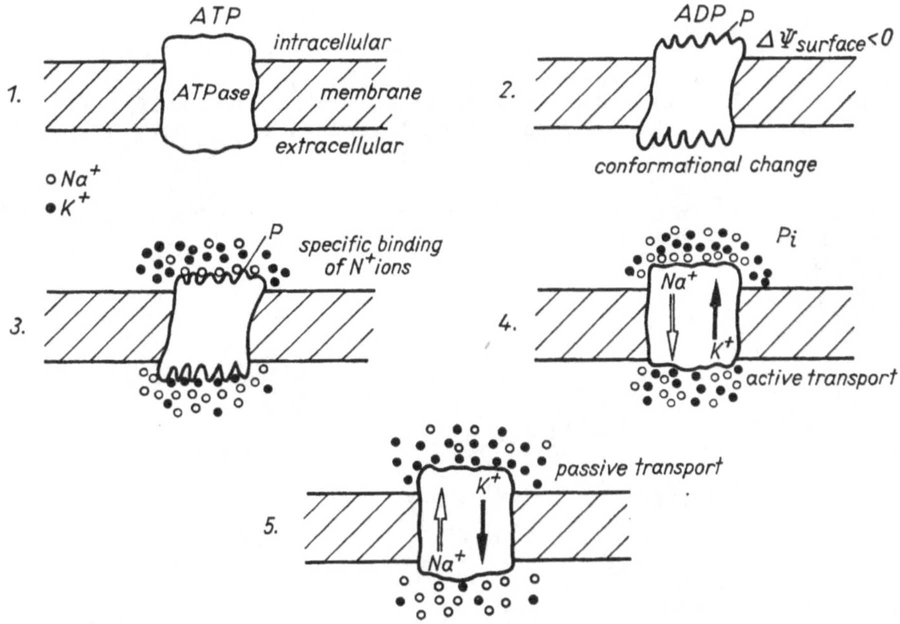

Fig. 3. Different states of an ATPase molecule in the membrane together with the respective steps of Na^+ and K^+ transport (see text)

This process cannot be interpreted with the negative surface potential only, because in this case the Li^+ ion, being smaller than the Na^+, should substitute and displace the Na^+ ion at the surface. Since the experimental results prove that Li^+ (and other cations) does not substitute for Na^+ in the active transport specific Na^+ binding sites have to be supposed.

On the other hand, the selectivity sequence of extracellular binding sites is: $Rb^+ > K^+ \gg Cs^+ \gg Na^+$ [4].

These facts, as well as the lack of K^+ ions at the inner surface result in a local inversion of K^+ electrochemical potential gradient.

After dephosphorylation of ATPase, presumably promoted by the adsorption of ions, its metastable state comes to an end, the energetically more stable state is restored and the specific binding sites of ATPase disappear. The ions having bound at the surface are released and ion flows start in the direction of local electrochemical potential gradients. This temporary and local Na^+ efflux and K^+ influx means the active ion transport of the membrane (Fig. 3/4).

Due to these ion fluxes the local concentration gradients, i.e. the driving forces of "active" Na^+ and K^+ transport decrease, then turn back corresponding to the bulk concentrations and this region of the membrane will join the passive transport (Fig. 3/5).

As regards the alkali ion selectivity of red cell membranes Figs. 1 and 2 clearly demonstrate that both factors (surface potential induced ion concentration at the intracellular interface and the conformational energy of binding sites with different sizes) depend on the ionic radius in such a way as to give a possibility for distinction among the cations of one group of the periodic system.

References

[1] BUFF, F. P., and F. H. STILLINGER, Jr.: J. chem. Phys. **39**, 1911 (1963)
[2] PLESNER, I. W., and I. MICHAELI: J. chem. Phys. **60**, 3016 (1974)
[3] SIMON, W., W. E. MORF, and P. CH. MEIER: Structure and Bonding **16**, 113 (1973)
[4] GYÖRGYI, S., and K. BLASKÓ: Acta Biochim. Biophys. Acad. Sci. hung. **9**, 97 (1974)

S. GYÖRGYI, Department of Biophysics, Semmelweis University of Medicine, H-1088 Budapest, Puskin utca 9., Hungary

Acta biol. med. germ., Band 36, Seite 913—918 (1977)

Department of Biochemistry, Kyushu University School of Medicine, Fukuoka 812, Japan

Uphill and selective transport of phosphoenolpyruvate through red cell membrane

N. Hamasaki, H. Harasaki*, A. Tomoda**, and S. Minakami

Summary

Accumulation of phosphoenolpyruvate and subsequent metabolism of the compound in human red cells were observed when the cells were incubated with the phosphate ester in isotonic sucrose media. Essentially no accumulation and metabolism were observed when the cells were incubated with other organic phosphate compounds such as 3-phosphoglycerate, fructose-1,6-bisphosphate and ATP. The accumulation of phosphoenolpyruvate seems to be an uphill transport of the compound. The rate of transport apparently followed a saturation kinetics and the intracellular concentration reached as high as ten times of the extracellular concentration. The transport was inhibited by an inhibitor of anion transport, 4-acetamide-4′-isothiocyano-2,2′-stilbene.

Introduction

Organic phosphate compounds have been generally regarded as impermeable to red cell membrane, which seems to be an indispensable condition for the cells to conserve the phosphate compounds within the cells. Similarly, organic phosphate compounds given exogenously are usually not metabolized by the cells.
We have previously reported that phosphoenolpyruvate (P-Prv) was accumulated in human red cells when the cells were incubated in an acidified isotonic sucrose solution with P-Prv and metabolic conversion of the ester inside the cells was observed [1]. P-Prv seems to have a unique property among organic phosphate compounds: the accumulation rates of other phosphate esters, e.g. 3-phosphoglycerate (3-P-Gly), fructose, 1,6-bisphosphate or ATP, were extremely slower than that of P-Prv in this condition. The rate of the accumulation apparently followed a saturation kinetics and the intracellular concentration became about ten times higher than the extracellular concentration in an incubation condition. It was inhibited by 4-acetamide-4′-isothiocyano-2,2′-stilbene (SITS), an inhibitor of anion transport. Thus the process of P-Prv accumulation seems to be an uphill and selective transport of the phosphate ester into the cells.

Results and discussion

Accumulation of P-Prv and its metabolism

When human red cells were incubated at 37 °C in an isotonic sucrose medium at *p*H 4.5 containing about 10 mM P-Prv, the P-Prv concentration in the cells increased rapidly to about 16 mM in 4 min with the concomitant decrease of the

* Department of Cardiac Surgery, Kyushu University School of Medicine
** Department of Biochemistry, Faculty of Medicine, Kanazawa University, Kanazawa, Japan

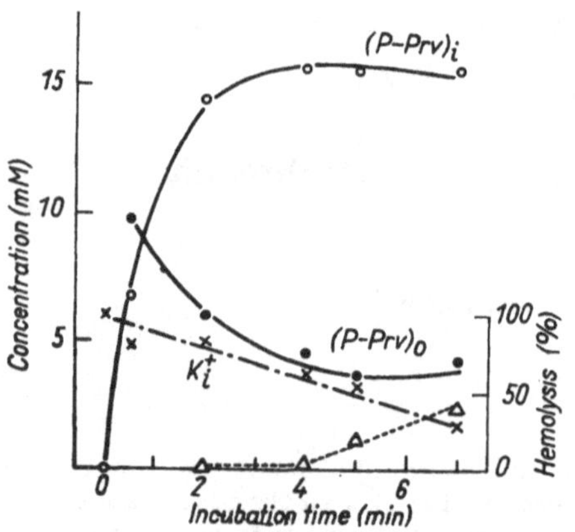

Fig. 1 Accumulation of P-Prv in red cells The cells were incubated in an isotonic sucrose medium at *p*H 4 5, 37 °C The hematocrit value of the suspension was about 30% at the beginning of the incubation The *p*H of the suspension was fixed by the addition of 0 1 M HCl

extracellular P-Prv concentration to 5 mM (Fig. 1). The P-Prv concentration of the cells suspended in the acidified sucrose medium without the addition of P-Prv was about 10 μM and it did not increase by the incubation. No appreciable hemolysis was observed until 4 min.

The cells loaded with P-Prv had also high concentration of monophosphoglycerates, which are supposed to be formed from P-Prv. The cells incubated for less than 4 min showed an apparently normal glycolytic activity when they were suspended in Ringer solution at *p*H 7.4 and incubated at 37 °C. The loaded P-Prv was metabolized rapidly to monophosphoglycerates, 2,3-bisphosphoglycerate-(2,3-P_2-Gly) and pyruvate (Fig. 2). The concentration of 3-phosphoglyceroylphosphate reached up to 90 nmol/g Hb, which is about two orders higher than the physiological concentration which is about 1 nmol/g Hb [2, 3]. The increment of 2,3-P_2-Gly during the incubation corresponded essentially to that of pyruvate, indicating that the formation of 3-phosphoglyceroylphosphate by the phosphoglycerate kinase reaction is coupled with the pyruvate kinase reaction. The reaction of phosphoglycerate kinase seems to be so rapid that the mass action ratio for the

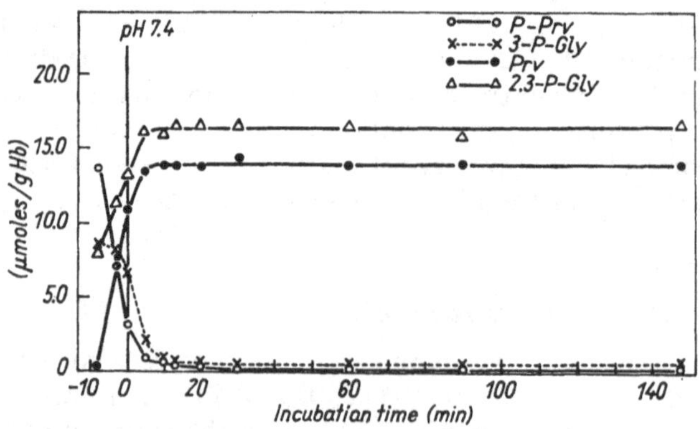

Fig 2. Changes of glycolytic intermediates in P-Prv loaded cells at neutral *p*H. After the treatment of red cells with P-Prv in the isotonic sucrose medium at *p*H 4 5, the *p*H of the suspensions was adjusted to 7 4 by dropwise addition of 0.1 NaOH with continuous stirring During incubation, the *p*H was kept constant under a *p*H-stat at 37 °C

reaction was 693 at 0 time and 641 at 10 min [1] which was close to the value obtained from fresh blood by Rose and Warms [4] but somewhat lower than that expected from the equilibrium of the reaction [2].

Accumulation of P-Prv in the presence of fluoride

In order to avoid the metabolic conversion of the accumulated P-Prv during the incubation, NaF as an inhibitor of both the enolase and ATPase was added to the incubation medium. The intracellular concentration of P-Prv increased linearly up to 43 mM in 15 min with the rate of about 2 μmol/ml cells per min, in the same conditions as in the incubation without NaF (Fig. 3). No accumulation of other glycolytic intermediates was observed. The addition of NaF is not only advantageous for quantitative estimation of P-Prv accumulation but also for the prevention of hemolysis. The uphill accumulation of P-Prv was also observed in an isotonic sucrose medium at *p*H 6.0 with the rate of about 0.4 μmol/ml cells per min (Fig. 4).

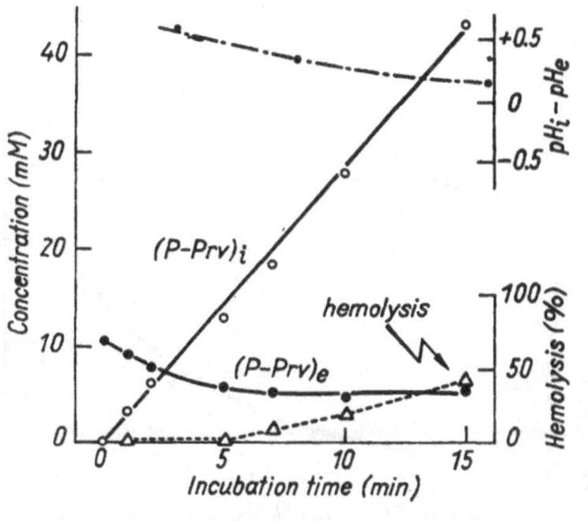

Fig. 3

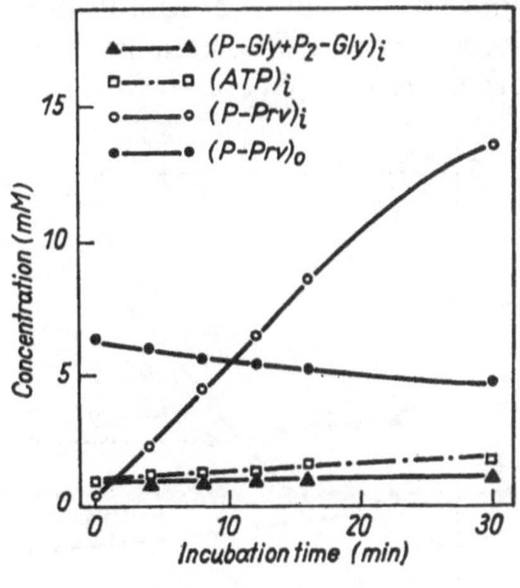

Fig. 4

Fig. 3. Accumulation of P-Prv in red cells incubated in the presence of NaF.
The cells were incubated in the sucrose medium with 10 mM NaF at *p*H 4.5, 37 °C. The hematocrit value of the suspension was about 30% at the beginning of the incubation

Fig 4. Accumulation of P-Prv in red cells at *p*H 6.0.
The cells were incubated in 0.25 M sucrose medium at *p*H 6.0, 37 °C. The hematocrit value of the suspension was about 10% at the beginning of the incubation. No hemolysis was observed

The intracellular *p*H of red cells incubated in an acidified sucrose medium (*p*H 4.5) was about 0.6 unit higher than the intracellular *p*H. This may lead to the uphill accumulation of P-Prv if the anion is freely permeable to the membrane. During the incubation with P-Prv, the intracellular *p*H decreased with the increase in the intracellular P-Prv concentration and the difference between the cellular and medium *p*H decreased from 0.6 unit to 0.16 unit (Fig. 3).

Accumulation of other organic phosphate compounds

The impermeability of organic phosphate compounds through cell membrane may be ascribed partly to the ionization of their phosphate residues and partly to their molecular size. At acidic *p*H, negative charges of phosphate compounds and the membrane are expected to decrease, which may facilitate the transport. If the transport is restricted solely by charges and molecular size, we could expect that other organic phosphate compounds, the molecular size and pK_a of which are similar to those of P-Prv may penetrate the membrane. We used 3-P-Gly to test the possibility. The molecular weight of the compound is 186 and pK_a's 5.98, 3.42 and 1.42 [5], which are quite similar to those of P-Prv (molecular weight 168, pK_a's 6.38, 3.5 [5] and probably around 1.5). Red cells were incubated in the sucrose medium containing NaF, P-Prv and 3-P-Gly at *p*H 4.5, taking advantage of the negligible interconversion of the phosphate esters in the presence of the inhibitor. The rate of 3-P-Gly accumulation was less than one tenth of the P-Prv accumulation rate (Fig. 5). Red cells incubated with organic phosphate compounds such as fructose-1,6-bisphosphate, glucose-6-phosphate and ATP showed negligible accumulation of the compounds in the same conditions. The reasons why P-Prv is more permeable to the cellular membrane compared with other organic phosphate compounds remain to be solved.

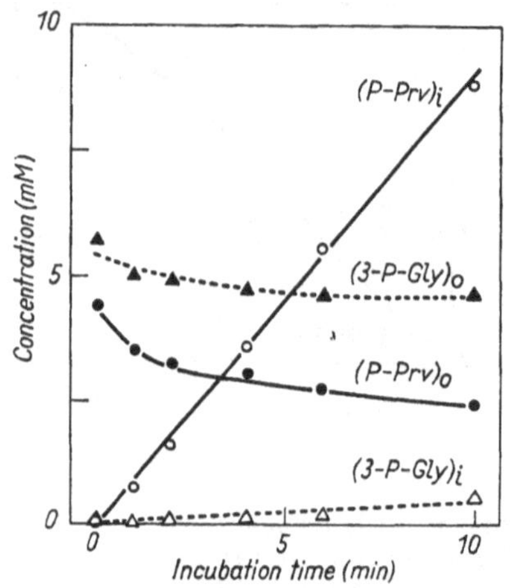

Fig. 5

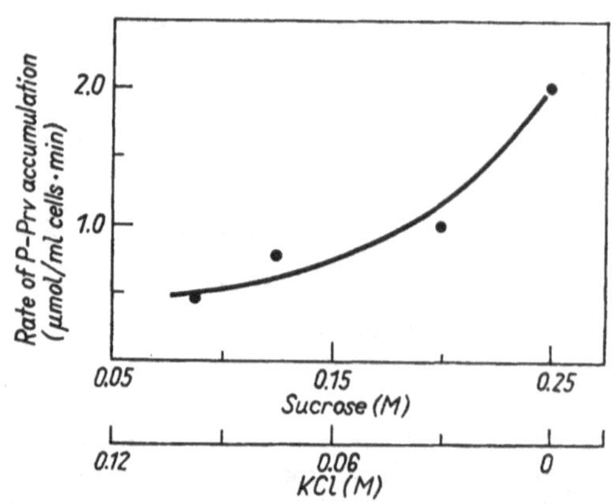

Fig. 6

Fig. 5. Comparison of the accumulation of P-Prv and 3-P-Gly.
Conditions of the incubation were the same as in Fig. 3. The hematocrit value of the suspension was about 20% at the beginning of the incubation

Fig. 6. Relationship between the sucrose concentration and the rate of the P-Prv accumulation

Some properties of P-Prv transporting system

The rate of P-Prv accumulation was decreased when sucrose was partly replaced by KCl or NaCl. The accumulation rate was decreased to about half when the cells were incubated in a medium consisting of 200 mM sucrose and 30 mM KCl

(Fig. 6). Further decrease in sucrose concentration in the medium caused decrease in the accumulation rate, but the experiment could not be carried out in an isotonic KCl medium because of the fragility of the cells at *p*H 4.5. The accumulation also depended on the *p*H of the medium as shown in Fig. 7. At neutral *p*H, the accumulation rate was very slow but the transport could be accelerated by increasing the incubation temperature up to 45 °C.

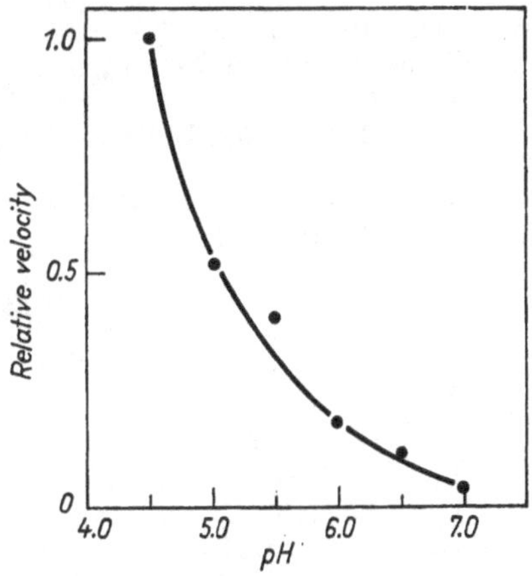

Fig. 7

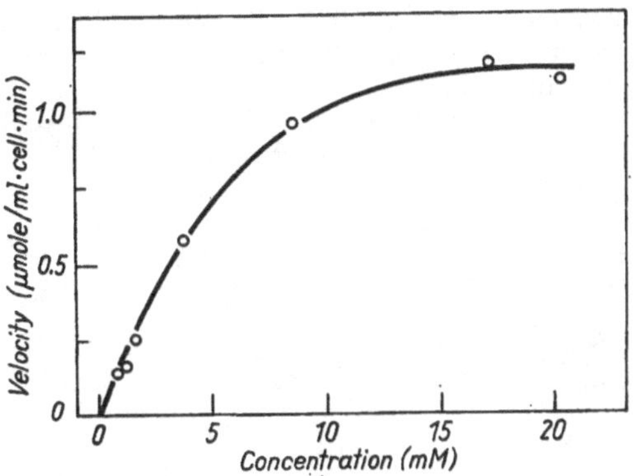

Fig. 8

Fig. 7. Relationship between the external *p*H and the relative velocity. The cells were incubated in 0.25 M sucrose medium with 7—8 mM P-Prv at 37 °C. The relative velocity was drawn from the decrease of external P-Prv

Fig. 8. Correlation between P-Prv flux and P-Prv concentration. The cells were incubated in 0.25 M sucrose at *p*H 6.0, 37 °C. The flux was calculated from the decrease of external P-Prv. The total phosphate concentration was fixed to 26 mM by the addition of potassium phosphate

By increasing the extracellular P-Prv concentration, the rate of the accumulation increased hyperbolically, reaching a limiting velocity at high concentrations of P-Prv. When red cells were incubated in 0.25 M sucrose solution with 10 mM NaF at *p*H 6.0, 37 °C, the K_m was about 7 mM and the V_{max} about 1.4 µmol/ml cells per min (Fig. 8). The accumulation rate depended on the incubation temperature and the activation energy was about 26 kcal/mol, which is similar to the activation energies for organic and inorganic anions [6]. The accumulation was completely inhibited by SITS at the concentration of 1 mM or 8 µmol/ml cells.

Scanning electron microscopic observation of the cells

A scanning electron microscopic observation of the red cells incubated at *p*H 4.5 for 2 min at 37 °C revealed that the cells adhered to each other forming clusters. The constituent cells of the cluster were spherocytic. This may reflect the decrease in negative charges on the cell surface at low *p*H.

Concluding remarks

Phosphoenolpyruvate seems to be unique among phosphorylated glycolytic metabolites that it can be transported through red cell membrane when the cells are suspended in an isotonic sucrose solution. This phenomenon may be used for the study on the mechanism of membrane transport as well for the modeling of glycolytic regulation in red cells.

References

[1] TOMODA, A., N. HAMASAKI, and S. MINAKAMI: Biochem. biophys. Res. Commun. **247**, 1127 (1975)
[2] MINAKAMI, S., and H. YOSHIKAWA: J. Biochem., Tokyo **59**, 139 (1966)
[3] SCHEWE, CH., G. JACOBASCH, and S. RAPOPORT: Acta biol. med. germ. **28**, 437 (1972)
[4] ROSE, I. A., and J. V. B. WARMS: J. biol. Chem. **245**, 4009 (1970)
[5] OESPER, P. in: Phosphorous Metabolism. **1**, 523. W. D. MCELROY and B. GLASS (ed.) Johns Hopkins, Baltimore 1951
[6] DEUTICKE, B. in: Erythrocytes, Thrombocytes, Leukocytes. E. GERLACH, K. MOSER, E. DEUTSCH and W. WILMANNS (ed.) Thieme, Stuttgart 1973, p. 81

Acta biol. med. germ., Band 36, Seite 919—920 (1977)

Biophysics Department, St. Mary's Hospital Medical School, London, W2 1PG. U.K.

The rate of reversible and irreversible sickling

M. W. Rampling

When erythrocytes containing haemoglobin (Hb)S are deoxygenated the internal haemoglobin polymerises and distorts the cells to produce sickle cells. On reoxygenation the haemoglobin polymers break up and the cells revert to the discoidal form. This is reversible sickling.
A constant-flow rapid reaction apparatus has been used to measure the rate of change of morphology of erythrocytes containing HbS after essentially instantaneous deoxygenation [1]. With this device suspension of cells in Ringer-Locke-solution, *p*H 7.4, was very rapidly deoxygenated by mixing with sodium dithionite ($Na_2S_2O_4$) solution. The cells were then allowed to undergo the morphological changes which result from the polymerisation of the reduced haemoglobin and they were fixed after various intervals of time by means of formalin solution.
The results obtained with this apparatus showed that sickling is an extremely rapid process. The first signs of morphological change occurred 30 to 40 ms after mixing with the dithionite solution, half the cell population was deformed after 80 to 90 ms and all were deformed after 150 to 160 ms. These results show that, under in vivo conditions, those cells that become sufficiently deoxygenated to induce sickling, e.g. as a result of passage through the liver where pO_2 can be as low as 20 mm Hg, have ample time to change their morphology. Therefore, during every circuit of the circulation a proportion of the cell population will sickle, only to revert back to the discoidal form on reoxygenation. Thus it is to be expected that during their lifetime the cells of a subject homozygous to HbS will cycle through the discoidal to sickled form and back again many times.
In a second series of experiments a comparison was made of the effect of prolonged in vitro incubation on the rate of degeneration of the membranes of cells containing either HbA or HbS. One index of degeneration was the rate of loss of membrane flexibility as measured by a centrifugal technique [2]. Another was the rate of echinozyte production. A third, which could only be applied to HbS cells, was the rate of loss of reversibility, i.e. the rate at which discoidal, oxygenated cells became unable to deform to the sickled shape on deoxygenation and vice versa. For each of these tests, 1 ml of heparinised whole blood was incubated at 37 °C, without added metabolic substrate, in an oxygenating (10 cm Hg pO_2, 4 cm Hg pCO_2 and N_2) or a deoxygenating (4 cm Hg pCO_2 and N_2) atmosphere. After various intervals of time samples were removed for flexibility or reversibility measurement, or were fixed and the proportion of echinocytes determined [3].
In the case of HbA cells, echinocyte production and loss of flexibility followed similar time courses. These degenerative changes began only after 8 to 10 hrs of incubation and continued progressively to completion at 40 to 50 hrs. The state of oxygenation of the cells made no difference to the time-course of the changes.

The similarity in the time-scales of these two phenomena suggests that they result from the same membrane degeneration and, from the work reported by WEED et al. [4] and SHOHET and HALEY [5], that they are related to ATP depletion and metabolic run-down.

Damage to the membranes of HbS cells, as determined by both echinocyte production and loss of reversibility, started only after 5 to 6 hrs of incubation and was complete after about 24 hrs. This too was independent of the state of oxygenation of the cells. The rate of loss of flexibility (which could only be measured on oxygenated cells, since deoxygenated cells were rigid due to the polymerisation of the haemoglobin) followed the same time-course. Since these changes are unaffected by the state of oxygenation, since the three indices of membrane damage have the same time-course and since it is known that HbS cells have a metabolic rate roughly double that of HbA cells [6]; it is concluded that these membrane changes observed in the HbS cells are, as in the case of HbA cells, due to metabolic depletion and are independent of the state of polymerisation of the cellular haemoglobin.

The blood of subjects who are homozygous to HbS usually contain a considerable proportion of circulating irreversibly sickled cells (ISC) i.e. cells which remain deformed irrespective of their state of oxygenation. Several hypotheses have been proposed to explain the production of ISC; the results reported above can be used to test some of these hypotheses. For example, EATON et al. [7] suggested that ISC result from the effect on the cell membrane of the increased ingress of Ca^{++} to cells in the deoxygenated, sickled state as compared to the oxygenated, discoidal state. Since no difference was observed between the rates of degeneration of oxygenated and deoxygenated HbS cells in the experiments reported above, this hypothesis is clearly no longer tenable. It is also difficult to see how the stagnation hypothesis of ISC production can be maintained. This proposes that the ISC are produced by cellular trapping in the microcirculation and consequent metabolic rundown and membrane damage. From the work reported here the stagnation would have to be in excess of 5 hrs and, for this to result in circulating ISC, subsequent release of the cells would have to occur. It is difficult to suggest a mechanism for this.

A more attractive hypothesis results from PADILLA et al's [8] finding that during the sickle/unsickle cycle some membrane loss and damage frequently occur. As described above, during its life time a cell can be expected to cycle many times through the sickled to the unsickled state and back again, so progressively more damage will take place. It is suggested that it is this cycling that leads to the in vivo production of circulating ISC and to the early destruction of cells of homozygous sicklers.

References

[1] RAMPLING, M. W., and J. A. SIRS: Clin. Sci. Mol. Med. **45**, 655 (1973)
[2] SIRS, J. A.: Phys. Med. Biol. **15**, (1970)
[3] RAMPLING, M. W.: Clin. Sci. Mol. Med. **48**, 367 (1975)
[4] WEED, R. I., P. L. LACELLE, and E. W. MERRILL: J. Clin. Invest. **48**, 795 (1969)
[5] SHOHET, S. B., and J. E. HALEY: Nouv. Rev. Franc. Hematol. **12**, 761 (1972)
[6] EYIMOFE-BOYO, A., and J. A. IKOMI-KUMM: Lancet **1972**/I, 1215
[7] EATON, J. W., T. D. SKELTON, H. S. SWOFFORD, C. E. KOLPIN, and H. S. JACOB: Nature, Lond. **246**, 105 (1975)
[8] PADILLA, F., P. A. BROMBERG, and W. N. JENSEN: Blood, **41**, 653 (1973)

Acta biol. med. germ., Band 36, Seite 921–924 (1977)

Humboldt-Universität zu Berlin, Sektion Biologie, Bereich Biophysik, 104 Berlin, DDR

Die Sedimentationsgeschwindigkeit von Erythrozyten als Indikator für Phasenübergänge in der Membran

U. Beutel und R. Glaser

Zusammenfassung

Aus temperaturabhängigen Sedimentationsuntersuchungen an Erythrozyten des Menschen ergab sich, daß die Sedimentationsgeschwindigkeit bei 21 °C in Krebs-Ringer-Lösung, *p*H 7,4, ein Minimum hat [1]. In weiteren Untersuchungen konnte gezeigt werden, daß hauptsächlich ein Phasenübergang der Membranlipide die Ursache dafür ist. Membranproteine beeinflussen das Sedimentationsverhalten der Erythrozyten nur über die Wechselwirkung mit den Membranlipiden.

Einleitung

Bei temperaturabhängigen Sedimentationsmessungen konnte gezeigt werden, daß die Sedimentationsgeschwindigkeit von Erythrozyten des Menschen in Krebs-Ringer-Lösung, *p*H 7.4, bei 21 °C ein Minimum hat. Als Ursache für dieses Minimum können Membranveränderungen angenommen werden, die unter anderem die Membranviskosität beeinflussen. Eine Korrelation mit Lipidphasenumwandlungen ist dabei möglich [1].
Dieser Gesichtspunkt und die Bedeutung von Membranproteinen standen in weiteren Untersuchungen zur temperaturabhängigen Erythrozytensedimentation im Vordergrund.
Es wurden Erythrozyten untersucht, bei denen durch die Erhöhung der Anzahl der Waschungen der Anteil auf der Membranoberfläche adsorbierten Proteine verringert wurde. Außerdem erfolgten Messungen in Suspensionsmitteln unterschiedlichen *p*H-Wertes und unterschiedlicher Ionenstärke.

Material und Methoden

Die Erythrozyten wurden aus frischen Blutkonserven durch Zentrifugation vom Plasma getrennt. Nach zwei- oder mehrmaligem Waschen und einstündiger Inkubation bei 37 °C erfolgte die Resuspension der Zellen in das für die Messung bestimmte Suspensionsmittel. Als Wasch- und Inkubationslösung diente Krebs-Ringer-Lösung *p*H 7,4. Als Suspensionsmittel kamen Krebs-Ringer-Lösungen mit *p*H-Werten von 6,0, 7,4 und 8,0 und isotonische NaCl-Saccharose-Lösungen mit Ionenstärken von 10^{-1} M, $5 \cdot 10^{-2}$ M und $3 \cdot 10^{-2}$ M und *p*H 7,4 zur Anwendung. Mit Erythrozyten, die bei *p*H 6,0, 7,4 bzw. 8,0 mit Glutaraldehyd fixiert wurden, erfolgten bei 20 °C in Krebs-Ringer-Lösung, *p*H 7,4, Vergleichsmessungen zum Einfluß der Form auf die Sedimentationsgeschwindigkeit. Meßkammer und Meßmethode wurden bereits kurz beschrieben [1].

Ergebnisse und Diskussion

Veränderungen der Zellmembran, die durch eine Verringerung des Anteils der auf der Membranoberfläche adsorbierten Proteine durch mehrmaliges Waschen entstehen können, haben bei gewaschenen Erythrozyten keinen signifikanten Einfluß

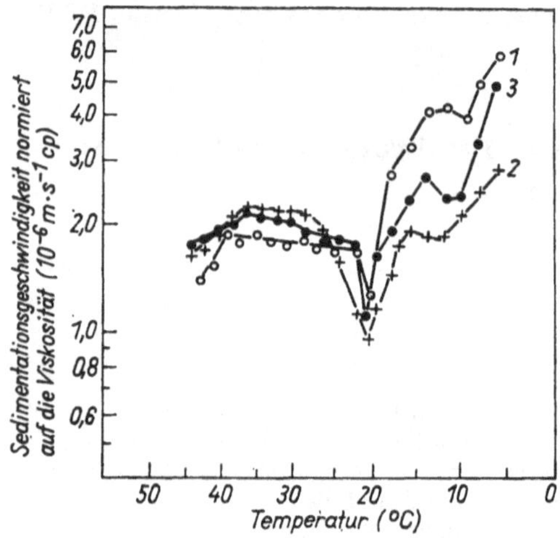

Abb. 1

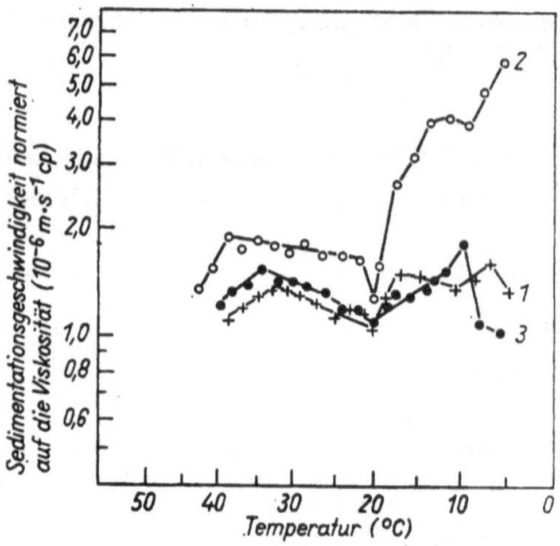

Abb. 2

Abb. 1. Einfluß des Waschens auf die Temperaturabhängigkeit der auf die Viskosität des Suspensionsmittels normierten Sedimentationsgeschwindigkeit: 1. nach 2maligem Waschen; 2. nach 4maligem Waschen; 3. nach 6maligem Waschen.
Darstellung im Arrhenius-Plot

Abb. 2. Einfluß des *p*H-Wertes auf die Temperaturabhängigkeit der auf die Viskosität des Suspensionsmittels normierten Sedimentationsgeschwindigkeit: 1. *p*H 6,0; 2. *p*H 7,4; 3. *p*H 8,0.
Darstellung im Arrhenius-Plot

auf das Sedimentationsverhalten der Zellen (Abb. 1). Das Minimum bei 21 °C bleibt erhalten. Der Kurvenverlauf bei Temperaturen über 21 °C weist keine bedeutenden Unterschiede auf. Die Differenzen bei Temperaturen unter 21 °C korrelieren nicht mit der Anzahl der Waschungen. Das bedeutet, daß die nach zwei- oder mehrmaligem Waschen noch auf der Membranoberfläche verbleibenden Proteine nur einen geringen Einfluß auf Parameter wie Membranviskosität oder Membranfestigkeit haben können. Weiterhin geht aus den Ergebnissen hervor, das diese Proteine nicht für die Entstehung des Minimums bei 21 °C verantwortlich sind.

Messungen bei unterschiedlichen *p*H-Werten zeigen, daß zumindest die Eiweiße, die die Membran teilweise oder vollständig durchdringen, das Sedimentationsverhalten des Erythrozyten mitbestimmen. Zum einen führen *p*H-Wertverschiebungen zu Konformationsänderungen und damit im Zusammenhang auch zu Veränderungen der Protein-Lipid-Wechselwirkungen [2]. Diese dürften auch die Hauptgründe für die Unterschiede im Sedimentationsverhalten der Zellen bei *p*H 7.4 und bei *p*H 6.0 bzw. 8.0 sein (Abb. 2). Von Bedeutung ist hierbei, daß das Minimum bei 21 °C erhalten bleibt. Das ist ein weiterer Hinweis dafür, daß die Ursache des Minimums ein Lipidphasenübergang ist. Die *p*H-Wertunabhängigkeit der Phasenübergangstemperatur in größeren *p*H-Bereichen wurde für eine Reihe von Lipiden nachgewiesen [3].

Die Unterschiede im Betrag der Geschwindigkeiten sind auch teilweise auf unterschiedliche Zellformen zurückführbar. Aus Kontrollmessungen an Erythrozyten,

Tabelle 1

Einfluß der Erythrozytenform auf die Sedimentationsgeschwindigkeit der Zelle

*p*H-Wert	Erythrozytenform	Sedimentationsgeschwindigkeit ± SD (10^{-6} m · s^{-1})
6,0	Stomatozyt	1,66 ± 0,32
7,4	Diskozyt	2,83 ± 0,26
8,0	Echinozyt	2,03 ± 0,44

die bei verschiedenen *p*H-Werten mit Glutaraldehyd fixiert wurden, ergab sich, daß Diskozyten schneller sedimentieren als Stomatozyten und Echinozyten (Tab. 1). Die Zellform wurde mit einem Mikroskop ermittelt. Die *p*H-Abhängigkeit der Erythrozytenform entsprach der aus der Literatur bekannten [4].

Weitere Hinweise über die Natur des Minimums bei 21 °C konnten aus Ionenstärkeabhängigen Messungen erhalten werden. Sie ergaben, daß sich das Minimum mit kleiner werdender Ionenstärke zu niedrigeren Temperaturen verschiebt (Abb. 3). Eine Ionenstärkeabhängigkeit der Phasenübergangstemperatur von Lipiden ist nachgewiesen worden [3]. Die Frage, welche Bedeutung die mit der Verschiebung der Ionenstärke verbundene Änderung des Membranpotentials für die Verlagerung des Minimums zu niedrigeren Temperaturen hat, bleibt ungeklärt.

Allgemein zeigen die Ergebnisse, daß die Sedimentationsgeschwindigkeit der Erythrozyten des Menschen als Funktion der Temperatur ein relativ empfindlicher Indikator für Lipidphasenübergänge und Veränderungen der Protein-Lipid-Wechselwirkungen innerhalb der Membran ist.

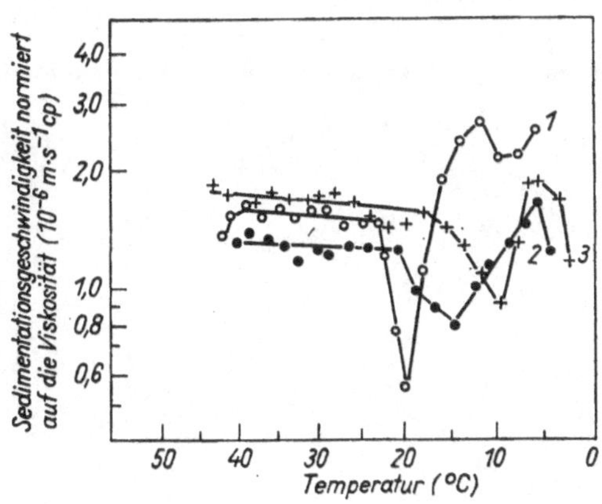

Abb. 3. Einfluß der Ionenstärke auf die Temperaturabhängigkeit der auf die Viskosität des Suspensionsmittels normierten Sedimentationsgeschwindigkeit: 1. $I = 10^{-1}$ M; 2. $I = 5 \cdot 10^{-2}$ M; 3. $I = 3 \cdot 10^{-2}$ M. Darstellung im Arrhenius-Plot

Literatur

[1] Beutel, U.: Acta biol. med. germ. (eingereicht)

[2] Bieri, V. G., u. D. F. Hoelzel-Wallach: Biochim. biophys. Acta **406**, 415 (1975)

[3] Träuble, H., u. H. Eibl: Proc. natn. Acad. Sci. U.S.A. **71**, 214 (1974)

[4] Bessis, M., R. I. Weed u. P. F. Leblond (Hrsg.): Red Cell Shape. Springer, Berlin, Heidelberg, New York 1973

Summary

U. BEUTEL and G. GLASER: Sedimentation rate of erythrocytes as an indicator for phase transitions in the membrane

Temperature dependent sedimentation meassurements with human erythrocytes showed that the sedimentation velocity at 21 °C in Krebs-Ringer-solution, *p*H 7.4, has a peak shaped minimum [1]. In further investigations it could be shown that a phase transition of membrane lipids is the main reason of this peak. Membrane proteins influence the sedimentation behavior of the erythrocytes only by changing the interaction with membrane lipids.

Acta biol. med. germ., Band 36, Seite 925—930 (1977)

Humboldt-Universität zu Berlin, Sektion Biologie, Bereich Biophysik, 104 Berlin, DDR

Der Einfluß von Hämatokritwert sowie oberflächenladungsverändernden Faktoren auf den ^{22}Na- und ^{86}Rb-Efflux menschlicher Erythrozyten

U. Kunter und R. Glaser

Zusammenfassung

An menschlichen Erythrozyten wurden Flux-Messungen mit dem Ziel durchgeführt, Aussagen über die Wechselwirkung von Zellen in Suspensionen sowie über die Rolle der Oberflächenstruktur zu treffen. Oberflächenmodifikationen wurden durch Neuraminidase- und La^{2+}-Aufarbeitung erreicht. Zur Klärung der Wechselwirkungen der Zellen untereinander wurden hämatokritabhängige Messungen des durch die (Na^+, K^+)-ATPase vermittelten aktiven ^{22}Na-Effluxes und des passiven ^{86}Rb-Effluxes durchgeführt. Die hämatokritabhängigen Messungen zeigen, daß die Zellen schon bei großen Abständen in Wechselwirkung treten.

Einleitung

Verschiedene Membrantransportprozesse sind eine Funktion der Zellkonzentration. Das wurde von mehreren Autoren für den Nichtelektrolyttransport an verschiedenen Versuchsobjekten nachgewiesen [1, 2]. Diese Phänomene werden zumeist durch Substanzen hervorgerufen, die von den Zellen in das Suspensionsmittel abgegeben werden, sich dort anreichern und als Hemmstoff auf Transportvorgänge einwirken [1]. Andere Autoren interpretieren ihre Ergebnisse dahingehend, daß die Zahl der Zellkollisionen mit Erhöhung der Zellkonzentration zunimmt [2]. Jedoch ist auch eine gegenseitige Beeinflussung der Zellen durch den sie umgebenden sog. „membrannahen Raum" möglich [3, 4]. Dies kann z. B. durch die Glykoproteine der Glykokalyx vermittelt werden, die durch ihre Struktur und ihre Ladung an der Barrierefunktion der Zellmembran beteiligt sind [5—7]. Wir wissen heute einiges über die Rolle von Ladungen innerhalb der Membran bei der Steuerung der Ionenpermeabilität und -selektivität (Überblick siehe [8]), wenig dagegen über den Beitrag, den die der Membran vorgelagerten Ladungen der Glykokalyx für die Beeinflussung aktiver und passiver Ionentransportprozesse leisten.

Das Ziel der vorliegenden Arbeit ist es

1. den Einfluß der Zellkonzentration auf den aktiven und passiven Ionentransport zu bestimmen,
2. einen Beitrag zur Klärung der Rolle der Oberflächenladungen zur Steuerung von Transportprozessen zu leisten.

Die Messungen erfolgten dabei in einer Meßkammer, die es erlaubte, zwischen direkten Wechselwirkungen zwischen den Zellen bei Erhöhung der Zellkonzentration und der Wirkung einer in der Suspension akkumulierten Substanz zu unterscheiden. Auf diesem Wege können Informationen über mögliche Wechselwirkungen zwischen Zellen in Suspension erhalten werden.

Material und Methodik

Verwendet wurden menschliche Erythrozyten aus maximal drei Tage alten Vollblut-Zitrat-Konserven. Die Erythrozyten wurden durch zweimaliges Zentrifugieren (1400 × *g*) mit der Suspensionslösung vom Blutplasma getrennt und bei einem Hämatokritwert Hk = 30% im Brutschrank bei 37 °C zur Einstellung eines stationären Zustandes 2 Std. inkubiert. Anschließend erfolgte die Inkubation mit radioaktiver Ringer-Lösung (1 Std. bei ^{86}Rb- und 2 Std. bei ^{22}Na-Inkubation) unter gleichen Bedingungen.
Die La^{3+}- und Neuraminidase-Behandlungen erfolgten vor der oben genannten Inkubation.
La^{3+}-Aufarbeitung: 1 ml gewaschener Erythrozyten wurde in 9 ml der entsprechenden $LaCl_3$-NaCl-Lösung aufgeschüttelt und nach 5 min abzentrifugiert. Die verwendeten La^{3+}-Konzentrationen sind der Abb. 4 zu entnehmen.
Neuraminidase-Aufarbeitung: Nach der Beseitigung des Blutplasmas wurden die Zellen in Anwesenheit von Neuraminidase (Serva, Heidelberg) 50 min bei einem Hämatokritwert Hk = 1% bei 37 °C inkubiert. Danach werden die Zellen bei 1400 × *g* abzentrifugiert und anschließend wie oben beschrieben weiterbehandelt.

Tabelle 1
Für die Efflux-Experimente verwendete Suspensionsmittel

	Lösung 1 (mM)	Lösung 2 (mM)	Lösung 3 (mM)
NaCl	142,4	141,3	141,3
KCl	5,7	5,7	5,7
NaH_2PO_4	0,3	1,2	5,8
$CaCl_2$	3,0	3,0	—
$MgSO_4$	1,4	1,4	—
Glukose	5,0	5,0	5,0

Meßanordnung: Die mit ^{22}Na bzw. ^{86}Rb inkubierten Erythrozyten wurden in eine Diffusionskammer zur kontinuierlichen Messung des Effluxes gefüllt. Diese Anordnung (Näheres siehe [9]) bestand aus einer 1 cm^2 großen und 1 mm tiefen Kammer, die durch ein Milliporfilter abgedeckt war und von inaktiver Ringer-Lösung kontinuierlich umspült wurde. Zur Verhinderung der Sedimentation war die Kammer mit der Filterfläche senkrecht auf eine langsam rotierende Achse montiert (12 U/min). Kontinuierlich wird mittels eines Szintillations-Meßkopfes die in der Diffusionskammer verbleibende Radioaktivität gemessen.
Die Kompartmentanalyse der Meßdaten ergab ein zwei-Kompartment-System in Reihenschaltung. Es läßt sich durch graphische Analyse die Austauschkonstante k des Zellkompartments direkt bestimmen. Das ist deshalb möglich, weil sich die Austauschkonstanten von Meßkammer (k_0) und Erythrozyten (k) um Größenordnungen unterscheiden und sich deshalb durch graphische Analyse trennen lassen [10].

Meßergebnisse

^{86}Rb- und ^{22}Na-Efflux als Funktion des Hämatokritwertes

Die Ergebnisse der Efflux-Experimente sind in Abb. 1 und 2 zusammengefaßt. Es zeigt sich, daß die ^{22}Na- und ^{86}Rb-Austauschkonstanten eine Funktion des Hämatokritwertes sind. Der ^{22}Na-Efflux wird durch die (Na^+, K^+)-ATPase vermittelt. Da diese stark *p*H-abhängig ist [11,12], wurden zur Klärung des Einflusses von eventuell auftretenden *p*H-Verschiebungen [13] Hämatokritabhängigkeiten bei unterschiedlichen Pufferkapazitäten aufgenommen. Um die auftretenden *p*H-Verschiebungen abschätzen zu können, wurde die *p*H-Abhängigkeit des Na-Effluxes gemessen (Abb. 3). In Übereinstimmung mit anderen Autoren zeigt die

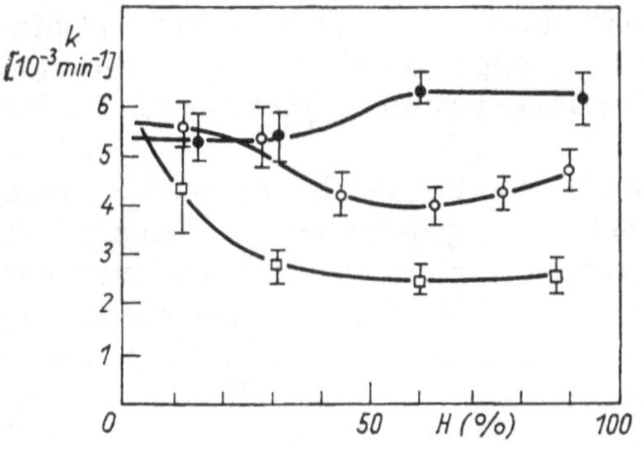

Abb. 1

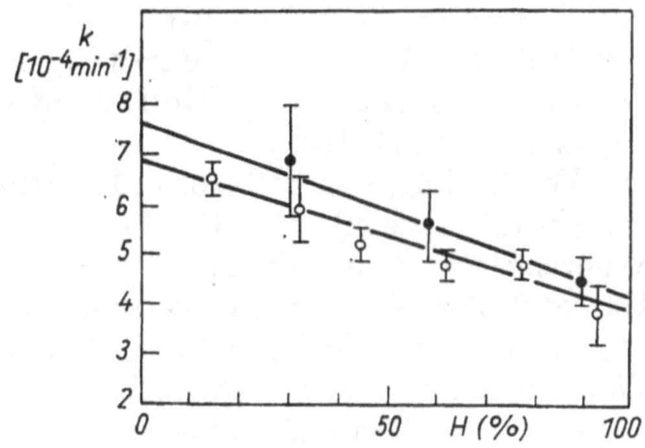

Abb. 2

Abb. 1. Abhängigkeit der Austauschkonstanten des [22]Na-Effluxes vom Hämatokritwert (H) *p*H 7,4; $T = (37 \pm 0{,}5)$ °C.
□: Suspensionsmedium 1; ○: Suspensionsmedium 2; ●: Suspensionsmedium 3 (s. Tab. 1)

Abb. 2. Abhängigkeit der Austauschkonstanten des ^{86}Rb-Effluxes vom Hämatokritwert (H). *p*H 7,4; $T = (37 \pm 0{,}5)$ °C.
●: Suspensionsmedium 3; ○: Suspensionsmedium 2 (s. Tab. 1)

Kurve ein ausgeprägtes Maximum bei *p*H 7,4 [11, 12]. Ist die Hämatokritabhängigkeit des Na-Effluxes tatsächlich auf *p*H-Effekte zurückzuführen, so müßte sich die *p*H-Verschiebung in der Suspension durch einen veränderten *p*H-Wert im an der Diffusionskammer vorbeiströmenden Suspensionsmittel gegentitrieren lassen und eine scheinbare Verschiebung des Maximums in der *p*H-Kurve bei höheren Hämatokritwerten bzw. geringeren Pufferkonzentrationen auftreten. Die Messungen dazu zeigt Abb. 3.

Der ^{86}Rb-Efflux ist im untersuchten Bereich (*p*H 6,0 bis 8,2) *p*H-unabhängig. Das stimmt mit Literaturdaten überein [14]. Weiterhin zeigt sich, daß die Pufferkonzentration im Suspensionsmedium ohne signifikanten Einfluß ist. Die Sus-

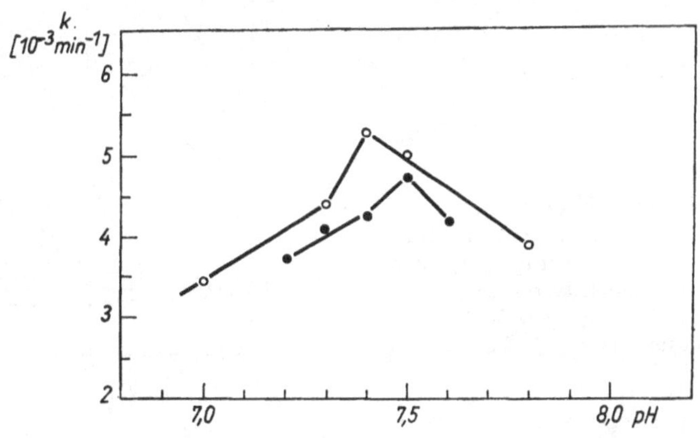

Abb. 3. Abhängigkeit der Austauschkonstanten des [22]Na-Effluxes vom *p*H-Wert im strömenden Suspensionsmedium.
$T = (37 \pm 0{,}5)$ °C; Standard-Abweichung $\pm 10\%$.
○: Suspensionsmedium 2 (s. Tab. 1), $H = (30 \pm 5)$ %; ●: Suspensionsmedium 2 (s. Tab. 1), $H = (60 \pm 5)\%$

pensionsmedien enthalten 2wertige Kationen in unterschiedlichen Konzentrationen (Tab. 1). Wie Literaturdaten ergeben, ist jedoch die Ca^{++}-Konzentration im Suspensionsmedium ohne Einfluß auf den Ionenaustausch metabolisch aktiver Erythrozyten [15, 16].

Die Verwendung von ^{86}Rb zur radioaktiven Markierung des K^+-Effluxes erscheint gerechtfertigt, da Rb^+ und K^+ ähnliche kinetische Eigenschaften aufweisen [17, 18] Während die Abhängigkeit der Austauschkonstanten des ^{22}Na-Effluxes vom Hämatokritwert durch *p*H-Verschiebungen innerhalb der Suspension erklärbar ist, kann jedoch die Hämatokritabhängigkeit des ^{86}Rb-Effluxes nicht auf diese Weise gedeutet werden. Da nicht auszuschließen ist, daß hier die Oberflächenstruktur der Zellen und ihre gegenseitige elektrostatische Beeinflussung eine Rolle spielen, wurden Untersuchungen mit oberflächenmodifizierten Erythrozyten durchgeführt.

Der Einfluß von Neuraminidase auf den ^{86}Rb- und ^{22}Na-Efflux

Neuraminidase bewirkt in den durchgeführten Experimenten eine 50%ige Reduktion der elektrophoretischen Beweglichkeit. Sie ist ohne Einfluß auf den aktiven Na-Transport bei *p*H 7,4. Jedoch ist eine Erhöhung der Austauschkonstanten für den ^{86}Rb-Efflux von $k = (5{,}9 \pm 0{,}6) \cdot 10^{-4}\ \text{min}^{-1}$ auf $k = (8{,}9 \pm 1{,}2) \cdot 10^{-4}\ \text{min}^{-1}$ feststellbar.

Der Einfluß von La^{3+} auf den ^{86}Rb- und ^{22}Na-Efflux

Abb. 4 zeigt die mit La^{3+}-Behandlung erhaltenen Ergebnisse. Alle Werte sind auf die Austauschkonstante ohne La^{3+}-Behandlung normiert. Es zeigt sich, daß der Na^+-Austausch wesentlich geringer beeinflußt wird als der ^{86}Rb-Austausch. Das steht in Einklang mit Angaben von QUIST und ROUFOGALIS [19], die nachweisen, daß die (Na^+, K^+)-ATPase im Gegensatz zur Ca^{++}-ATPase kaum von La^{3+} im Suspensionsmedium beeinflußt wird.

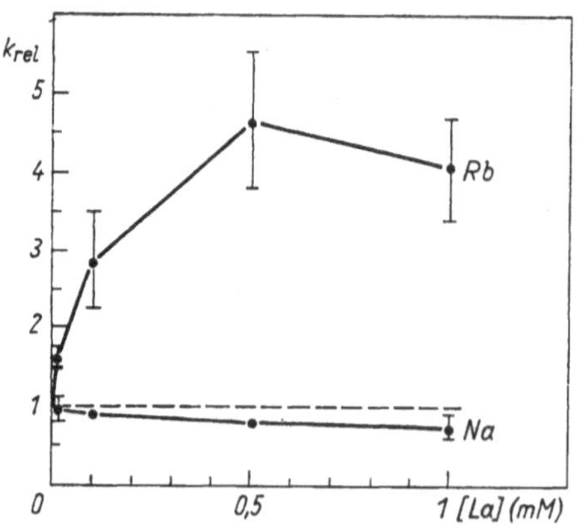

Abb. 4. Abhängigkeit der relativen Austauschkonstanten k_{rel} (Austauschkonstante für Messungen ohne La^{3+}-Behandlung = 1) von der La^{3+}-Konzentration $c_{La^{3+}}$. *p*H 7,4; $T = (37 \pm 0{,}5)$ °C. Suspensionsmittel 3 (s. Tab. 1)

Diskussion

Die experimentellen Ergebnisse weisen auf die Existenz von Wechselwirkungen zwischen Erythrozyten hin, die durch folgende Faktoren realisiert sind:

1. *p*H-Verschiebungen in Suspensionen hohen Hämatokritwertes,
2. direkte elektrostatische Wechselwirkungen.

Veränderungen der Struktur und der elektrostatischen Ladung der Zelloberfläche haben einen signifikanten Einfluß auf die Effektivität der Membran als Diffusionsbarriere. Der Einfluß der Oberflächenladungen kann einerseits auf einer direkten elektrostatischen Wechselwirkung zwischen den Festladungen und den betrachteten penetrierenden Ionen beruhen, andererseits auf der Beeinflussung der Struktur membranaler Makromoleküle und damit der Membranstruktur durch die Oberflächenladungen [5]. Experimente mit ^{86}Rb-inkubierten neuraminidasebehandelten Zellen zeigen, daß eine Verringerung der negativen Oberflächenladung den Ionenaustausch zwischen Zellen und Suspensionsmittel erhöht und somit die Effektivität der Membran als Diffusionsbarriere sinkt. Das gleiche gilt für La^{3+}-behandelte Zellen. Hier können jedoch durch unterschiedliche La^{3+}-Konzentrationen unterschiedliche Oberflächenladungszahlen eingestellt und der graduelle Einfluß der Oberflächenladung gezeigt werden (Abb. 4).
Vergrößert sich mit sinkender Oberflächenladung die Austauschkonstante für den passiven ^{86}Rb-Efflux, so kann andererseits geschlußfolgert werden, daß eine Vergrößerung der negativen Oberflächenladung den gegensätzlichen Effekt haben sollte. Nach experimentellen [3] und theoretischen [13] Untersuchungen steigt mit abnehmendem Zellabstand und damit zunehmender Wahrscheinlichkeit für die Annäherung zweier Zellen die effektive Oberflächenladung. Es ist deshalb die in Abb. 2 gezeigte Abhängigkeit zu erwarten. Dabei muß festgestellt werden, daß die gegenseitige Beeinflussung der Zellen schon bei geringen Hämatokritwerten resp. großen Zellabständen meßbar wird. Die so erhaltenen Aussagen zeigen die Rolle des membrannahen Raumes bzw. der Strukturen der Zelloberfläche für die Steuerung von Ionentransportprozessen über Erythrozytenmembranen. Die experimentellen Daten, besonders die nach Neuraminidase- und La^{3+}-Behandlung erhaltenen, zeigen den Zusammenhang zwischen der elektrostatischen Struktur der Zelloberfläche und der passiven Diffusion. Das gilt jedoch nicht in gleichem Maße für den aktiven Ionentransport.
Die Ursachen für die *p*H-Verschiebungen in Suspensionen hoher Zelldichte, die als Ursache für die Hämatokritabhängigkeit des ^{22}Na-Effluxes festgestellt wurden (Abb. 3), sind noch nicht klar. Es muß jedoch hier darauf hingewiesen werden, daß es sich um *p*H-Verschiebungen am Wirkort der (Na^+, K^+)-ATPase handeln muß, die aber durchaus extrazellulären Ursprung haben können.
Die für andere Zellen gefundenen spezifischen Transporthemmstoffe, die sich in Abhängigkeit von der Zelldichte im Suspensionsmittel anreichern [1], können für die dargestellten Ergebnisse auf Grund der Wahl der Meßanordnung ausgeschlossen werden. Das die Erythrozyten zurückhaltende Membranfilter ist großporig genug, um für derartige Moleküle kein Diffusionshindernis darzustellen.

Für technische Assistenz bei der Durchführung der Versuche danken die Autoren Frau I. Dichtl.

Literatur

[1] Bhargava, P. M., M. A. Siddiqui, H. Krauti-Kumar u. K. S. N. Prasad: J. Membrane Biol. **22**, 357 (1975)
[2] Bladé, E., L. Harel u. N. Hanania: Expl Cell Res. **41**, 473 (1966)
[3] Wolf, H.: Promotion B der Humboldt-Universität Berlin, Biowiss. Fakultät 1975
[4] Glaser, R., H. Wolf, u. St. Blottner: Bioelektrochem. Bioenerget. **1**, 343 (1974)

[5] EMMELOT, P., u. C. J. BOS: Biochim. biophys. Acta **99**, 578 (1965)
[6] WEISS, L., u. C. LEVINSON: J. cell. Physiol. **73**, 31 (1969)
[7] GLICK, J. L., u. S. GITHENS: Nature, Lond. **208**, 88 (1965)
[8] PASSOW, H.: Prog. Biophys. molec. Biol. **19**, 424 (1969)
[9] BLOTTNER, ST.: Studia biophys. **42**, 25 (1974)
[10] SOLOMAN, A. K. in: Mineral Metabolism. C. L. COMAR u. F. BONNER (Hrsg.) Academic Press, New York, London 1960
[11] BEAUGÉ, L. A., u. N. ANDRAGNA: Biochim. biophys. Acta **352**, 441 (1974)
[12] DUNHAM, E. T., u. J. M. GLYNN: J. Physiol., Lond. **156**, 274 (1961)
[13] GINGELL, D.: J. theor. Biol. **17**, 451 (1967)
[14] La CELLE, P. L., u. A. ROTHSTEIN: J. gen. Physiol. **50**, 171 (1966)
[15] RIORDAN, J. R., u. H. PASSOW: Biochim. biophys. Acta **249**, 601 (1971)
[16] ROMERO, P. J., u. R. WHITTAM: J. Physiol., Lond. **214**, 481 (1971)
[17] DIAMOND, J. M., u. P. M. WRIGHT: A. Rev. Physiol. **31**, 581 (1969)
[18] BEAUGÉ, L. A., u. O. ORTIZ: J. Physiol., Lond. **210**, 519 (1970)
[19] QUIST, E. E., u. B. D. ROUFOGALIS: FEBS Lett. **50**, 135 (1975)

Acta biol. med. germ., Band 36, Seite 931—934 (1977)

Humboldt-Universität zu Berlin, Sektion Biologie, Bereich Biophysik und
[1] Sektion Chemie, Bereich Magnetochemie, 104 Berlin, DDR

Changes in the shape of human erythrocytes under the influence of a static homogeneous magnetic field

A. Leitmannová, R. Stösser[1], and R. Glaser

Summary

The influence of a homogeneous magnetic field on the shape of erythrocytes was investigated. The specific influence of the magnetic field probably changes the energetic balance of the red blood cell resulting in its transformation. The cells were observed and photographically registered in hanging drops. Under the influence of a magnetic field only echinocytes changed their shapes. In the range of the magnetic field strength from 1 to 11.1 kOe a transformation of echinocytes into discocytes, stomatocytes and hemolyzing cells took place. The number of transformed echinocytes depends both on time and on the magnetic field strength. Also the ratio of newly formed discocytes to the newly formed stomatocytes is dependent on the field strength. The results are discussed in three ranges of the magnetic field strength: 1—4.6 kOe, 4.6—8.9 kOe and 8.9—11.1 kOe. In each range of the field strength an analogous behavior between the components of an erythrocyte membrane and fluid-crystalline systems as well as macromolecules with typical diamagnetic and paramagnetic properties, respectively, may be observed. The results are interpreted in terms of the fluid-mosaic theory and a hypothesis previously established by us on changes of the erythrocyte shape.

Introduction

The shape of human erythrocytes may be regarded as a multistable energetic state where energy-dependent membrane states and the volume regulated by the osmotic equilibrium play an important role [1, 3, 4]. Continous changes in such parameters obviously lead to discrete transitions in the shape of erythrocytes. For this reason the human erythrocyte may be suited as an indicator of the influence of magnetic fields on biological objects which undoubtedly also directly manifests itself in the energy balance.

Materials and methods

A suspension of erythrocytes was prepared in the following way: Blood stored up to 5 days in an ACD medium was centrifuged at $500 \times g$ and washed twice with an isotonic Krebs-Ringer solution at $2000 \times g$. After that the hematocrit value amounted to 40%, the erythrocyte suspension was incubated for 1 hr at 37 °C and then diluted with Krebs-Ringer solution up to a hematocrit value of 0.1%. The erythrocyte shapes of this suspension were observed at 37 °C in a microscope with 160fold magnification as hanging drops. The chamber was thermostated at 37 °C. The microscope was situated between the coils of an electromagnet according to Weiss. The chamber containing the probe was located in the center of the homogeneous magnetic field. By means of this experimental procedure magnetic fields with field strengths between 0.1 and 13 kOe at 37 °C could be produced at the locus of the probe. At each field strength (1.0, 2.5, 4.0, 5.4, 6.9, 8.4, 9.5, 10.2 and 11.1 kOe) in about 10 probes under physiological conditions were investigated. The photographic registration took place 1 and 10 min after switching on the field. Control experiments were performed after 5, 10, 15 and 20 min in the absence of a magnetic field.

Results and discussion

From Fig. 1, 2 and 3 it can be seen that the process of echinocyte transformation in a static magnetic field is dependent on the time and on the field strength applied. Both the value of the field strength and the time interval required for the trans-

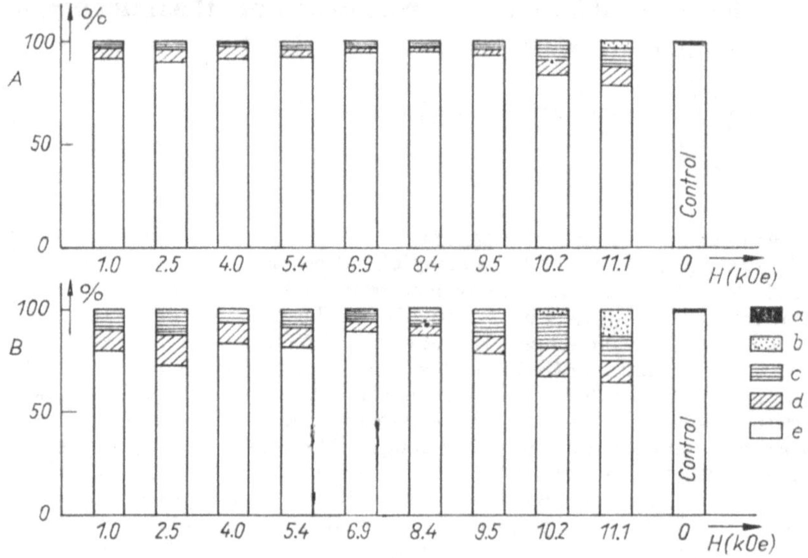

Fig. 1. Distribution of red blood cells between different shapes at several constant values of the magnetic field strength.
A) After 1 min B) after 10 min in the field.
Symbols: a) total number of transformed echinocytes at 37 °C after 5—20 min without magnetic field (control); b) echinocytes transformed mostly into stomatocytes and then hemolyzed; c) echinocytes transformed into stomatocytes; d) echinocytes transformed into discocytes; e) unchanged echinocytes

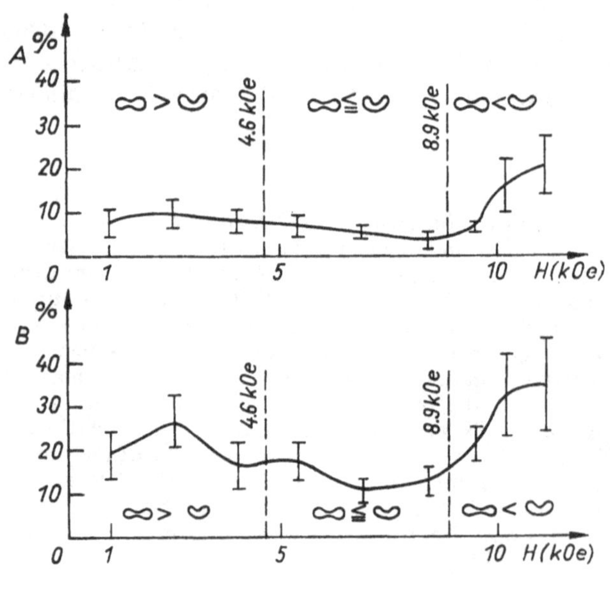

Fig. 2. Dependence of the total number of transformed echinocytes (sum of discocytes, stomatocytes and hemolyzed cells, respectively) on the magnetic field strength.
A) after 1 min. B) after 10 min in the magnetic field. In the figure is also given the ratio of the discocytes to the stomatocytes both formed from echinocytes at different magnetic field strengths. All values are weighted means with the standard deviation of the mean

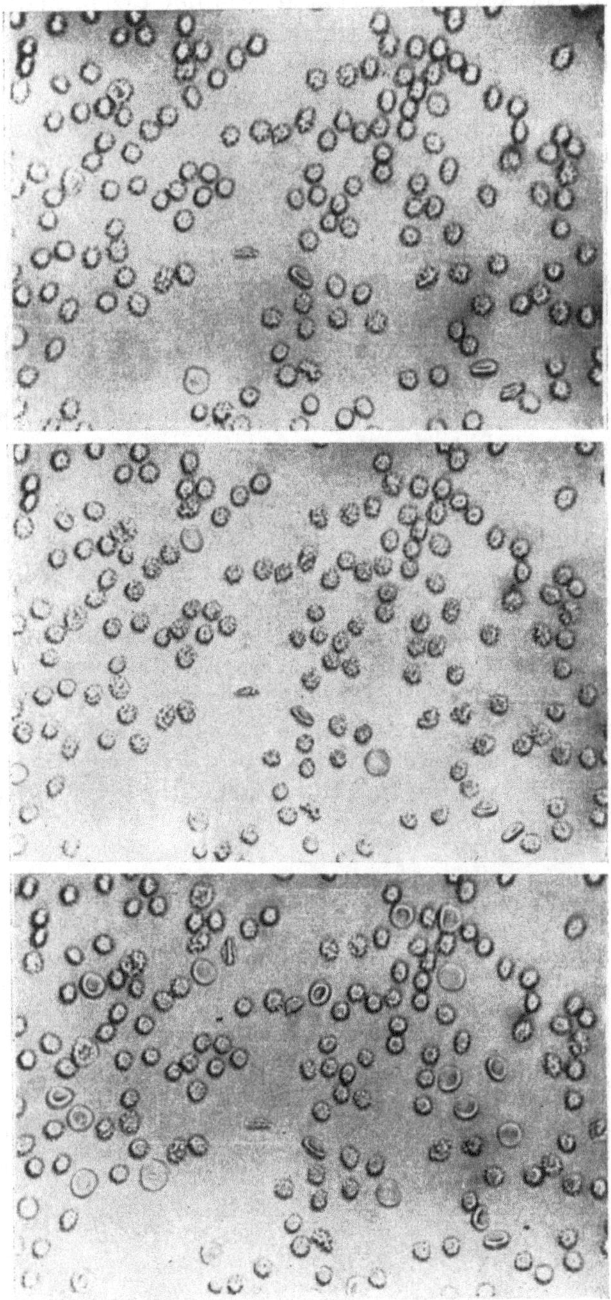

A

B

C

Fig. 3. Dependence of the transformation of echinocytes into discocytes, stomatocytes and hemolyzed cells, respectively, on the time at 2.5 kOe which is typical for the phase transitions in fluid-crystalline systems.

A) 1 min without the field; B) 1 min in the field; C) 10 min in the field

formation lead to the assumption that it was not the primary influence of the magnetic field on the erythrocytes which has been observed. Rather the shape of human erythrocytes must be regarded as a metastable state which is determined by the surface, volume and plane inhomogenities of certain mechanic membrane parameters such as bending elasticity and viscosity. While the membrane surface may be assumed to be nearly constant, the two other parameters are determined by a variety of molecular and cell-physiological conditions. The results reported here about shape transformations of erythrocytes permit a division of the effects in three ranges of the magnetic field strength (Fig. 2A and 2B). The first range extends from about 1—4.6 kOe with a maximum occurring at 2.5 kOe. This range is typical for fields causing phase transitions in fluid-crystalline systems and analoguous structured biological model membranes [2, 5]. In the second range there is a minimum in the number of erythrocytes transformations. The third range (8.9—11.1 kOe) is characterized by a strong increase in the number of transformed cells. In this range the field begins mainly to influence macromolecular components with a large anisotropy of the magnetic susceptibility. Furthermore, small inhomogenities in the magnetic field may cause diffusion processes of paramagnetic — and to a lesser extent also diamagnetic — substances. These effects may counteract the formation of field-induced ordered states. The results obtained in the mean field range support this assumption.

The application of high magnetic field strengths can lead to a superposition of all the effects mentioned which in its turn in combination with sequential reactions may cause a destruction of the membrane. This has been really observed at magnetic field strengths higher than 9 kOe. In the range between 20 and 40 kOe. this effect was even more clearly observed where a fragmentation of the main part of cells took place. It is assumed that the action of the magnetic field is not limited to only one cell parameter.

References

[1] CANHAM, P.: J. theor. Biol. **26**, 61 (1970)
[2] GAFFNEY, B. J., and H. M. MCCONNELL: Chem. Phys. Lett. **24**, 310 (1974)
[3] GLASER, R., and A. LEITMANNOVÁ: Studia biophys. **48**, 219 (1975)
[4] GLASER, R.: Acta biol. med. germ. **35**, 715 (1976)
[5] PERPLIES, E., A. RINGSDORF, and J. H. WENDORF: Makromolek. Chem. **175**, 553 (1974)

Acta biol. med. germ., Band 36, Seite 935–936 (1977)

Department of Biochemistry, Tokyo Medical and Dental University Yushima, Bunkyo-ku, Tokyo, Japan

Complete separation of red cells from whole blood with SE-cellulose

M. Nakao and T. Nakayama

Introduction

In an experiment for the separation of whole cells from the medium [1, 2], we noticed that the white cells were eliminated completely through a column of Sepharose 6B. The phenomenon was closely reproducible only in the case of one lot of Sepharose 6B. When other lots were used, the results were quite different. This prompted us to find a new technique for the separation of white cells from red cells. To examine the enzyme activities and minor components in red cells, the red cells must first be separated from the white cells. The repeated centrifugations in various media usually used for this purpose are time-consuming, and there is also the disadvantage of an accompanying loss of the youngest population of red cells, which frequently show stronger enzyme activities than older red cells.

Material and method

1.0 ml of wet SE-cellulose and 1.0 ml of wet Sephadex G 25 (medium) were mixed, packed in a column (3 cm × 1 cm) and equilibrated with a solution containing 0.138 M NaCl, 0.005 M $MgCl_2$, 10 units/ml of heparin and 0.01 M potassium phosphate buffer, *p*H 7.5. 1 ml of heparinized blood was added to the column and eluted with 5 ml of the NaCl-$MgCl_2$-heparin mixture.

Results and discussion

By this technique, the elution of erythrocytes and the elimination of leucocytes were complete as indicated in Table 1. The effects of column size and blood volumes and various eluents were examined and a great latitude was allowed. The presence of magnesium ion yielded a better separation. Even in the case of SE-cellulose, the results obtained in the presence of magnesium seemed to be better. Various ion exchange celluloses, or resins were also effective, though to a lesser extent.

Table 1
Effect of various amounts of Sephadex G-25 on SE-Cellulose column for removal of leucocytes

SE-cellulose (ml)	Sephadex G-25 (ml)	Leucocytes removed (%)	Elution rate
0.5	0	> 99	with pressure
0.5	0.5	> 99	slow
0.5	1.0	> 98	moderate
0.5	1.5	> 98	fast

The principle of this procedure may be chiefly ionic binding of white cells with SE-cellulose, because P-cellulose, Sephadex (prepared by us), and ionic exchange resins were also, but less, effective. Agar gels showed a good separation, too, but the best results were obtained using SE-cellulose (various lots and various commercial products showed a good separation).

For the examination of cell species remaining in the eluates, the second best material, Dowex 1, was used, because with SE-cellulose practically no count of cells was obtained. No difference was seen before and after the column passage. Gel shape, reticulocyte number, acetylcholine-esterase, osmotic hemolysis of the eluent cells showed no changes. Activities of various enzymes were compared between red cells obtained by SE-cellulose or centrifugation procedure; no striking differences were noted.

The procedure is very easy and the separation is almost complete. It takes about 20 min at room temperature. The red cells obtained appear to be normal as far as the various examinations stated above are concerned. The technique is particularly useful for comparing enzyme activities and minor components in red cells from different individuals when white cells may have some effect. The comparison with other separation methods is also shown. A preliminary result was already published [3].

References

[1] KANKURA, T., S. KURASHINA, and M. NAKAO: J. Lab. Clin. Med. **83**, 840 (1974)
[2] KANKURA, T., W. NAKAMURA, H. ETO, and M. NAKAO: J. Radiat. Biol. **15**, 125 (1968)
[3] NAKAO, M., T. NAKAYAMA, and T. KANKURA: Nature New Biol. **246**, 94 (1973)

Acta biol. med. germ., Band 36, Seite 937 (1977)

Discussion

The discussion of the report of PASSOW dealt mostly with the band 3 protein fraction. In contrast to a literature report liposomes with incorporated band 3 protein fail to show anion exchange (PASSOW). There is no [^{3}H]-lysine binding to H_2DIDS-treated cells, which may indicate formation of intramolecular bridges. With respect to the report of GÁRDOS the identity of Ca^{2+} and La^{3+} binding sites was questioned. It was suggested to test the competition of Ca^{2+} and Na^+ on active Na^+ outward transport (ERDMANN). In reply to a variety of questions WOLF pointed out that there are 1200 copies of Ca-ATPase per cell, assuming a dimer $\alpha_2\beta_2\gamma_2$. The enzyme does not appear to be identical with that of AVISSAR. Both the high and the low affinity enzyme are significant for ATP consumption. The high-affinity enzyme is responsible for the Ca-transport. The low affinity enzyme is probably connected or in part identical with spectrin. The mixed micelles were prepared from crude phosphatidylcholine from hen egg yolk. Phospholipid extracts from human red cell membranes may also be used.

The report of RAMPLING occasioned an extensive discussion. It was pointed out that the sedimentation rate depends not only on the flexibility but also on the shape and size of the cells (PASSOW). The kinetics of packing is probably determined by the extrusion of intracellular fluid, since the force for the deformation calculated from the suction experiments of RAND is much smaller than that applied by centrifugation (GLASER). There is no significant relaxation at the end of the spinning period (RAMPLING). The effect of EDTA to reduce the flexibility is unclarified, possibly there is an interaction with the actomyosin-like protein of erythrocytes. *p*H influences strongly the flexibility. Normal and sickle cells do not differ in their flexibility in the oxygenated cells. After deoxygenation interpretation is difficult owing to the bizarre shapes of sickle cells. The question was raised whether the asymmetry of the frequency curves is related to the asymmetrical distribution of the cells (SYLLM-RAPOPORT).

With respect to the report of GLASER the following questions were discussed. The biconcave form of erythrocytes may be due to 2 factors: firstly, tension related to the state of actomyosin-like protein with properties of a Mg-ATPase and secondly the asymmetrical distribution of membrane lipids. It is assumed that with a decrease of ATP the membrane tension declines and echinocytes are formed (MIRCEVOVA). An immediate explanation of the form of elliptocytes is not available.

Acta biol. med. germ., Band 36, Seite 939—940 (1977)

Workshop

Die Diskussion berührte folgende Fragenkomplexe:

1. Formveränderung und Flexibilität der Erythrozyten im Zusammenhang mit dem metabolischen Zustand der Zellen unter dem Einfluß verschiedener Agenzien:

Die Formveränderungen der Erythrozyten spiegeln verschiedene Membranreaktionen wider. Dabei könnte der Zustand der Spektrine von Bedeutung sein. Struktur und Funktion der Spektrine sind jedoch noch weitgehend unklar, da die bisherigen Spektrinpräparationen heterogen sind. Auch auf der äußeren Membranfläche können evtl. Spektrine auftreten (Nakao). Es ist jedoch zweifelhaft, daß sie hier eine kontraktile Funktion haben und ob überhaupt eine Spektrin-ATPase-Aktivität in situ vorkommt.
Ausführlich wurde die ATP-Abhängigkeit der Flexibilitätsänderungen diskutiert. Die Funktion des ATP, aufgrund seiner Eigenschaft als Chelatbildner, die Konzentration des freien Kalziums, das mit dem Spektrin in Wechselwirkung treten kann, zu regulieren, ist z. Z. noch umstritten. Eine Konkurrenz von Ca^{2+}- und Mg^{2+}-Ionen um den Chelatbildner ist auf jeden Fall zu beachten. Es ist jedoch unwahrscheinlich, daß die Ca-ATP-Komplexbildung unter den Bedingungen physiologischer Ca- und ATP-Konzentrationen überhaupt von Bedeutung für Flexibilitätsänderungen der Zellen ist. Die Ca-ATPase hat eine hohe Kapazität für den Ca-Transport und eine hohe Affinität zu Ca^{2+} ($K_m \approx 1\ \mu M$ bzw. $100\ \mu M$). Bei experimentell veränderter intrazellulärer Kalzium-Konzentration verlieren die Zellen sowohl bei starker Erhöhung als auch bei extremer Erniedrigung ihre Diskozyten-Form (Frunder).
Bei künstlicher Erniedrigung der intrazellulären ATP-Konzentration beobachtet man erst um 0,5 mM ATP Sphärozyten, obwohl Membranveränderungen schon bei 1 mM ATP auftreten.
Reversible Sphärisierung der Zellen tritt auch bei Temperaturerhöhungen auf 47,5 °C ein; ab 48 °C werden irreversible Membranveränderungen, die mit Mikrosphärulation verbunden sind, beobachtet (Jung). Es ist noch unklar, welcher Mechanismus der Auslösung dieser Erscheinung zugrunde liegt. Der Prozeß der Mikrosphärulation, induziert durch Temperaturerhöhung oder andere Agenzien ist wahrscheinlich als laterale Phasenseparation aufzufassen (Glaser).

Sphärisierung und Mikrosphärulation wurden auch bei Patienten mit Verbrennungen beobachtet. Die Tatsache, daß auch hier eine erhöhte Glykolyserate der Erythrozyten gemessen wurde, deutet auf eine im Detail noch unklare Beziehung zwischen Formveränderung und Stoffwechsel der Zellen hin. Bei Membranveränderungen könnte demzufolge bei u. U. konstanter ATP-Konzentration der ATP-Umsatz aktiviert sein (ATPase-Aktivierung). In diesem Zusammenhang wurde darauf hingewiesen, daß auch bei Sichelzellen eine Verdopplung der Glykolyserate gefunden wird (Rampling).

Die relative Formkonstanz und die Flexibilität der Erythrozyten scheinen weiterhin auch abhängig von der Ladung der Außenmembranoberfläche zu sein, da Sialidasebehandlung der Zellen zu starken Formveränderungen führt (Nakao).

2. Zur Membranfluidität:

Die Diskussionsteilnehmer kritisierten den Begriff der „Membranfluidität", der aus zweierlei Gründen unexakt erscheint: 1.) Er stammt aus dem Bereich der Kontinuumsmechanik und ist folglich für molekulare Mosaikstrukturen, wie es die Membran ist, ungeeignet. 2.) Er ist als Eigenschaft eines 3-dimensionalen Systems definiert, wird jedoch im Falle der Membran für einen 2-dimensionalen Film verwendet (GLASER).

Zur Debatte stand weiterhin die Frage, ob ein Protein innerhalb der Membran um eine Achse rotieren kann, die in der Membranfläche liegt, d. h. im Sinne eines Transportproteins, wie es von einigen Autoren angenommen wird. Viele Experimente sprechen gegen eine solche Möglichkeit. Auch thermodynamisch erscheint ein solches Verhalten höchst unwahrscheinlich (GLASER).

3. Zur Reinigung und Charakterisierung von Membranproteinen:

Es wurde diskutiert, ob es möglich ist, durch Anwendung neuer Solubilisierungsverfahren reine Membranproteine mit nativer Konformation zu erhalten, bzw. durch Zugabe von Phospholipiden native Strukturen rekombinieren zu können, z. B. durch Verwendung von Lösungsmitteln. Der Kritik des Einsatzes von Detergenzien wurde entgegengehalten, daß diese aufgrund ihrer hydrophil-hydrophoben Eigenschaften beim Solubilisierungsprozeß so an die Proteine binden könnten, daß deren native Konformation erhalten bleibt (PASSOW, SCHÖN).

CONTENTS

Molecular biology of the erythron

Regulation of energy metabolism

Enzymopathies and metabolic defects

Membrane Processes

СОДЕРЖАНИЕ

Энзимопатии и метаболические эффекты

Мембранные процессы

INHALT

Enzymopathien und Stoffwechseldefekte

Membranprozesse

Bezugsmöglichkeiten

Bestellungen sind zu richten:

- **in der DDR** an eine Buchhandlung oder an den Akademie-Verlag
  DDR-108 Berlin, Leipziger Straße 3–4
- **im sozialistischen Ausland** an eine Buchhandlung für fremdsprachige Literatur oder an den zuständigen Postzeitungsvertrieb
- **in der BRD und Westberlin** an eine Buchhandlung oder an die Auslieferungsstelle
  KUNST UND WISSEN, Erich Bieber, 7 Stuttgart 1, Wilhelmstraße 4–6
- **in Österreich** an den Globus-Buchvertrieb,
  1201 Wien, Höchstädtplatz 3
- **im übrigen Ausland** an den Internationalen Buch- und Zeitschriftenhandel; den Buchexport, Volkseigener Außenhandelsbetrieb der Deutschen Demokratischen Republik,
  DDR-701, Leipzig, Postfach 160, oder an den Akademie-Verlag,
  DDR-108, Berlin, Leipziger Straße 3–4.

MIX
Papier | Fördert gute Waldnutzung
FSC® C083411

Zeitfracht Medien GmbH
Ferdinand-Jühlke-Straße 7
99095 Erfurt, Deutschland
produktsicherheit@kolibri360.de